Latin Eugenics in Comparative Perspective

Latin Eugenics in Comparative Perspective

Marius Turda and Aaron Gillette

Bloomsbury Academic
An imprint of Bloomsbury Publishing Plc

BLOOMSBURY

LONDON · NEW DELHI · NEW YORK · SYDNEY

Bloomsbury Academic
An imprint of Bloomsbury Publishing Plc

50 Bedford Square
London
WC1B 3DP
UK

1385 Broadway
New York
NY 10018
USA

www.bloomsbury.com

BLOOMSBURY and the Diana logo are trademarks of Bloomsbury Publishing Plc

First published 2014

British Library Cataloguing-in-Publication Data
A catalogue record for this book is available from the British Library.

ISBN: HB: 978-1-4725-3140-7
ePDF: 978-1-4725-2369-3
ePub: 978-1-4725-2210-8

Library of Congress Cataloging-in-Publication Data
Turda, Marius, author.
Latin eugenics in comparative perspective / Marius Turda and Aaron Gillette.
p. ; cm.
Includes bibliographical references and index.
ISBN 978-1-4725-3140-7 (HB) – ISBN 978-1-4725-2210-8 (ePub) –
ISBN 978-1-4725-2369-3 (ePDF)
I. Gillette, Aaron, 1964- author. II. Title.
[DNLM: 1. Eugenics–history–Europe. 2. Eugenics–history–Latin America.
3. Cross-Cultural Comparison–Europe. 4. Cross-Cultural Comparison–Latin America.
5. History, 20th Century–Europe. 6. History, 20th Century–Latin America.
7. Public Policy–history–Europe. 8. Public Policy–history–Latin America.
9. Social Change–history–Europe. 10. Social Change–history–Latin America. HQ 751]
HQ751
363.9'2–dc23
2014017404

Typeset by Integra Software Services Pvt. Ltd.
Printed and bound in Great Britain

Contents

List of Illustrations and Tables vi
Acknowledgements viii

Introduction 1
1 Precursors 15
2 Early Latin Eugenics 41
3 Latin Eugenics in Interwar Europe 67
4 Latin Eugenics, Sterilization, and Catholicism 103
5 Eugenics in Interwar Latin America 129
6 The Latin Eugenics Federation 165
7 Latin Eugenics and Scientific Racism 199

Conclusion 237
Epilogue: Latin Eugenics after 1945 242
Notes 250
Bibliography 252
Index 293

List of Illustrations and Tables

Fig. 1.1	Adolphe Pinard (*Source*: Wellcome Library, London. Image reproduced under the Creative Common Licence.)	34
Fig. 2.1	Lucien March (*Source*: Collections École Polytechnique, Palaiseau, France.)	46
Fig. 2.2	Birth rates and mortality in various European countries (*Source*: Lucien March, 'Depopulation and Eugenics' (part 1), *The Eugenics Review* 5, 3 (1913): 237. Courtesy of the Galton Institute, London.)	50
Fig. 2.3	French–Spanish–American Medical Alliance (*Source*: Dr. Dartigues, ed., *Livre d'or et Annuaire de L'Union Médicale Franco-Ibéro-Américaine* (Paris: Laboratoires Darrase, 1926), front page. Courtesy of Bibliothèque Interuniversitaire de Santé, Paris.)	58
Fig. 3.1	Iuliu Moldovan (centre) together with his collaborators (*Source*: Library of the Institute of Medicine, Institute of Public Health, Bucharest.)	71
Fig. 3.2	Gregorio Marañón (*Source*: Fundación Ortega-Marañón, Madrid.)	75
Fig. 3.3	Corrado Gini (*Source*: Photo Division, Ministry of Information & Broadcasting, Government of India.)	90
Fig. 3.4	Nicola Pende (*Source*: Library of the Institute of Medicine, Institute of Public Health, Bucharest.)	92
Fig. 3.5	Silvio Canevari, Il Littore, Stadio dei Marmi, Rome (*Source*: Aaron Gillette's personal collection.)	99
Fig. 4.1	Antonio Vallejo-Nágera (*Source*: Archivo General de la Administración, Alcalá de Henares, Spain.)	111
Fig. 4.2	Agostino Gemelli (*Source*: Archivio generale per la storia dell'Università Cattolica del Sacro Cuore, Milan.)	118
Fig. 5.1	Arturo R. Rossi (*Source*: *Anales de Biotipología, Eugenesia y Medicina Social*, vol. 3 no. 42, 15 April 1935, p. 2. Courtesy of the National Library of Argentina.)	132
Fig. 5.2	Renato Kehl (front row seated, second left) (*Source*: Fundação Oswaldo Cruz, Departamento de Arquivo e Documentação, Rio de Janeiro.)	144

Fig. 5.3 Eugenic Baby Prize Winner, São Paulo, 1929 (*Source*: Maria Antonieta de Castro, 'A influência da educação sanitária na redução da mortalidade infantil'. Arquivo de Antropologia Física, Museu Nacional/ Universidade Federal do Rio de Janeiro.) 148

Fig. 5.4 Domingo F. Ramos, on the right (*Source*: Courtesy of Claude Moore Health Sciences Library, University of Virginia.) 152

Fig. 6.1 Lucien March (first on the left) and Corrado Gini (centre) at the 1927 World Population Congress in Geneva (*Source*: Wellcome Library London. Image reproduced by permission of Alexander C. Sanger.) 169

Fig. 6.2 Gheorghe Marinescu (*Source*: Library of the Institute of Medicine, Institute of Public Health, Bucharest.) 178

Fig. 6.3 The Programme of the 4th Congress of the Latin Medical Press (*Source*: Marius Turda's personal collection.) 182

Fig. 6.4 Participants at the 17th Congress of Anthropology and Prehistoric Archaeology in Bucharest, 4 September 1937. Manuel López Rey (third from the left), Georges Paul-Boncour (third from the right) and Henri Berr (second from the right) are seated on the first row. (*Source*: *În amintirea profesorului Fr. J. Rainer (1874–1944)*, Bucharest, 1945, p. 16.) 192

Fig. 7.1 António Mendes Correia (*Source*: Patrícia Ferraz de Matos, *Mendes Correia e a Escola de Antropologia do Porto*, Lisbon 2012, p. xxxiii.) 203

Fig. 7.2 Réne Martial (*Source*: Bibliothèque de l'Académie nationale de Médecine, Paris.) 208

Fig. 7.3 Journée des mères, 31 Mai 1942 (*Source*: Archives Départementales du Loiret, Orléans.) 220

Fig. 7.4 Gheorghe Banu (*Source*: Library of the Institute of Medicine, Institute of Public Health, Bucharest.) 225

Fig. 7.5 Iordache Făcăoaru (standing up) and Sabin Manuilă (on Făcăoaru's left) (*Source*: Iordache Făcăoaru's Personal File, Consiliul Național pentru Studierea Arhivelor Securității, Bucharest.) 231

Table 6.1 The Latin International Federation of Eugenic Societies (1935–39) 186

Table 6.2 Presidents of the Latin International Federation of Eugenic Societies (1935–39) 196

Acknowledgements

I owe a great deal of debt to a number of friends and colleagues who have helped me in more ways than just one during the writing of this book: first and foremost, Francesco Cassata, Maria-Sophia Quine, Richard Cleminson, Andrés Reggiani, and Chiara Beccalossi. Their comments, suggestions, and assistance in obtaining primary sources have been invaluable.

Equally important, I want to thank indviduals and institutions for helping with information, and for granting us the permission to use the illustrations included in this book: in the UK, John C. Aldrich, Bette Nixon, and Crestina Forcina; the Galton Institute and the Wellcome Library; in France, Ștefan Lemny, Michel Prum, André Lebrun, Olivier Azzola, Damien Blanchard, and Stéphanie Charreaux; École Polytechnique; Bibliothèque de l'Académie nationale de Médecine, Bibliothèque interuniversitaire de Santé and Bibliothèque Nationale de France; in Spain, Miguel Ángel del Arco Blanco, Juan José Villar Lijarcio, and Ramón Castejón Bolea; Archivo General de la Administración and Fundación Ortega-Marañón; in Portugal, Patrícia Ferraz de Matos; in Italy, Susanna Cimmino and Maurizio Romano; Biblioteca Museo Galileo, Florence and Ufficio Archivio storico d'Ateneo, Università Cattolica del Sacro Cuore, Milan; in Brazil, Ricardo Ventura Santos and Claudio Arcoverde; Museu Nacional Universidade Federal do Rio de Janeiro; Fundação Oswaldo Cruz, Rio de Janeiro; in Argentina, Gustavo Vallejo; in the USA, Alexander C. Sanger, Cristina Bejan, and Radu Ioanid; United States Holocaust Memorial Museum, Washington, DC; in Romania, Răzvan Pârâianu, Laura Bădescu, Ștefan Ionescu, Adrian Majuru and Mioara Georgescu; Biblioteca Institutului de Igienă şi Sănătate Publică.

Both Oxford Brookes University, in particular the Centre for Health, Medicine and Society, and the Wellcome Trust (grant no: 082808 and grant no: 096580) deserve special thanks for providing financial and moral support for this work. I also owe a debt of gratitude to Camelia Cazac, who was always understanding of my work, while diligently looking after my daughter Ariadne, when I was unable to do so.

Most significantly, I want to thank my wife, Aliki, and my daughter, Ariadne, to whom both I dedicate this book. I owe each of them my daily moments of happiness.

Marius Turda

This project was first conceived during a National Endowment for the Humanities Summer Seminar summer course on the Italian Risorgimento, taught by David Kertzer and John Davis at the American Academy of Rome in 2003. I would especially like to thank John for discussing this project with me. Over the years I have used a number of libraries and archives for this project. I would especially like to thank the librarians of Truman State University, who did an unprecedented amount of photocopying for me, virtually without charge; Archivio Centrale dello Stato in Rome; the Smithsonian Institute of Anthropology Archives; the National Academy of Sciences Archives; the Carnegie Institute of Washington Archives; the Cold Spring Harbor Laboratory Archives; and the Department of Anthropology of the University of Rome 'La Sapienza'. Michael Gelb, assistant editor at the Center for Advanced Studies at the United States Holocaust Memorial Museum, spent many hours discussing an earlier version of this book with me.

Various archivists, librarians, and photographers have permitted us to use the images in this book. In particular, I would like to thank Dan Cavanaugh, Historical Collections Specialist at Claude Moore Health Sciences Library, University of Virginia, for the photograph of Domingo F. Ramos; Laura N. Braga, librarian at the Hemeroteca of the Biblioteca Nacional de la República Argentina, for the image of Arturo R. Rossi; and Frédérique Hamm at Archives Départementales du Loiret, Orléans for the 1941 French poster.

The University of Houston-Downtown has given me tremendous support for this project, including a 2006 faculty research grant. Adolfo Santos, my department chair in 2008, gave me a grant to conduct research at Stanford University. My good friend and colleague David Ryden provided substantial advice regarding the manuscript.

This work would not have been possible were it not for the help I received from many members of my family. My mother, Roxane Price, graciously assisted in funding my visit to the Bancroft Library at Berkeley. My mother-in-law, Isabel López Panisello, one of the nicest, most patient and tolerant persons I have ever known, endured my idiosyncrasies for several summers. Anthony and Rosa Barilla, my cousins, were no less generous, and treated me like a son for the long months I spent with them in Washington, DC. Above all, I would like to thank my wife, Montse Feu, who took on innumerable chores and delayed publication of one of her books, to allow me time to complete mine. It is to her, with my eternal love, that this book is dedicated.

Aaron Gillette

***

Both Aaron and Marius would like to thank Emma Goode and Claire Lipscomb at Bloomsbury. It has been a real pleasure to work on this book with you.

Finally, we offer the disclaimer that none of those mentioned should be held responsible for any of the views expressed in this book, which are entirely those of the authors. Both of them contributed to the writing of this book equally.

London/Rome 2014

Introduction

This book offers a history of Latin eugenics in comparative context; that is, of a particular strand of eugenics that became an aggregate of ideas and practices espoused by many eugenicists who considered their countries to belong to an international Latin cultural community during the late nineteenth and the first half of the twentieth centuries. As they understood it, the culture of Latin countries was based on a synthesis of ancient Roman civilization, linguistic and cultural commonality, and Roman Catholicism (in the Romanian case, Christian Orthodoxy). The following countries in Europe (France, Italy, Spain, Belgium, Romance-speaking Switzerland, Portugal and Romania), and in the Americas (particularly Argentina, Mexico, Cuba, Brazil, Paraguay, Peru, Venezuela, and Chile) considered themselves to be Latin. Some observers even considered modern *Latins* as a distinct ethnic or racial group. At the same time, these countries (most of them predominantly rural) were struggling to find their place in a rapidly changing world, and enjoy the benefits of modern medicine, science and technology.

This book is thus concerned with the questions about individual and collective improvement promoted by eugenicists in Latin countries, which have traditionally been ignored or treated in isolation. Our purpose here is to reconstitute what eugenicists at the time defined as Latin eugenics, while at the same time to explore the bases upon which this cluster of ideas formed, developed, and ultimately defined itself *in opposition* to other theories of human betterment, most notably Nordic and Anglo-Saxon eugenics. While we acknowledge that this is a contested term in the historiography on eugenics, we believe it nevertheless to be a valid one. Latin eugenics was one of central concepts of a scientific, social, and political vocabulary developed after the mid-nineteenth century, whose purpose was to improve society. Contrary to Nordic and Anglo-Saxon eugenics, Latin eugenics relied less on race and class and more on individual and the national community.

This history of Latin eugenics involves manifold and often contradictory historical experiences. From the 1880s through the 1950s, eugenics consisted of a cluster of scientific methodologies devised to control heredity and the environment in such a way as to improve the biological and social quality of human populations. In most cases, this also implied a powerful state, guided by scientific elites, intent on controlling human reproductive patterns for social and biological purposes (Broberg and Roll-Hansen 1997). A number of critical questions then emerge: is this degree of state empowerment compatible with the preservation of a role for the Catholic (or other Christian) Church, especially given the Church's centuries-long involvement in the social and reproductive functions of the family and the individual? Could eugenicists succeed in convincing the state to grant them powers over the population? How could their ambitious modernizing goals be achieved in countries which were relatively impoverished and less industrialized? And how could nationalist sentiments, increasing throughout the period, be harmonized with notions of an international 'Latin sisterhood', with shared values, aspirations, and common enemies? These are some of the questions we hope to answer in this book.

For the past two decades the history of Latin eugenics has been defined by Nancy Leys Stepan's work on Latin America (1991). Stepan traced a series of eugenic developments, often overlapping one another, in such countries as Brazil, Mexico, and Argentina. According to Alexandra Minna Stern, Stepan has *Latinized* eugenics, showing 'decisively how Latin American countries produced and not only replicated or reengineered new biological knowledge' (Stern 2011: 432). More recent work has appeared by Armando García González and Raquel Álvarez Peláez on Cuba and Spain; Vanderlei Sebastião de Souza and Robert Wegner on Brazil; Arturo Orbegoso on Peru; Michael Richards and Ramón Castejón Bolea on Spain; Maria Sophia Quine and Francesco Cassata on Italy; Jean-Paul Gaudillière, Paul-André Rosental and Andrés H. Reggiani on France; Gustavo Vallejo, Marisa Miranda and Yolanda Eraso on Argentina; and Patrícia Ferraz de Matos and Richard Cleminson on Portugal. While incorporating elements from Stepan's interpretation, these authors seek primarily to identify idiosyncratic national developments, and to place them into the context of extensive changes in twentieth-century demography, medicine and genetics, politics and culture.

First and foremost, the history of Latin eugenics explored here establishes the intra-European dimension of Latin eugenics, a dimension always implicitly assumed in the scholarship but never properly studied, unlike the case of Latin American eugenics, which has long been an object of study. Further, rather than focusing on the administrative history of eugenic organizations in various Latin countries, we concentrate on examining Latin eugenicists' attempts to achieve the social and political goals of the modern welfare state. By so doing, we see the impact of the Latin eugenics movement on many countries' population and family policies, maternal and infant health, preventive medicine, and social hygienic and public health campaigns against tuberculosis, alcoholism, venereal diseases, and prostitution.

After 1900, many Latin eugenicists believed that multiplying signs of social and biological degeneration threatened their nations' future. These included increased urban crime, uncontrolled immigration, the breakdown of family structures in rural communities, poor living conditions for the majority of the population, underdeveloped systems of social hygiene and medicine, vagabondage, alcoholism, widespread disease, decreased birth rates, sexual deviancy, and anti-maternal feminism. Eugenics was expected to cure the national body of these and other social and biological ills. Ironically, to the extent that Latin eugenicists were actually able to shape society according to their programmes, their actions only served to intensify modernization, which it appeared was driving this supposed biological degeneracy.

Until the outbreak of the Second World War, Europe was at the apex of its global power, reflected also in its cultural and scientific domination. These facts are essential in explaining how Latin eugenics came into existence, and the crucial roles played in formulating and promoting Latin eugenics by France and Italy. Italian and French eugenicists led the Latin nations in their quest to construct one broad eugenic theory that promised to encourage national unity, modernization, and economic progress. For a century, these two countries maintained the network of loyalties, alliances and patterns of subordination and domination that made up the Latin world.

Building on the work of a number of scholars, and combining our own familiarity with primary sources from these and many other countries, we have attempted to avoid simplistic reductionism in exploring the multiplicities and varieties of ways in which eugenics was expressed across geographical and national boundaries. We are, however, aware of the daunting task of writing an integrated history of Latin eugenics' ever-changing causal matrix as well as of its developing general doctrine and associated politics. In this sense, Latin eugenics was the *historical* expression of both scientific refinement and situated political interests. Therefore, to engage with a subject of such international scope and conceptual range also requires sensitivity to national variations. Although sharing commonalities, the so-called Latin nations – represented here by over a dozen states spread across three continents – were actually very diverse; political relationships between them were unstable and sometimes even belligerent (such as the short Italian–French War of 1940). Moreover, these countries were not a cultural monolith: Romania, for instance – in many ways as Latin as any of the others – was predominantly not Roman Catholic. Furthermore, due to their history and geography, the Iberian states were more focused on their transatlantic relationships with their former colonies in Latin America than they were on links with their Mediterranean neighbours. Elsewhere, Portugal, Spain, France, Belgium, and Italy had colonial empires that absorbed considerable national effort and political determination, redefining at the same time ideas of race and eugenics in the metropole.

The history of Latin eugenics begins with the scientific and cultural developments of the mid-nineteenth century. Positivist intellectuals asserted that science could understand – and manage – human societies. Through Romanticism and nationalism, they claimed to have discovered the 'essential nature' of various peoples, and gave sharper form to previously existing, vague notions of national identity and ethnicity. These developments led to a re-interpretation of historical and contemporary events as racial struggles for supremacy, power and ultimately, survival. It was commonly accepted that certain nations demonstrated the ability to dominate others through a combination of innate racial qualities and technological achievements.

Some aspects of this synchronized process of racialization and modernization seemed heretical to the religious authorities, who rejected the modern world's secular utopianism and its racism. However, the cultural and scientific elites of many predominantly Catholic countries were attracted to modernizing ideologies, exacerbating tensions between science and religion. Latin countries in Europe and the Americas struggled to reconcile secular liberal with more conservative religious ideologies. In this context, Latin eugenicists hoped that their 'largely preventive eugenics with a strong flavor of social hygiene' (Rosental 2012: 542) would become a 'middle way' to the future, a synthesis of positivist modernization in harmony with national traditions and spiritual values.

By the late nineteenth century, few observers contested the fact that modernization, and its corollary industrialization, was essential for producing national wealth and power as well as a numerous population. An industrialized country could at once confidently assert itself in international politics, defeat its less-developed adversaries in a military conflict, and bask in economic prosperity. The modern nation-state began regulating *laissez-faire* economies and enacted new labour laws and work regulations aimed at improving conditions for the working class, increasing national economic efficiency, and decreasing the potential for revolutionary action that tended to arise from severe worker discontent. The management of national populations to foster social integration and centralization became one of the most important goals of the modernizing nation states.

The spread of compulsory mass education at this time is an excellent example of the interconnection of industrialization, ethnic homogenization, and nationalization. In the second half of the nineteenth century, mass education became imperative for all industrializing nations. However, education was not seen simply as a means to facilitate economic development. With growing number of children increasingly under the state's direct control, no opportunity was missed to indoctrinate them with nationalistic sentiments. Such attitudes included reverence for the nation's past and a desire to emulate its glorious heroes, hatred towards its enemies, support for its military, and loyalty to the state, coupled with an appreciation of the benefits the state conferred upon its citizens and the family (Weber 1976; Anderson 1983; de Luca 2005: 11–35). To achieve an idealized national community, then, the state extended its power over the family and the individual. As an Italian socialist physician, Gaetano Pieraccini, put it in 1898, 'society is everything, the individual nothing' (quoted in Mantovani 2003: 115).

Social commentators and health reformers argued that as a result of urbanization, industrialization and the breakup of traditional family life, alcoholism, prostitution, and venereal diseases soared in the growing cities (Harsin 1985; Guy 1990; Caulfield 2000; Vallejo 2005; Sippial 2013). Increased urban population densities also prompted incidents of communicable diseases and overwhelmed poor urban sanitation systems. Thus, medical authorities felt empowered to employ more dramatic measures to control the burgeoning population, such as compulsory vaccinations, quarantine and hospitalization. The idea that these communities' biological condition could also be improved upon with the help of external factors such as education and a controllable environment, including the prevention and eradication of contagious diseases, as well as modern sanitation and housing, was central to medical theories developed by social hygienists and public health reformers (Zylberman 2001; Baldwin 2005; Rodriquez 2006; Zulawski 2007; Mckiernan-González 2012; Turda 2012: 125–40).

By the early twentieth century, eugenicists added foreign and domestic immigration to the list of forces counteracting the work of medical and health professionals in curing the nation's body. It was assumed that the biological human variability represented by immigrants and ethnic minorities *deviated* from the national statistical norm. In light of this, as national capitals became more cosmopolitan and ethnically diverse, eugenicists saw them generally not as socially beneficial and worthy of celebration, but nefarious, criminal and racially harmful. Indeed, how could one cure the population of a city when so many poor, diseased people were flooding into it? Furthermore, many elite groups felt threatened by the perceived racial and cultural changes implied by the immigration of different peoples seeking economic opportunity and political change. The new social scientists of the nineteenth century duly studied these disturbing changes, and often denounced them in their writings.

Reflecting such growing feelings of discontent, reports in national mass media on crime, poverty, prostitution, floods of poor immigrants and ethnic minorities, disorder and disease caused grave concerns to state and health officials. These problems, and the discomforting changes accompanying them, further augmented fears of biological degeneration and racial decline. During the Second World War, the greater acceptance of state intervention, including medical control of military and civilian populations, only strengthened the work of eugenicists, public hygienists and health reformers (Turda 2011: 325–50 and 2014). The fascist regimes that emerged during the interwar period epitomized the expectation that the biopolitical purpose of eugenics was to mold the family and the individual in the shapes dictated by the regime and in turn scientifically manage the national population as a whole (Quine 2002; Reggiani 2010; Cassata 2011).

Though the expansion of cities and increased immigration caused disturbing signs of crisis, population growth as a whole was encouraged in order to both expand national markets as well as lists of military conscripts. In the mid-1850s, however, demographers in France noticed a countervailing trend: the birth rate of the French population was appreciably slowing. Demographic decline had military and economic consequences (Quinlan 2007). In the second half of the nineteenth century, the Latin countries suffered a number of military defeats, beginning most dramatically with the stunning rout of the French armies by Prussia in 1870, resulting in a revolution in European political relations: the defeated French were at the mercy of the newly established German Empire, whose peace terms demanded the French provinces of Alsace and Lorraine and much more besides. France's defeat in the war augmented not only domestic fears of national decline but also pessimistic predictions about the future of the 'Latin race' (Manias 2009: 733–57).

The 1890s were another disastrous decade for the Latin powers. The British Ultimatum of January 1890 abruptly ended Portuguese expansion in Africa, triggering wide-ranging debates on Portugal's racial decline and the waning of its former imperial glory (Cleminson 2014). In 1896, Italy's attempt to expand its African colonial empire into Ethiopia suffered a humiliating setback in the battle of Adwa (Jonas 2011). Finally, in 1898, the emerging United States quickly and easily stripped Spain of the remnants of its once-vast American and Pacific empire (Pérez 1998). These military and diplomatic defeats were also interpreted as signs of the 'decline of the Latin race'. Given the widespread belief in the 'struggle for existence', Latin intellectuals worried that the ultimate Darwinian fate of the defeated might apply to their nations as well. As we will see, anxious observers suggested a torrent of causes for Latin racial decline, and proposed various solutions. Many of these proposals directly or indirectly contained eugenic elements, and in some cases, the Latin countries implemented them.

By the first decades of the twentieth century, eugenicists were claiming the ability to accurately quantify the desirability of human traits, and classify individuals (and even entire races) according to perceived worth (Bashford and Levine 2010). For instance, just as economic rationalization and systems engineering had improved economic efficiency, so too would the biological engineering of humans – through eugenics – improve workforce productivity. The eugenic engineering of national communities would therefore advance social and economic improvement. If one considers that eugenics was born out of this process of modernization, it is hardly surprising that Latin countries intent on rapid self-empowerment embraced eugenics with a sense of urgency.

During the first half of the twentieth century, European and Latin American states considered a wide range of eugenic proposals. Developing eugenics programmes for an entire nation was, however, a complex process. Though national homogenization and centralization was desirable, distinctive national and regional traits had to be respected and preserved. In accordance with the hereditary and statistical study of human biology in this period, nationalists often regarded the nation as an enduring biological unit. A biologization of national belonging occurred widely during the interwar period (Turda 2007: 413–41 and 2010: 6–8). 'Biopolitics', for example, was a biological theory of the nation common in fascist and quasi-fascist states, such as Italy and Vichy France. In Romania, eugenicists like Iuliu Moldovan conceptualized 'biopolitics' in exclusively national terms, connecting it to state interventionism and radical measures to regulate the health of individuals. These biopolitical theories projected the relationship of cells in a complex multi-cellular organism onto the nation. The most basic cellular unit was the individual and all individual cells were bound together in families. Although individual cells may die, their descendants would replace them, preserving their hereditary heritage. This meant that families were, in a sense, eternal. Since the aggregate characteristics of these families provided the state with its biological identity, the survival of the family, as an institution, was essential for the survival of the nation. By this logic, eugenicists envisioned the development of idealized, healthy, and patriotic citizens and nations.

During this period of modernization, biological racism gave the question of national identity even greater urgency. Certainly before the Second World War, sweeping generalizations about the 'rise' or 'decline' of various races thrived as many nationalist writers distorted historical facts to support the racial hierarchies that were popular at the time. For many, uneven rates of economic development and military prowess in different nations could be explained by racial qualities. Few Europeans questioned the assumption that they were endowed with special qualities that had allowed them to create the various wonders of the modern world. It appeared to many writers that biological racism directly justified European imperialism and thus allowed for Europe's self-definition as 'superior' and 'civilized'. Furthermore, many anthropologists and social scientists asserted that there were important differences still within the 'European races' and even proposed a number of categories to label these differences, such as 'Latin', 'Mediterranean', 'Anglo-Saxon', 'Germanic', and 'Slavic'.

The casual use of such terms by almost all educated Europeans, North Americans, and South Americans was ubiquitous. However, it is not always clear what an author meant when using the term 'Latin race', for instance. Few anthropologists of the time believed that a distinct 'Latin race' existed, arguing instead that there was something akin to a Latin ethnicity or culture shared by peoples of different races who had either inhabited regions once a part of the Roman Empire, or whose ancestors were from these regions. In terms of racial categorization, various segments of these populations were classified as Mediterraneans, Celts, Ligurians, Iberians, Italics, or even as Aryans. Indeed, quite a few argued that the attempt to limit Latin cultural identity to a particular race, such as the 'Mediterranean race', was contrary to the assimilationist principles of Latinity inherited from ancient Rome.[1]

Engaging with these issues, this book demonstrates that an allegiance to Latinity – defined culturally and linguistically – existed within the international Latin eugenics movement, even though the individual countries also aimed to create their *own national* eugenics, infused with their own specific cultural values. This is one of the most distinctive characteristics of Latin eugenics: it had a determination to modernize the nation state while preserving its traditional cultural heritage. At the beginning of the twentieth century, Latin identity in countries like France, Italy, Romania, Spain, Portugal, and a host of others in the Americas provided a platform upon which eugenicists, demographers, social hygienists, and child welfare activists built their theories of a distinctive, Latin eugenics, serving the cultural particularities of their own societies. A humanistic ideal also survived in the notion of a Latin identity based on the argument that a better society could be achieved through positive improvements in the population's hereditary health and living conditions, public sanitation, education, and child welfare.

'Pan-Latinism' – a concept that grew in popularity after the 1840s and was especially favoured by the Romantic nationalists of the late nineteenth century – embraced the notion of a Latin community with its roots in the ancient Roman Empire, but with its modern reflection in French and Italian culture and civilization. The Latin community thus defined was transnational: French, Italians, Spanish, Portuguese, French-speaking Swiss, Romanians as well as Mexicans, Argentinians, Cubans, Brazilians, and Chileans were included within the definition. The 'brotherhood of Latin' nations, which blossomed on the barricades of Paris and Milan during the 1848 Revolutions, was further enhanced by France's growing cultural prestige, the unification of Italy, and the liberation of the Romanian principalities from the Ottoman Empire during the 1860s and 1870s. These latter events, for instance, prompted renewed enthusiasm about the future of the modern Italians and Romanians and re-energized their claim to be the direct descendants of the ancient Romans.

In terms of international cooperation, eugenicists' desires for personal aggrandizement, their narrowness of vision, professional jealousy, and national pride all played an important role in delaying the establishment of a unified international eugenic movement (Kühl 2013). The resurrection of an aggressive Germany during the 1930s determined to assert its racial superiority also contributed to this process of fragmentation, ultimately provoking the advocates of Latin eugenics to break away from the increasingly German-dominated International Federation of Eugenic Organizations and create their own, *Latin* eugenic organization (*Federazione Latina fra le Società di Eugenica; Fédération Internationale Latine des Sociétés d'Eugénique*; *Federațiunea Societăților Latine de Eugenie*) in 1935. By that time, Latin eugenicists had come to see their movement as a major defender of their cultural heritage in an increasingly racialized and antagonistic world. It is, therefore, important to emphasize both *cooperation* and *competition* when discussing international eugenics in general and Latin eugenics in particular.

Some of the scientific elements of the eugenics debate also reflected cultural differences. Virtually all eugenicists emphasized the need to 'weed out' those with 'undesirable traits', inhibit their reproduction in some fashion, while simultaneously encouraging the reproduction of those with 'superior traits', and thus guide national improvement. It was also agreed that eugenicists should have some role in guiding human reproduction. The concern for healthy and numerous populations had, however, clear nationalistic and cultural overtones. Reproducing the next generation was a national duty which the common man and (especially) woman were asked to fulfil.

During most of the twentieth century, eugenicists had identified, corrected, isolated, and even eliminated alleged degenerates, criminals, sexual deviants, prostitutes, the mentally deficient or unbalanced, revolutionaries, alcoholics, those with chronic health problems, or those who were reproductively sterile. Yet contrary to eugenicists in Germany, Britain, the USA, and Scandinavia, Latin eugenicists prompted the state to exert control over national reproduction not through sterilization and segregation, but by developing various health and hygiene programmes, together with the establishment of institutions designed to encourage the desired family development. Financial and symbolic incentives to bring up healthy children and families, the establishment of mother and child health clinics, and eugenics-themed educational curricula became common in all Latin countries.

Latin eugenics was also intrinsically political. For example, the Italian dictator Benito Mussolini embraced eugenics and biotypology in the earlier years of his regime, and supporters of Latin eugenics in Italy such as Nicola Pende and Corrado Gini fed the Duce's desire to rapidly modernize and empower Italy, as well as to project its power internationally. With the ability to concentrate and direct resources as he wished, Mussolini adopted a full-fledged Latin eugenic programme in the late 1920s. He sought to quickly advance Italy as the leader of the international Latin eugenics movement, and sent out the country's foremost eugenicists to advance the cause. In so doing, Italy consolidated and expanded its influence in the Latin American world. It offered Latin Americans the cultural and scientific means of national development vis-à-vis the dominant power of the United States in the Western hemisphere.

This eugenic vision, which focused less on heredity and more on environment, became one of most recognizable features of Latin eugenics. The widespread assumption is that most Latin eugenicists preferred neo-Lamarckist theories of inheritance over Mendelian explanations. Neo-Lamarckism claimed that biological organisms, through their interaction with their environment, tended to develop particular inheritable physical traits over the course of their lives. Conversely, the followers of Mendelian genetics believed that the hereditary genetic material was not modified by an organism's life environment or experiences. Rather, genetic changes occurred based upon the variable combination of genes inherited from the parents; species change over time was the result of certain genetic types being more fit to survive and reproduce than were other types (Bowler 2003; Gillette 2007). Significantly, from the neo-Lamarckian perspective, the individual with so-called 'defective genes' could take measures to cure their otherwise inheritable defects. Moreover, better living and working conditions, maternal and child care, improved diet, and physical exercise, were but a few measures deemed to encourage the nation's eugenic improvement. Neo-Lamarckism emphasized the transformative power of education, assimilation, and spiritual renovation; interference in reproduction was not necessary to evolve a better race.

To some extent, these values were also reflected in Catholic tradition in the Latin countries. The relationship between eugenics and religion is of crucial importance when examining Latin eugenics. During the 1920s and 1930s, the practice of voluntary and compulsory eugenic sterilization, as well as the acceptance of birth control methods, spread throughout the Protestant world. The Catholic Church, concerned about these developments, unequivocally condemned all such interventions in the natural process of the creation of life. The promulgation of the papal encyclical *Casti connubii* ('On Christian Marriage') on 31 December 1930 further clarified the Catholic Church's position on abortion and eugenics.

Other historical developments contributed to the crystallization of a Latin eugenic movement. The transformation of Germany into a militant, aggressive 'racial' state after 1933 provided an incentive for many eugenicists in France, Italy, Spain, Portugal, Romania, and Latin America, to define their work and ideas in opposition to German racism and militarism. In the process, these eugenicists sought to clarify the still rather abstract concept of Latin eugenics and use it to attack an expansionist German racial hygiene. However, this common front was difficult to maintain, and Latin eugenicists were soon forced to accept the political realities of the time. By 1938, Mussolini decided that Italians were 'Aryans' and thus related to the Germans rather than the French, and hence had to suitably reorient their political alliances (De Donno 2006; Wolff 2013; Livingstone 2014). By 1940, France and Belgium were both under German occupation. During the Second World War, Latin eugenics lingered on in France, Italy, Spain, and Romania (by then all allies, to varying degrees, of Nazi Germany), but bereft of its rejection of German racial hygiene. Yet eugenics grew in importance during the war period. Many eugenicists in Latin countries became government officials and once placed in strategic institutional positions were entrusted with the nation's racial protection.

In some European countries, such as France and Romania, the importance of Latin eugenics was reinforced by a series of eugenic measures introduced in the early 1940s. Latin eugenicists adopted various theories of social and biological improvement tailored to their own national context, and in some cases their eugenic concerns took a distinctively racial turn: attention was now fixed on the alleged source of national degeneration posed by the Jews, the Roma or other ethnic minorities. New laws promoted the social, economic, and political power of the dominant racial group. The ethnic state became the central trope of eugenic imagination during this period. Reflecting this racial turn, as in French and Romanian cases, eugenicists promoted legal decrees introducing new ideas of public health and hygiene, medical protection for mothers and children, and medical screening and mandatory premarital examinations. These laws further testify to the changes in eugenic thinking and practice which occurred during severe upheavals of the Second World War. Maintaining the nation's racial survival had become of prime political importance, irrespective of which eugenic strategy was adopted. The programme of the ethnic state, imbued with an all-encompassing eugenic and biopolitical vision of national survival, and first developed in Germany after 1933, was eventually adopted by the Latin countries as well.

Latin eugenics was at the same time a scientific, political, and cultural programme based on identifying hereditary traits, and assigning them *individual* value in order to increase the social and biological quality of both *current* and *future* generations. As the twentieth century progressed, new discoveries in human heredity and genetics made it increasingly clear that Latin eugenic goals could not be reached without a more rigorous scientific underpinning or without the state's more intensive interference in family and individual life. Eugenicists in Latin countries – as their counterparts elsewhere – redefined the nation's body politic and the role of the state in protecting it (Turda 2009: 77–104). Thereafter, they introduced substantial innovations in individual and family health, as well as in social control and population management. Yet many of these eugenicists have been largely ignored in the scholarship on international eugenics. In this and other respects, this book reinstates their roles in the history of eugenics, and in shaping modern medical and social welfare in their own countries.

Finally, more research is needed to unveil the political ramifications of post-war eugenics in the Latin and Catholic countries. As such, this book engages with more recent scholarship on the post-war history of eugenics, which stresses conceptual continuities with the pre-war period rather than abrupt ruptures in 1945 (Comfort 2012; Cassata 2013: 217–28; Hansen and King 2013; Spektorowski and Ireni-Saban 2013). One revealing indicator of the extent to which the history of Latin eugenics is instructive to current debates on genetic demography, population policies, medical genetics, and so on, is how its fate has differed from that of German racial hygiene, particularly after 1945. Although Latin eugenics was fluid and sometimes even contradictory, it was also held together by strong elements of continuity and coherence, owing to a commonality of purpose: the ideal of a healthy nation (Turda 2013: 109–26). During the period discussed in this book, Latin eugenics promoted not only the improvement of living conditions and health standards, but also a longing for healthy and numerous families. This ideal remains recognizable in post-war population and family planning policies in democratic countries like France and Italy, as well as in communist Romania and in the authoritarian states of Latin America. This was a eugenic project of national engineering that transcended ideological differences established during the Cold War, and whose troubled epistemology and long-term consequences continue to be felt to this day.

1

Precursors

On 10 May 1871, French representatives signed the Treaty of Frankfurt, which dramatically highlighted the military decline of France from its predominant position on the Continent several generations before. That title was figuratively ceded to Germany, along with the French provinces of Alsace and most of Lorraine. The French were stunned by their defeat and humiliation, often describing this event in apocalyptic terms. Because of the fascination with race and evolutionary theories at the time, some writers attributed the decline of France to the decline of the 'Latin race' in general. For the rest of the century, numerous books were published suggesting that France was suffering from terminal degeneracy, and many of France's leading intellectual and scientific authorities fretted over France's racial decline. Gustave Flaubert set the tone with his lamentation to Madame Règnier on 11 March 1871 that: '*Nous assistons à la fin du monde latin*' (Flaubert 1975: 624). Revelling in the Decadent literary movement, novelists entertained their readers by highlighting the sexual depravity, debauched luxury, and moral corruption of Imperial Rome and the Byzantine Empire, with obvious analogies to France in the decades after the Franco-Prussian War. Émile Zola attributed degeneration to inheritable depravity. In his novel *Nana*, written several years after the Franco-Prussian War, degeneration was present in the guise of prostitution and uncontrollable sexual desire. Nana Coupeau, the novel's protagonist, was the product of neo-Lamarckian degeneracy, contaminating the Parisian elite with her diseased body. Zola described her as 'the offspring of four or five generations of alcoholics, her blood tainted by a long heredity of deep poverty and drink' (Zola 1992: 190).

France's humiliation echoed widely in the Latin world. The Spanish writer José de Castro y Serrano, for instance, feared that, with the fall of France, 'the greatest calamities' would befall the Latin peoples because the 'barbarians of Teutonic civilization' would overwhelm them. Serrano dreaded that German advanced military power had given rise to a new 'racial law', the 'brute force' that seemed destined to 'lead to the subjugation of the Latin peoples, their becoming *Genízaros* (Native American slaves)' (De Castro y Serrano 1871: 3–6).[1] Meanwhile, the Italian anthropologist Giuseppe Sergi did not hesitate to call the French defeat of 1870: 'The most grandiose disaster, that rips away the veil under which the decadence of the Latin nations hid' (Sergi 1900: 88).

Uniting against such pessimistic predictions, other Latin authors stressed the cultural and religious qualities of their *raza latina*, insisting not only on past achievements but equally on future greatness. The journal *La Raza Latina*, published throughout the 1870, illustrates both the malleability of transnational solidarity and some of the enduring themes of the Latin movement. Its editors – the prominent French and Spanish politicians Léon Gambetta and Cánovas de Castillo, respectively – while accepting the growing power of the German and Anglo-Saxon races, produced a regenerative narrative, in the hope to draw together the members of a reborn Latin race (Goode 2009: 30–1). Physicians, anthropologists, and eugenicists, directly concerned with the alleged biological inferiority of the Latin and Mediterranean race, gradually endorsed this cultural and political vision of Latinism.

Causes of racial decline

This process of racial cartography, which favoured the Aryan–Nordic elements within European races, was based on the assumption that such categorization was scientific, and therefore it was seen as legitimate to represent European nations in terms of their racial strength (D'Agostino 2002: 319–43). There was no doubt – the American anthroposociologist Carlos C. Closson argued – that the Aryan *Homo Europaeus* was superior to *Homo Mediterranaeus* (Closson 1897: 314–27). The Anglo-German writer Houston Stewart Chamberlain laid out what became the 'orthodox' deprecation of contemporary Italians and Spaniards in his 1899 book, *Die Grundlagen des neunzehnten Jahrhunderts* (*The Foundations of the Nineteenth Century*). He explained that the Italians of his day were not the descendants of 'the sturdy Romans' of ancient times, nor the 'artistic geniuses of the Renaissance', but were instead the spawn of innumerable African and Asian slaves brought to the peninsula during late Roman Empire. The resulting 'Völkerchaos' (Chaos of Peoples) had ultimately overwhelmed the 'true' Romans who, of course, were Aryan. It was only with the appearance of the Teutonic people that the 'Völkerchaos', which caused the downfall of the Roman Empire and threatened to destroy Western civilization, was brought to an end. The Teutons preserved the achievements of Roman civilization, transforming Northern Italians through a new racial synthesis, which served as foundation for both the Renaissance and modern European culture (Chamberlain 1912).

Chamberlain's ideas about the cultural supremacy of the Germanic (Teutonic) race were further developed by Ludwig Woltmann, one of the co-founders with Alfred Ploetz of the *Archiv für Rassen und Gesellschaftsbiologie* (*Journal of Racial and Social Biology*). In a number of disputed publications, including the 1905 *Die Germanen in Italien* (*Germans in Italy*), *Die Germanen und die Renaissance von Italien* (*Germans and the Italian Renaissance*), and the 1907 *Die Germanen in Frankreich* (*Germans in France*), Woltmann argued that Germanic creativity was responsible for Italy's and France's cultural achievements. This influx of Germanic blood rescued these countries' latent Aryanism and, with it, modern Italians and French from racial degeneration.

This steady stream of pejorative German propaganda profoundly influenced many Latin intellectuals, who seemed to accept its basic premises. The Italian anthropologist Giuseppe Sergi, for instance, remarked: 'The Latin nations run, thus, toward the abyss of decadence; and following the old ways, and continuing to live and direct their residual energies with the same methods that had served them in times past, doubtlessly precipitate towards the bottom of this abyss' (Sergi 1900: 157). His compatriot, the historian Guglielmo Ferrero, in an equally dramatic tone, quipped that: 'The world will perhaps be flooded in the next century by a Germanic flood, in the midst of which will stick out the isolated Latins, laughing in their pleasures and joys, but miniscule in the midst of a grey sea' (Ferrero 1897: 216). In contrast, the French writer Léon Balzagette claimed that the addiction to 'verbal excess', a sign of the 'the impotence to act', was the most 'fatal of the maladies which disturb the Latin soul' (Balzagette 1903: 96).

Some observers were, however, more pragmatic in their judgement. The Portuguese anthropologist António Augusto Mendes Correia, for instance, denied the scientific validity of most arguments put forward by the German school of anthroposociology, describing some of them as mere 'pan-Germanic fiction' (Mendes Correia 1919). Other Latin authors looked inward, and blamed the antiquated and inadequate educational systems in their own countries, for the lack of technological and logistical sophistication that led to defeat at the hands of a better-educated army. The popular French playwright Etienne Rey, for instance, remembered that, after the obvious decline of French geopolitical power, 'even we ourselves were dupes and victims of [the] opinion [...] that to regenerate ourselves, we had to learn from the German schoolmaster' (Rey 1912: 197–8).

As the nineteenth century drew to a close, learning from the Germans meant bemoaning France's relatively slow rate of industrialization. The loss of Alsace-Lorraine to Germany was particularly severe, given that the region was one of France's most industrialized areas. The economic calculus had the worst possible result: what France had lost, Germany had gained. Furthermore, economic growth was possible only with population expansion. All signs indicated that the French were in the midst of a disastrous population decline, due in part to the widespread use of birth control that often accompanied urbanization. Between 1850 and 1910, the population of France increased from 35.7 to 39.1 million, while during the same period the population of Germany and Great Britain grew from 33.4 million to 58.4 million and from 17.9 million to 36.0 million (Offen 1984: 651). Causes for this demographic decline ranged from infant mortality to voluntary sterility and movements of population from the countryside to the city, prompting pessimistic predictions about France's national future (Piot 1900 and 1902), and eventually determining the senate to appoint a Commission on Depopulation (*Commission de la dépopulation*) in 1901.

European observers reacted to the French demographic decline according to their own national interests. In 1875, the Prussian statistician Arthur Freiherr von Fircks concluded that France's low birth rate and poor-quality conscripts were signs that the French people 'were becoming old and decrepit' (quoted in Spengler 1938: 121). Similarly, the French geographer Émile Levasseur felt that 'it is truly humiliating to think that a nation of thirty-eight million souls, [...] one of the wealthiest of the globe [...] one of the most capable of enlightening the world [...] is a nation which, according to the statistics, is destined to disappear' (quoted in Garner 1914: 259).

Although racial degeneration in late-nineteenth-century Italy was not attributed to population decline (this concern would come a generation later), the future of Italy seemed no less ominous. The new state, created in 1861, was almost overwhelmed by the problems it faced, and the inability of the central government to solve them. Cultural and economic differences between the provinces, political corruption, and conflicts with the Catholic Church were endemic. Observers at the time often turned to racial explanations for Italy's disappointing development. Many concluded that the cause of Italy's misery lurked in the South. The Southerners had sunk so deeply into racial putrescence that they had almost reverted to barbarism. Furthermore, the heterogeneity of the population promoted regional and racial conflict.

Anthropologists and positivist-minded social scientists produced a plethora of statistics to chart Italy's Southern degeneracy (Gibson 1998: 99–116). Foremost among them, perhaps, was Alfredo Niceforo, a social scientist who argued that, by virtually every measure, the South proved inferior to the North. This held true in terms of mortality, crime, literacy, capital accumulation, industrialization, and agricultural development. Italians in the South seemed inclined to anarchy, rebelled against legitimate government, and allowed organized crime to flourish. The stifling hand of tradition and lack of enterprise seemed the complete antithesis of what modern Italians so desperately needed. Niceforo blamed contemporary lethargy on the South of Italy's racial decline (Niceforo 1898). Similar debates occurred in Portugal at the end of the nineteenth century, positing the superiority of the North, and its connection to the foundation of Portugal and its empire. As in Italy, it was assumed that the North of Portugal was dominated by the influence of Aryan and Germanic peoples, contrasted with the South's Mediterranean and Berber populations (Sobral 2008: 205–24).

It was not only modern Latin countries in Europe that, at the end of the nineteenth century, suffered from this accumulated historical process of decline and degeneration. For example, the Argentinian educator and president of the republic from 1868 to 1874, Domingo Faustino Sarmiento, attributed Spain's decline to Catholic repression. The Spanish brain, he believed, had become atrophied by 500 years of Inquisitorial domination. He feared that Spanish decadence had contaminated Latin Americans as well (quoted in Helg 1990: 40). Sarmiento's attempt to explain Argentina's struggle between 'civilization' and 'barbarity' (Rodriguez 2006: 14) in terms of Spain's racial history was not unusual. In 1912, an article written by Eduardo Acevedo Díaz for the Argentinian Industrial Union complained that southern Italians were 'an inferior race, a mix of races that were even more inferior' and looked forward to restrictive immigration laws to lessen this flood of lowly people (quoted in Barbero and Felder 1988: 157).

Even liberals among the white elite of Latin America believed that the Continent's degeneration was due, in part, to the persistence of native and African populations, and their intermixing with the Spanish settlers. A chorus of Argentinian notables accepted this assessment, bemoaning 'the negative effects of the country's modernization on its spiritual life', while emphasizing the 'cultural incompatibilities of certain "races" and the Argentine Indo-Hispanic heritage' (Zimmermann 1992: 25). In one typical example, in his 1914 book *La anarquía argentina y el caudillismo* (*Argentinian Anarchy and Caudillismo*) the physician Lucas Ayarragaray described the Indian, the caudillo, and the gaucho as indolent and 'mainly alcoholics for the most part, with a rudimentary sense of modesty and morality, living in bestial promiscuity, many of them epileptic, cretins and idiots, and most mentally weak, in a word, antisocial and degenerates due to ethnic causes' (quoted in Di Liscia 2005: 46).

Brazilian and Cuban intellectuals, too, expressed anxieties over their countries' composite racial identity (Petruccelli 1993: 251–62). Both countries had a large population of African-origin and mixed-race peoples, and slavery ended particularly late, at the end of the nineteenth century. In line with their Brazilian counterparts, Cuban leaders worried that the increase in the number of black people in the country would cause its 'slow decadence, its certain intellectual ruin' (De la Fuente 2001: 49). This perceived Latin American degeneration had much to do with economic transformation in the late nineteenth and early twentieth centuries. Once industrialization became a driving force in the Western European and North American economies, Latin America felt the pull of European and North American demand for raw materials and agricultural products.

By the end of the nineteenth century, a neo-colonial economic structure was solidly in place in Latin America. Frequently, European and US businesses controlled Latin American industrialization and absorbed most of the profits (Burns 1983). Moreover, prosperous Latin American agriculture and natural wealth attracted millions of European immigrants, whose numbers stimulated production for the growing internal market. There is no doubt, for instance, that the relatively highly skilled newcomers contributed to making Argentina the most advanced industrialized state in Latin America. Not surprisingly then, an Italian writer who had visited Argentina in 1896, Angelo de Gubernatis, rhapsodized on the wonders of migration to that country. He believed that contact with the 'new virgin earth' of Argentina would transform the 'tired, infirm, and exhausted' old Latin race into a 'new, simpler, freer, purer life' from which would pour forth a new progeny. 'As in ancient Rome', de Gubernatis claimed, 'the Latin peoples of Argentina' would merge to create 'a new, higher harmony [...] a new type of beauty' (De Gubernatis 1898: 34).

Population growth was essential not only for economic development but was equally important for racial survival. In many Latin American countries, such as Argentina and Cuba, academics and politicians were concerned that their nations were under-populated. As a result, many Latin American countries encouraged immigration, particularly from Europe, both to 'whiten' their populations and to fuel economic growth. Better public sanitation aided in bolstering the economy through facilitating trade with more developed countries, and enticing desirable European immigrants. Overall, from 1820 to 1930, sixty-two million people moved to the Americas from Europe and Asia. In Latin America,

Argentina proved to be the most popular destination for these immigrants (about six million), with Brazil second (four and a half million) (Thorp 1998: 49).

At the beginning of the twentieth century, Latin American cities grew because of immigration, coupled with internal movement of population from rural areas. Industrialization, transportation, construction, port development, and commodity processing drew rural people to settle in urban centres. Latin America's rural poor, many of whom were indigenous peoples, were also thrown off the land as commodity production became more prevalent. This further encouraged urban migration and racial heterogeneity in the cities. Although this urban growth was a sign of accelerating modernization, urbanization also brought enormous problems: for instance, the rapid influx of the poor created large slum areas (Burns 1983: 134–5) Even in the relatively wealthy Buenos Aires, one third of the working class population lived in severely overcrowded but exorbitantly expensive slum tenements (Bethell 1998: 156 and Table 1; 193) Various social ills, such as alcoholism, prostitution, crime, juvenile delinquency, and racial antagonisms rapidly worsened. Public health systems found it very difficult to keep pace with urban growth. Mortality rates, due to poverty, war, and endemic disease, were also on the rise. All these factors seemed certain to inhibit national development in Latin America (Stern 2003: 187–210; Birn 2005: 72–100).

The persistently high rate of infant mortality was especially worrisome. To Argentinian demographers, a decline in birth rates was almost as harmful as high infant-mortality rates: in the country's most developed areas, such as Buenos Aires, birth rates were falling. Argentina began the twentieth century with a fertility rate of 44.3 annual births per thousand women; this figure steady dropped over the years. By the time of the Second World War, the analogous figure was 26.1. Here, as in France, demographers worried that this decline was an indication of the physical and moral degeneration of the race (Sánchez-Albornoz 1974: 171, Table 5.12).

Biological explanations, and corresponding solutions, were eagerly sought after. In Italy, the criminal anthropologist Cesare Lombroso laid the groundwork for an understanding of biological degeneration. Lombroso was a late-nineteenth-century Italian professor of psychiatry who developed a theory of criminal anthropology that built upon theories of evolution, Social Darwinism, and forensic psychiatry. Lombroso alleged that crime was often the product of a biologically degenerate mind. Criminal degenerates generally displayed specific physical and behavioural features, conveniently catalogued by Lombroso, which allowed criminologists to identify and scientifically analyse these alleged 'born criminals'. Commonly, visible physical deformities were outward signs of deviant, innately primitive minds (Lombroso 2004 and 2006). Degeneracy was inherited through natural 'aberrations' as well as neo-Lamarckian 'use and disuse of organs' (Gibson 2002: 98–9).

Anthropologists Giuseppe Sergi in Italy and António Mendes Correia in Portugal were among the first Latin eugenicists to connect eugenics with Lombrosian degeneration theory and neo-Lamarckism (Cassata 2011: 16–8; Tedesco 2011: 51–65; Henriques 2012: 44–6). Both were concerned with identifying, isolating, and (if possible) curing degenerates. In a book published in 1913, entitled *Os Criminosos Portugueses* (*Portuguese Criminals*), Mendes Correia presented a list of degenerates that corresponded closely with that of the Lombrosian school: criminals, prostitutes, the insane, vagabonds, beggars, parasites, 'servile' people, and suicides. He may not have agreed with all of Lombroso's claims, but Mendes Correia fully accepted the notion of inherited criminal predisposition (Mendes Correia 1914).

Like Mendes Correia, Sergi had a complex relationship with hereditary theories (Verdicchio 1997: 29; Lucamante 2008: 219–20). Sergi was not averse to Mendelism (Sergi 1912: 22), but attributed 'the role of prime motor in the modification of the germ plasm to environmental conditions' (Cassata 2011: 17) As a neo-Lamarckist, he was convinced that poor environmental conditions induced social and biological degeneration. To impede the birth of degenerates, he recommended the improvement of living conditions as well as increasing employment opportunities and the provision of healthy environments, including good nutrition, rest, and recreation. Sergi also emphasized the benefits of education for those deemed worthy, which would encourage civic mindedness, good work habits and, ultimately, a healthy, eugenic lifestyle. Conversely, Sergi worried that such benefits would not simply prevent degenerate children from being born to 'at-risk' parents, but would also allow those who were already degenerate to survive and reproduce further. Italy's widespread social and biological degeneration required efficient eugenic treatment, Sergi maintained. Degenerate criminals or mental defectives could be medically and eugenically treated; more severe defectives, including children with tuberculosis, rickets, and scrofula, had to be prevented from marrying, thus inhibiting their reproduction. Sergi's attempts to prevent degeneration were not, however, based on sterilization, seen as a dangerous and overly radical medical measure, given the lack of knowledge that still surrounded human heredity. Instead, he advocated the segregation of degenerates from the national body (Mantovani 2004: 29–31; Tedesco 2011: 51–65; Quine 2012: 98–103).

Lombroso's theories reverberated throughout the Latin world (for Romania see Ionescu-Muscel 1929). In Argentina, for example, Lombrosian criminology thoroughly dominated the activities of the Society of Criminal Anthropology of Buenos Aires (*Sociedad de Antropología Jurídica de Buenos Aires*) – founded in 1888 by José María Ramos Mejía – and of the most important Argentinian criminologists (Zimmermann 1992: 33–4; Rodriguez 2006: 39). Italian influence in Argentine criminal anthropology was furthered strengthened in 1907 with the establishment of the Criminology Institute at the National Penitentiary. José Ingenieros, the country's foremost psychiatrist and the Institute's founding director, hoped that the new institution would 'solve our own problems of prevention and repression' (quoted in Rodriguez 2006: 165). The inauguration of the Institute was an expression of the strong Latin connection existing between Argentina and Europe, with experts from France, Spain, and Italy (including such luminaries as Guglielmo Ferrero and Enrico Ferri) attending the event (Rodriguez 2006: 166).

Dr. Juan Santos Fernández, president of the Cuban Academy of Sciences, similarly engaged with Italian anthropology when complaining that Cuba's racial mixing caused genetic mutation and the appearance of atavistic, animalistic features in the population: the 'regressive savage type, the cruel, thieving, hypocritical atavist' (quoted in Guerra 2002: 153) The belief that normal human types and 'deviants' could be identified through physical and mental tests was especially popular at the time (Harris 1989; Mucchielli 2000: 57–89).

Eugenicists, together with social and health reformers, often argued that criminality was a source of biological degeneration, and that isolating the criminals from the population would permanently improve society. This was one strategy envisioned to reaffirm the critical importance of eugenic regeneration. Equally important, it was recognized that a scientific programme of national improvement could only succeed when the economic, social, and political causes of non-degenerative crime had been eliminated through such means as public works, better sanitation, vaccination, hygiene education, and a restoration of family morality. Rescuing the race from its alleged social and biological decline and restoring its deserved historical destiny were deemed essential to the future of any nation. In the case of the Latin countries, the widespread importance this vision of racial renewal was accorded reinforced the tendency to subvert the opposing myth of the ascending Anglo-Saxon and Germanic races by adopting its main ontological foundations. In the words of one Argentinian commentator, it was this 'fusion of the Latin genius with the Anglo-Saxon energy' that signified the nation's racial prestige and its civilizing mission (quoted in Zimmermann 1992: 30). By popularizing the idea of synthesis between two 'superior races', the Latin and the Anglo-Saxon, these authors provided a new basis for the study of their own national community.

Proposals for revival

With the onset of the twentieth century, the eugenic tropes of rejuvenation and improvement were often used to provide solutions to racial and social degeneration. Moreover, emphasis on the power of the biological and medical sciences, including eugenics, to revitalize the nation's body became more common (Murard and Zylberman 1996). 'Artificial selection', Sergi claimed, was 'regeneration' (Sergi 1889: 228). The prominent Romanian neurologist Gheorghe Marinescu further asserted that racial regeneration was a central task of modern medicine. Science, and medicine in particular, was the key to the nation's social and biological renewal (Marinescu 1906: 3–34). A similar commitment to the regenerative power of medicine characterized the work of the Spanish doctor and eugenicist Enrique Diego Madrazo. In *¿El pueblo español ha muerto? (Have the Spanish People Died?)*, published in 1903, Madrazo envisioned a programme of national renewal based on eugenics and racial mixing, and refuted allegations of Spain's racial decline (Cleminson 2006: 229–33). A year later, in his book *Cultivo de la Especie Humana. Herencia y Educación* (*Cultivation of the Human Species: Heredity and Education*), Madrazo proposed the creation of a Centre for the Promotion of the Race (*Centro para la Promoción de la Raza*), an institution combining research on heredity with exploring the possibility of introducing various 'positive' and 'negative' eugenic measures (Cleminson 2000: 82–3).

Napoleone Colajanni, a widely respected Italian sociologist and statistician, held similar views (Frétigné 2002). Like Marinescu and Madrazo, Colajanni rejected Lombrosian criminal anthropology and racial determinism, albeit from a sociological rather than medical, perspective (Colajanni 1890; D'Agostino 2002: 328). He denied that laws of natural sciences were applicable to social and racial development. Degeneration, he further argued, was exclusively the result of social, rather than biological, factors. Colajanni also believed that a certain degree of racial mixing was beneficial to the social progress of the nation. Biologists had shown that 'racially pure' marriages were essentially consanguineous, and actually would lead to racial degeneration, constitutional weakness, infertility, and serious illnesses. In fact, crossbreeding augmented racial development,

Colajanni asserted. Even racial crosses between whites and blacks or between blacks and Asians were 'not always bad' (Colajanni 1906: 144–5).

Social Darwinists treated the 'inferior' and 'defective' members of society with a similar moral crudeness, Colajanni believed. As examples, he cited Herbert Spencer's lamentation that charitable societies aided the weak, allowing them to reproduce (Spencer 1897) and John Haycraft, who suggested that medicine should *not* focus its attention on epidemiology, but rather should allow microbes to continue their beneficial work of culling the weak from human populations (Haycraft 1895: 51–7). American eugenicists, moreover, wanted to sterilize delinquents, epileptics, alcoholics, the insane, people suffering from tuberculosis, and so on. Other authors came up with equally extreme means to encourage those with 'superior racial qualities' to reproduce, Colajanni argued. For instance, the French racial eugenicist Georges Vacher de Lapouge desired 'a small number of males of absolute perfection to impregnate all women worthy of perpetuating the race' (Lapouge 1896: 472). In this, as in other instances, Colajanni drew selectively on racial theories and positivist anthropology to ridicule racial theories of de Lapouge and others (Taguieff 1991: 23–45).

Biological engineering based on race and hereditarianism had to be avoided, Colajanni asserted. Inheritance in the Darwinian sense was not 'a fatal, unchangeable force that condemns individuals, peoples, humanity, to never be able to leave heavily worn paths and to forever repeat the same moral views'. Rather, neo-Lamarckian inheritance and Darwinian evolution together explained 'the progressive moral and intellectual evolution which we have witnessed across the centuries' (Colajanni 1906: 141–6). In light of this, races were malleable. Though each race had its own 'psychological characteristics' change and adaptability were much more flexible than August Weismann and his followers alleged. This was just one more reason why Italians were not doomed to poverty, but could improve themselves. Therefore, by appealing to neo-Lamarckism, Colajanni was able to save his strongly held positivist and socialist convictions. For him, eugenic manipulation of human reproduction was not necessary as long as environmental improvements were effective.

The historian of ancient Rome, Guglielmo Ferrero, also argued that the modern Latins were more intelligent and artistically brilliant than the Germanic peoples (which included the English as well as the Germans). The Germanic character suffered from 'a ponderous intellect, cerebral poverty, and a lack of spiritual flexibility'. Yet, these Germanic 'brutes with a hard head' had become the creators of the most marvellous civilization that had yet existed. Ferrero set out to explain this paradox: he concluded that the modern Latins were oversexed.

Unfortunately, the Latin race had a natural predisposition to unusually early and vigorous sexual activity; sex for them was an obsession that their culture enhanced through its glorification of sensuality. While the Latin races dissipated their energies in sexual excess, the Germanic race showed much less interest in sex. Their formative years were devoted to the love of moral and intellectual beauty, rather than a lustful obsession with physical attractiveness (Ferrero 1897).

In an 1897 article entitled 'Le cagioni della effeminatezza latina' ('The Causes of Latin Effeminacy') another Italian anthropologist, Angelo Mosso, conceded that the Latin races were decadent in relation to the Germanic peoples. However, Mosso sounded a note of hope that certain changes in Italian culture would lead to a rebirth of the modern Latins, a theme that would become more common in the decades ahead. Mosso further argued that Ferrero was wrong to claim that the Latins were hereditarily condemned to racial inferiority and Germanic domination. What was needed was a better educational system. Unlike the Italian system, British and American education sought to train students to think independently, to work hard, and to study practical subjects; the Latin countries, however, trained students to indulge in aesthetic sensuality and pursue dry intellectualism, then turn them out to seek easy government jobs.

According to Mosso, history and culture also had a deleterious effect on the Italian (and especially Southern Italian) people. After the Germanic invasions, Italian culture had turned to religion and effete 'intellectualism'. Catholicism had encouraged contemplation rather than action. The most intelligent men were not allowed to reproduce, but were consecrated as clergy and forced into sexual abstinence. Since according to neo-Lamarckist dogma 'habits, if they persist for many generations, tend to become hereditary', this growing fear of action, and intellectual abstractionism, had caused the Latin race's mental degeneration (Mosso 1897: 249–65).

Theories of degeneration were also widely debated in Spain and Portugal at the end of the nineteenth century, influencing the emergence of both the hygienist movement and a wide range of scientific disciplines, including sociology and criminology (Peláez 1985a: 95–122 and 1985b: 622–9; Marín and Huertas 2001: 171–87; Richards 2004: 827–8; De Matos 2013). Moreover, as was rather common in the Latin world of the time, a certain fatalism was apparent in some Spanish writers, who could do little more than complain that they were not Germans or Anglo-Saxons – admirable peoples who were ambitious, intelligent, practical, and 'tenacious'. For one such writer, Emilia Pardo Bazán, this was 'a race that rules the world'. The Spaniards, by contrast, were a fusion of 'African personality and Latin Epicureanism', which to some extent explained their hedonistic and idle nature (quoted in Dendle 1970: 21). One of the leading proponents of the Regenerationist (*Regeneracionismo*) movement, Lucas Mallada, forlornly suggested that Spanish degeneration could only be overcome with a fresh injection of Northern European blood into the Spanish population (Ayala-Carcedo and Driever 1998).

The philosopher José Ortega y Gasset concurred. In his youth, Ortega read Ernest Renan's *Intellectual and Moral Reform* and accepted its principle thesis: that the Latin nations must incorporate German scientific thinking into their culture in order to rise from their historical passivity. Ortega also agreed that Spanish Catholicism was an obstacle to modernization. The Mediterranean peoples were irredeemably superficial. In his 1914 book *Meditaciones del Quijote* (*Meditations on Quixote*), Ortega claimed that his soul could hear 'intimate voices' from the German forests of the north. 'Do not try to turn my being against itself, in civil war', he commanded, 'do not encourage the Iberian with me, with his rough, wild passions, to struggle against the blond, meditative, sentimental German who breathes in the twilight zone of my soul. I aspire to find peace between my inner selves and force them into concord' (Ortega y Gasset 1975: 154–8).

Latin American intellectuals echoed this vision of a Latinity resurrected through its fusion with Aryanism. 'Cuban Aryanism' is a particularly illuminating example. By 1862 around 43 per cent of Cuba's population was black; most of the rest of the population was of predominantly Castilian and Catalonian ancestry. To retain European dominance in Cuba, the Secretary of Public Works in Havana, General Rafael Montalvo, demanded, in 1888, that measures be taken to ensure that 'the descendants of Aryans keep forever [their] material and political superiority' (quoted in De la Fuente 2001: 44). Much of the demographic and eugenic agitation from Cuba's white elite, at least until World War II, was focused on increasing the percentage of whites in the Cuban population. This elite remained in power after the Spanish–American War, and worked with the US government in setting up a new government for Cuba. This included the creation, in 1899, of a modern Department of Sanitation, charged with applying the latest (North American) knowledge to improving public hygiene in Cuba (González and Peláez 1999; Speckman Guerra 2005). Cuba's first immigration law, passed in 1902, followed the USA in incorporating clauses prohibiting Chinese immigration and dissuading non-whites from entering the country (Helg 1990: 56–7).

Cuba's white elite thus envisioned a racial nation enriched by encouraging Aryan immigrants from Europe to settle in the country. The 1906 law on immigration and colonization was meant to accomplish this (De La Fuente 2001: 46). Juan Santos Fernández, President of the Cuban Academy of Sciences, argued that the new law was designed to end Cuba's social and cultural stagnation by devoting attention to the physical characteristics of its population, much as had ancient Greece, modern Germany and the USA. Supporters of the law in the Cuban Congress explained that Cuba had to avoid the mistake of some of the other Latin nations in not managing the development of their racial stock, and thus had remained stationary, backward, poor, and even descended into near barbarity (Speckman Guerra 2002). Masses of Germans, English, and Scandinavians did not, however, move to Cuba, but nearly one million Spaniards did, thus significantly increasing the 'white', albeit Latin, character of the country.

This cursory discussion of ideas of degeneration and racial decline across Europe and Latin America reveals the ambivalent feelings Latin intellectuals expressed towards their common racial inheritance. Many of them valued, however, the unique qualities of their own national culture and its proud heritage, while promoting eugenic ideas of national revival. It was this constant tension between traditionalism and modernism that gave Latin eugenics its impetus. However, cultural traditions and linguistic ties alone could not create a Latin eugenics. Such an edifice had to rest also on a scientific theory, and it was neo-Lamarckism that served this function.

Neo-Lamarckism

In his 1809 *Philosophie zoologique* (*Zoological Philosophy*), the French biologist Jean-Baptiste Lamarck provided a teleological interpretation of hereditary change and its relationship to the environment (Lamarck 1914). Half a century later, the English biologist Charles Darwin published *The Origin of Species*, in which he attempted a monumental explanation of the basic concepts of evolution. These included the 'struggle for existence', 'natural selection' (survival of the fittest), 'sexual selection' (selection of reproductive partners), microevolution (the slow hereditary change of a species), and macroevolution (the formation of new species as a culmination of microevolution). However, Darwin did not offer a satisfying explanation for how these hereditary changes occurred. Thus, the Lamarckian theory of adaptive and environmentally driven hereditary change remained viable. The synthesis of Darwinism and Lamarckism became known as 'neo-Lamarckism' (Ward 1891; Le Dantec 1899) – a cluster of ideas centred on two premises: the inheritance of acquired characteristics and progressive adaptation (Mayr 1982: 626–7).

The viability of neo-Lamarckian theory was severely put to the test as studies of inheritance progressed. With the publication of *Das Keimplasma: eine Theorie der Vererbung* (*The Germ-Plasm: A Theory of Heredity*) in 1892, the German biologist August Weismann persuaded many of his colleagues that the mechanism of inheritance was located exclusively in the germ cells (sex cells or gametes), and the other cells of the body could in no way influence the hereditary matter of the sex cells (Weismann 1893). According to this argument, the organism's interaction with its environment could not directly alter its offspring. Weismann's germ-plasm theory was further validated by the turn-of-the-twentieth-century discovery of the laws of Mendelian inheritance (Punnett 1911).

Despite this challenge, neo-Lamarckism enjoyed a particularly long life, especially in France (Conry 1974; Persell 1999). Many of the leading French eugenicists endorsed it: Adolphe Pinard, Frédéric Houssay, Lucien March, Eugène Apert, Georges Papillaut, Edmond Perrier, among others. As late as 1938, the French eugenicist Henri Briand reported that French scholarly journals and societies only rarely discussed Mendelism (Briand 1938b: 307–14). According to William H. Schneider, neo-Lamarckism 'permitted French eugenicists to argue that improving the quality of the French population would not only permit the birth and survival of more offspring, but that the superior qualities would be passed along to subsequent generations' (Schneider 1982: 271).

Neo-Lamarckism was at home in other Latin countries as well, partly due to the international influence of French medicine (for Portugal see Almaça 2000: 85–98; for Uruguay, for example, see Birn 2008: 311–54; for Argentina see Novoa and Levine 2010). The proliferation of educated professionals in the late nineteenth century, and their employment in government positions, prompted many physicians to equate their nation's racial improvement with modern sanitation, hygiene education, and various other public health programmes. Naturally, physicians formed the majority of eugenicists in many countries; but many other professionals were also intimately involved: anthropologists, sociologists, and statisticians.

Significantly, as the Portuguese physician, anthropologist, and eugenicist, António Mendes Correia, noted in his 1915 *Antropologia* (*Anthropology*), neo-Lamarckism provided 'anthropology with a criterion of the first order for determining some of its most important issues, especially those relating to the origin of Man, heredity, upbringing and eugenics'. And furthermore, 'Adaption, the inheritance of acquired characteristics, the influence of environment and diet – these are the terms from the Lamarckian vocabulary that should be constantly in the forefront of the minds of those who study the evolution of Man and the human races, and who aim to perfect it' (quoted in Almaça 2000: 97. See also De Matos 2012: 92–7).

Neo-Lamarckian eugenicists insisted that society as a whole would become healthier, more peaceful, and more productive if a number of prescriptions were followed, including the improvement of living conditions in urban areas; nationwide programmes of vaccination; the criminalization of prostitution and pornography, and so on. In Latin America, the transformation of Rio de Janeiro at the end of the nineteenth century stood as a triumphant symbol of neo-Lamarckist principles, emphasizing social, educational, and medical reform, rather than the invasive management of reproduction associated with Mendelian eugenics (Borges 1993: 248–9). Other neo-Lamarckian eugenicists, however, worried that the ease of modern urban life was weakening the racial quality of future generations, particularly in the predominantly rural countries of Mediterranean and South-eastern Europe. The much-desired eugenic improvement of the nation was primarily not due to heredity but to social, education, and medical reform. For the Romanian hygienist Iacob Felix, for instance, this was an ongoing process of self-improvement, alongside measures to prevent depopulation, the increase in infant mortality, and the future degeneration of the Romanian peasantry (Felix 1897). Consistent with the degeneration terminology of his time, Felix problematized the generalized anxiety about urban development and its resulting chronic poverty and social unrest. Public health reforms were, he insisted, to be tailored to scientific principles and, equally important, to reflect a new understanding of health and its role in protecting the Romanian population.

While Latin eugenics primarily identified itself at the confluence of medical and biological sciences, it included a number of fundamental cultural elements. Catholicism was arguably the most important. The Catholic Church continued to exercise a deep influence in most Latin countries, unlike its Protestant counterparts. Neo-Lamarckism's apparent concordance with Catholicism's religious paradigm is one reason neo-Lamarckism had such a surprisingly long and successful life. Unlike Protestant determinism, the Catholic relationship to God allowed for free will – the ability to improve one's life in the eyes of God. Because of this, the destitute or fallen might always find inspiration through religion and reform. Like Saint Paul, the sinner could become a saint. Those who showed mercy towards unfortunates or were charitable were engaging in good deeds and made their own salvation more likely. Therefore, it is not surprising that Catholic eugenicists were sympathetic to neo-Lamarckian attempts to help 'cure' those with genetic defects; that all, in essence, could be 'saved'.

Until the second half of the twentieth century, the Catholic Church remained responsible for guiding men and women in their decisions on whom to marry, how to raise their children, and how to promote a morally healthy family. The Catholic Church saw women as mothers devoted to the care of the family, whose appropriate model was the Virgin Mary. The Catholic Church and Latin eugenicists were in accord with these aspects of a woman's essential and proper function in society. In Catholic countries, this was an important bridge between religion and eugenics. As we will discuss later, many Latin eugenicists reinforced traditional and religious views on the appropriate social and biological role of women.

However, there was a palpable tension between Catholic and eugenic views regarding the state control of reproduction. Eugenicists believed that it was critical that women's gendered biological imperative – their duty as mothers – be honed and facilitated, thus making them efficient reproducers of the next generation. As mothers, moreover, women were the first and most important influences on their children's biological qualities, as well as their moral upbringing. A healthy mother, devoted to her children's proper upbringing and care, would nurture healthy and productive citizens. Physical or mental infirmities, promiscuity, contagious illnesses (especially venereal disease), unhealthy behaviour, and undisciplined family life were the weak links that threatened this evolutionary chain.

In view of their neo-Lamarckian convictions, the social and biological management of women and children became a central concern of Latin eugenicists. In France, most parties of whatever political persuasion could agree on the need for women to reproduce prolifically. French conservatives and Catholics were never in doubt that birth control and various neo-Malthusian trends were morally degenerative. Even the liberals, who worried about demographic decline, came to define the production of large families as women's patriotic duty. Beginning in the 1890s, moderate feminists sought to ally themselves with the more liberal parties using this issue. They argued that women should be granted a greater role in the political life of the nation because of their importance as mothers (Offen 1984: 648–76; Cole 1996: 639–72; Camiscioli 2001: 593–621).

Latin eugenics drew sustenance from the neo-Lamarckist interpretation of the environment and its impact on the racial welfare of the individual and the community. Eugenics, according to this view, was both preventive and corrective. Take prostitution, for example: both eugenicists and the Catholic

Church shared much common ground in their fight against this social problem. Instead of contributing to the welfare of their nation as wives and mothers, many prostitutes were single mothers and had illegitimate children, whose place in society was unclear and unstable. Moreover, prostitution was also closely connected with the spread of venereal diseases, which threatened the health of the individuals and families concerned, and undermined the nation's racial fecundity. Venereal diseases could be controlled, the eugenicists hoped, through the state inspection of prostitutes and police control of brothels, the proscription of prostitution, and even the criminalization of sexual intercourse if a person knew he or she was infected with venereal disease.

If prostitution was associated primarily with women, alcoholism was seen as principally a male dysgenic disease. Drunken men were more prone to have sex with prostitutes or other women outside of marriage, commit crimes, become violent, and squander money. Alcoholism seemed to be a disease that could be passed on to children through neo-Lamarckian heredity, and contribute to an increase in hereditary defects and high infant-mortality rates. It was generally accepted that chronic alcoholism could be inherited or transmitted to descendants as a morbid nervous predisposition. The Peruvian medico-legalist Leonidas Avendaño believed that alcoholic fathers were liable to have epileptic descendants, while the Chilean physician Jaoquín Castro Soffia added insanity and alcoholism to the list of 'diseases' suffered by the children of alcoholics (Stepan 1985: 133–5). In 1901, the Mexican criminologist Carlos Díaz Infante associated alcoholism with (not surprisingly) criminality and vagrancy (Speckman Guerra 2005: 240).

Together with tuberculosis and syphilis, alcoholism was regarded as a major cause of degeneration, and as such defined as a public threat that should be curbed by public health measures. In the first decade of the twentieth century, France for example, had the highest consumption of alcohol per capita of all the Western European countries. The degree of alcohol consumption was also substantial in Romania: the average Romanian consumed 10 litres of alcohol per year, compared to the 2.5 litres by a Norwegian (Banu 1927: 99).

Eugenicists decried the effects of alcohol on the French and Romanian nations, lessening their fertility and their overall well-being. The French Anti-Alcohol Union warned of dire consequences: 'the fatherland is in danger', it announced (quoted in Nye 1984: 157). In 1912, the Klotz Commission agreed with these general observations, while renewing the government's surveillance of abortion and birth control (Garner 1914: 250; Pedersen 2003: 185). Once again, it was argued that preventive and corrective social and public health policies could prevent demographic decline, infant mortality, the spread of venereal diseases, and the effect of alcoholism on current and future generations (Drouard 1992a: 435–59). Reiterating their support for a healthy nation, expressed both in the quality and the quantity of its members, French eugenicists of the early twentieth century remained devoted to their own perspective on what facilitated social and biological improvement. This meant an emphasis on maternity and childcare – an environmental eugenic philosophy known as puériculture.

Puériculture

A remarkable dedication to improving the conditions of mothers, newborns, and children was crucial to the dissemination of Latin eugenics. First proposed by the French physician Alfred Caron in 1865 (Pinard 1908: 4–5), puériculture was revived and refined by the obstetrician Adolphe Pinard in the 1890s (Lefaucheur 1992: 413–35). As Professor of Clinical Obstetrics at the Paris Medical School and the leading physician of the renowned Baudelocque Clinic, Pinard (see Figure 1.1) was one of the world's foremost specialists in intrauterine paediatric care (Dunn 2006: 231–2).

In 1895, Pinard presented his research on the 'intra-uterine puériculture' to the Academy of Medicine, in which he demonstrated that infants whose mothers had rested for at least ten days prior to labour weighed (on average) more than those infants whose mothers worked until the onset of labour (Pinard 1895: 593–7). In true neo-Lamarckian fashion, Pinard argued that the protection and the supervision of the mother during pregnancy were critical to the welfare of the foetus (*puériculture intra-utérine*) and subsequently of the newborn baby (*puériculture extra-utérine*).

Yet Pinard wanted puériculture to be more than just maternal prenatal and postnatal care. In his 1899 text on the 'Preservation and Improvement of the Species' ('De la conservation et de l'amélioration de l'espèce') he proposed an all-encompassing programme of medical eugenics based on 'puériculture before procreation' (Pinard 1899: 141–6 and 1908: 25–6). This strong emphasis on the health and hereditary value of the parents was couched in traditional pronatalist and neo-Lamarckist language. The task of puériculture was to determine the means to conserve and improve the human species equally through parent selection prior to conception, and through proper maternity and childcare (Schneider 1986: 269–71).

Figure 1.1 Adolphe Pinard
Source: Wellcome Library, London. Image reproduced under the Creative Common Licence.

Pinard repeatedly stressed the importance of a healthy mother in the successful production of children as well as the importance of the medical professionals in supporting the women's role as mothers. In the embattled field of demographic decline, Pinard maintained, procreation, pregnancy, and birth provided a source of biological power. Reproduction was thus defined within a biological and social framework. The ability to control prenatal influences on the future child, therefore, began not with conception but with marriage. In a 1908 text devoted to marriage, Pinard connected puériculture with social selection and with the celebration of a eugenic family ideal. He chastised those women and men who did not subscribe to a eugenic agenda and denied themselves children. It was essential, he propounded, that women's reproductive role was acknowledged financially by measures meant to compensate for her nurturing work. The delicate nature of pregnancy made pregnant women unusually sensitive to the environment. To this effect, Pinard maintained that women who worked before or during pregnancy subjected themselves to debilitation, industrial disease, and unhealthy conditions in general, all of which could harm their future children. Indeed, since mothers should devote their time to raising their children, Pinard judged it better for women not to work in the last stages of pregnancy, lest it result in a debilitated child who could not fully contribute to society (Pinard 1908).

Pinard's ideas of puériculture dovetailed not only with ideas of eugenic regeneration but also with wider concerns about natalism. One of the leaders of the natalist movement, Paul Strauss, directly invoked puériculture in his work on depopulation (Strauss 1901). In this and other writings, Strauss promoted the introduction of nationwide welfare and social assistance programmes to protect mothers and children, urging the state to elaborate new laws to assure the propagation of future generations. As Rachel G. Fuchs remarked, 'Strauss's political philosophy and activities capture the decline in the traditional, religious view of the family and the emergence of a new morality: practicing proper modern hygiene under medical guidance in order to save the children' (Fuchs 1992: 72).

Pinard's appointment in 1902 to the Commission on Depopulation confirmed the growing importance afforded to puériculture by state officials. Medical expertise – as illustrated by Pinard's subsequent report – was thus enrolled to communicate official concerns about demographic decline (Pinard and Richet 1903: 15–24). As Mary Lynn McDougall pointed out, 'obstetricians, demographers, and charitable organizations defined infant mortality as a major social problem, one that could be solved by maternity leaves. Furthermore, they devised a "scientific" rationale for intervention in childbearing, which policy makers could use to create a consensus' (McDougall 1983: 80). Building upon these arguments, the demographer and deputy to the National Assembly and the National Alliance's vice-president, Adolphe Landry, was instrumental in passing the Strauss Law of 1913, which stipulated a daily allowance for two months, before and after birth, for 'all women of French nationality who habitually work for wages outside the home, whether as a worker, an employee, or a domestic' (quoted in McDougall 1983: 103–4).

As these examples illustrated, in the years preceding the First World War, puériculture thoroughly suffused natalist and eugenic rhetoric in France. Confident in their ability to improve the quality of future generations through pre- and post-natal care as well as the protection of mothers and infants, *puériculteurs* in France (and their followers elsewhere) promoted a distinct eugenic programme based not on the elimination of the 'unfit' but on the promotion of sanitary and public health measures destined to improve the health of the population (Variot 1908). To this effect, the Institute of Puériculture (*Institut de Puériculture*) was inaugurated on 8 June 1911. As outlined by its first director, the celebrated child reformer Gaston Variot, the purpose of the new institution was to become 'a centre of study and scientific research for students and physicians, and a school for the popularization of infantile hygiene' ('Inauguration de l'Institute de Puériculture', *Paris Médical* 1911: 81).

William H. Schneider pondered whether it made a difference 'that French eugenics grew out of puériculture rather than biology and anthropology?' (Schneider 1990: 82). At a general level, he answered, 'it meant that [eugenics] enjoyed [a] potentially broad base of support in French society'. In terms of institutional support, 'The identification with puériculture also gave French eugenicists alliances with those in social hygiene organizations fighting alcoholism, tuberculosis, and venereal diseases.' Crucially important, it also assured the support of the Catholic Church, 'who might otherwise have been opposed to a eugenics stressing the elimination of certain elements from the population' (Schneider 1990: 82–3). For Schneider, however, puériculture was in a politically subaltern and conciliatory position to eugenics and was destined to be replaced by the latter. Once the French Eugenics Society was formally established in 1912, he claimed, 'Puériculture had served its purpose and was officially left behind' (Schneider 1990: 83).

Yet puériculture (and maternal and childcare in general) did not need the French Eugenics Society to survive and thrive; the opposite was the case in France and other Latin countries. The combination of social protection and maternal and child care through improvements in living and working conditions remained a constant source of much eugenic activity, and not only in France. By the 1910s, the practices of puériculture became widespread among other Latin nations (for Romania see Alexandresco 1907: 474–9). Thus in Chile an Institute of Puériculture (*Instituto de Puericultura*) was established in 1906 in Santiago, at the initiative of physician Alcibíades Vicencio, followed by a school for obstetrics and puériculture in 1913 (Illanes 2007: 122–8). In Spain, the National Institute for Maternology and Puériculture (*Instituto Nacional de Maternología y Puericultura*) was established in 1910 (Orzaes 2009: 164; also Gómez 2005: 641–64).

By 1912 there were similar *Institutos de Puericultura* in Argentina (Lavrin 1995: 103). As Yolanda Eraso noted, Argentinian obstetricians 'anticipated the functionality of the French principles of puériculture through a comprehensive social and medical system of three-state assistance: prenatal, natal, and postnatal' (Eraso 2013: 30). One of the obstetricians, Alberto Peralta Ramos, was remarkably influential, both as director of the maternity service of the Hospital 'Bernardino Rivadavia' in Buenos Aires and as chair of obstetrics at the Faculty of Medicine. Like other supporters of puériculture, Alberto Ramos' eugenic rhetoric was explicitly pronatalist, envisioning a central role for women and their reproductive duties. As Enrique Feinmann, a leader in Argentinian puériculture, put it in his 1915 textbook *La ciencia del niño* (*The Science of the Child*): 'Woman will be the good fairy of the new era. Her nursery of human beings will be an immense blooming garden, and the children, instead of going to heaven as angels, will populate the earth as men, to make it better and more beautiful' (quoted in Rodriguez 2006: 120).

It was quite logical to extend the concept of puériculture to encompass eugenically beneficial changes to the care of individuals throughout their lifetime, rather than simply in childhood. The idea of 'lifelong puériculture' varied in name and to some extent in focus among eugenicists. Cuban eugenicists, for instance, popularized the science of 'homicultura' – a term coined by the eugenicist Eusebio Hernández. Homiculture was not a substitute for puériculture but a complex system of social and biological ideas pertaining to human improvement. It encompassed the following areas of research and eugenic concern: '*progonocultura* (care of the gonads), *patrimatricultura* (culture of the parents), *matrifeticultura* (care of the pregnant mother and the fetus together), *matrinacultura* (care of the mother and baby together), *puericultura* (care of the baby), and *post-genitocultura* (care of the individual after birth)' (Stepan 1991: 79; Bronfman 2004: 118–9; González and Peláez 2007: 33–4). This holistic view of human improvement was a fundamental feature of the transformation of Cuban society according to the scientific planning of family and reproduction. What this transformation would entail was described by Manuel Varona Suárez, the secretary of hygiene and welfare, in a document presented to the Cuban president José Miguel Gómez in 1910. Homiculture, he wrote, was 'a subject of such transcendence for the future of our nation, which sees to the physical and mental vigor of present and future generations and the development of citizens who can help themselves and contribute to the nation' (quoted in Bronfman 2004: 119). President Gómez was certainly convinced: a Division of Homiculture (*Negociado de Homicultura*) was created within the Ministry of Hygiene and Welfare (*Secretaría de Sanidad y Beneficencia*) on 22 September 1910. This set the stage for a more systematic dissemination of ideas of puériculture and homiculture to the Cuban population, a course of action adopted in 1914 at the Third National Medical Congress (*Tercer Congreso Médico Nacional*). A direct expression of the growing popularity of such ideas was the National League of Homiculture (*Liga Nacional de Homicultura*) founded by Hernández and another prominent Cuban eugenicist, Domingo F. Ramos, and the establishment of the Infant Hygiene Service (*Servicio de Higiene Infantil*) by the physician Enrique Núñez – both in Havana in 1913 (González and Peláez 1999: 121–32; Bronfman 2004: 119; Schell 2010: 477).

Nancy Stepan aptly summarized the relationship between homiculture and eugenics and its impact on the development of Latin eugenics: 'Homiculture represented, as it were, one stage on the road from the puériculture of the late nineteenth century to the more radical and innovative neo-Lamarckian eugenics of the early twentieth, and reflected the growing hereditarianism of medical thought in the period' (Stepan 1991: 79). This observation should be placed within the broader context of the global circulation of eugenic ideas. The incipient Latin eugenics was embedded in a wide set of transnational relationships. While Latin American eugenicists drew upon French ideas of neo-Lamarckism and puériculture as well as Italian theories of criminal anthropology, European eugenicists also benefitted from the relationship.

As will be discussed in the next chapters, Latin eugenics continued to absorb its intellectual inspiration from French and Italian traditions of puériculture, social hygiene, demography, and criminal anthropology, but it now relied on work carried out in various national contexts in places as far apart as Latin America and Eastern Europe. However, a self-conscious pan-Latinism remained an essential ingredient of Latin eugenics. Pan-Latinism was the political expression of 'Latinity', or Latin culture. This conceptual entity was not necessary a hoped-for new super-state of all Latin peoples, although the most radical pan-Latinists did envision such a federation. More commonly, pan-Latin enthusiasts hoped to retain the current nation-states, but build between them special economic, political, and cultural ties. Indeed, pan-Latinism was at its most realistic, and thus most influential, when it sought to bolster an existing nation-state through 'natural' alliances against common enemies. It was in this sense that pan-Latinism intersected with Latin eugenics.

To be sure, Latinity and pan-Latinism were limited forces. The nation-states that emerged in the late nineteenth century inevitably had conflicting agendas. States had powerful means at their disposal to inculcate loyalty in the masses, such as schools, armies, propaganda, national rituals, churches, and so on. This was also an age of imperialism, and the lands lusted after by the colonizing powers invariably overlapped. While pan-Latinism often attracted first-rate intellectuals, it seldom gained the allegiance of more than second-rate politicians. No significant pan-Latin institutions were created before 1914, although pan-Latinist sentiments did influence political events of the First World War more deeply than is generally recognized.

As will be discussed in the next chapter, pan-Latinist ideas also revealed a deep, festering anxiety that the Latin nations were incapable of modernizing rapidly enough to ensure their survival. Military defeat was the most visible and disastrous consequence of the inability to modernize. The antagonists inevitably belonged to countries of Anglo-Saxon and Nordic culture: Germany, Great Britain, and the USA. This division between 'Latins' and 'Germanic or Aryan people' also fed into the racial-linguistic stereotypes of the nineteenth century, especially with reference to the idea of a historical, millennial 'Roman versus Teutonic' antagonism (Chadwick 1945).

Given the desire of almost all Latin countries to increase their populations, radical pan-Latinists argued that there was one way to drastically increase population size almost immediately: for the Latin nations to establish some form of a common political and economic unit. The new populationist rhetoric caught on immediately after the Franco-Prussian War. The Spanish newspaper *Igualdad* (*Equality*) suggested that, in the wake of the French defeat, the Latins of Europe 'form a federation; remember our common ancestry and again unite as we did fifteen hundred years ago against these modern Vandals'. This time, the Latins' chief weapon was their numbers: 'eighty million men which compromise our [Latin] family' (quoted in López-Cordón 1975: 395).

In the years immediately before the First World War, some pan-Latinists were relieved to detect a new stirring of pride among the Latin people, most notably in France. In 1912, the French playwright Étienne Rey looked back with chagrin on the past decades, when the entire world had accepted the 'legend' of the decadence of the Latin race. 'Even we ourselves were dupes and victims of this opinion', Rey lamented, 'and [believed] that to regenerate ourselves, we had to learn from the German schoolmaster'. However, the Latins were now awakening; the French, for example, had embraced 'the cult of heroism' and 'the taste for action' (Rey 1912: 197–9). This revival of Latinity was also marked in these years by a proliferation of pan-Latinist organizations and journals which acted as lobbyists for pan-Latin causes through meetings, quasi-official government committees, and in the press. It is likely that Latin eugenics would never have become a self-conscious movement were it not for the influence of the growing pan-Latinist movement. World War I would bring pan-Latinism to its pinnacle, and strongly influence some of the most important Latin eugenicists of the succeeding decades.

2

Early Latin Eugenics

Early Latin eugenics consisted mainly of contributions from Italy, France, and Spain, although by the 1910s articles on eugenics appeared in Mexico, Romania, and Portugal. In Latin America, many eugenic programmes were presented at the Pan-American Sanitary Conferences (*Conferencias Sanitarias Panamericanas*), held periodically from 1902. As elsewhere, participants at these conferences saw their efforts as laying the foundation for nation building, modernization, and national empowerment. Furthermore, these conferences had explicitly activist agendas, which included control of disease; protective labour laws and agencies for women and children; assistance for abandoned children and single young mothers with children; the dissemination of puériculture to medical professionals; basic hygienic instruction in the schools; and separate juvenile and adult court systems. Sections also discussed the relationship between eugenics and education, demography, social hygiene, and public medicine.

However, by the end of the 1910s growing scientific interest in heredity indicated new directions of research for many eugenicists. As demonstrated by the 4th International Conference of Genetics (*IVe Conférence Internationale de Génétique*) organized in Paris in 1911 by the French National Society of Horticulture (*Société Nationale d'Horticulture de France*), Mendelian genetics was already widely discussed with respect to animal and plant breeding. As will be discussed in this chapter, prior to 1914 eugenicists in Europe and the USA articulated new social and biological interpretations of national health, while at the same time linking heredity and genetics to sociology, history, and statistics. In this context, eugenics was formulated strongly enough not only to provide a new vocabulary of national improvement but also to command widespread political and scientific attention. This dual process of scientification and politicization constituted the basis for the First International Eugenics Congress to convene in London in July 1912.

The impact of the First International Eugenics Congress

The history of this congress has often been told and needs no repetition here. Our focus will only be on the contributions offered by the Latin eugenicists. The French delegation included Georges Schreiber, Frédéric Houssay, Lucien March, Eugène Apert, Adolphe Landry, and Georges Papillaut. Due to a severe illness, Adolphe Pinard could not attend but his disciple, the physician Georges Schreiber, advocated his mentor's theories at the Congress. Achille Loria, Alfredo Niceforo, Roberto Michels, Corrado Gini, Vincenzo Giuffrida-Ruggeri, Antonio Marro, and Enrico Morselli represented Italy. Two Belgian eugenicists, Norbert Ensch and Louis Querton, also attended.

While all participants agreed that eugenics had to address the growing social and biological degeneration in modern society, disagreement persisted over the most appropriate methods. In general, the French and the Belgian participants approached eugenics through puériculture, education and theories of neo-Lamarckian heredity, but some of them insisted on greater government intervention in population management and even agreed on the importance of negative eugenic measures. Pinard, for instance, described Galtonian eugenics – namely 'the science having for its object the study of the causes subject to social control which can improve or impair the racial qualities of future generations, whether physical or mental' – as 'nothing else' but puériculture. According to Pinard, French puériculture promoted the health of the parents, conception, pregnancy and infant care, with the goal of improving 'the reproduction, preservation, and improvement of the human species' (Pinard 1912: 458). Pinard's description of puériculture as non-Galtonian eugenics represents, in fact, the first programmatic articulation of a different intellectual tradition of theories of social and biological improvement, one from which Latin eugenics would draw sustenance.

Belgian sociologist and educationalist Louis Querton offered a more nuanced interpretation of eugenics, one indebted to both Galton and Pinard. His programme of practical eugenics was guided by the twin principles of nature and nurture. He thus prioritized the social factors 'capable of improving or impairing the racial qualities of future generations' and 'the knowledge of the facts of heredity' together with 'the action of social institutions on the development of the race' (Querton 1912: 146). Querton then spelled out the main instrument of his eugenic programme: the 'systematic observation of the child', which – on the one hand – would document 'the absolute or relative insufficiency of the reproductive fitness of the parents', while on the other, would 'facilitate the spread of the elementary ideas of eugenics'. It is revealing here that even though

Querton spoke of approaching 'the eugenic ideal anticipated by Galton', he framed it in explicitly neo-Lamarckist terms, asserting the importance of family, education and social environment (Querton 1912: 147–9). Child welfare was therefore an essential component of practical eugenics. '[T]o be efficient and to really favour the perfecting of the individual, and the amelioration of the race,' Querton believed, 'the control of the development ought to be extended to all children and to be prolonged during the whole period of their development' (Querton 1912: 150). What Querton offered here was the formulation of a practical strategy of biological improvement through the promotion of education and social protection.

Other Latin eugenicists endorsed, in some form or another, this neo-Lamarckist view of racial improvement. The Italian psychiatrist Antonio Marro, for instance, looked at the social environment and the age of the parents in order to explain certain behavioural patterns in children (Marro 1912: 118–36). The French biologist Frédéric Houssay, however, offered a detailed critique of eugenic sterilization, which he described as a 'one-sided' method of 'artificial selection'. What was needed, Houssay insisted, was 'to enlighten ourselves on the origin and perpetuation of defects by heredity; and with this object in mind we must cling to the principles on which rest the Lamarckian doctrines' (Houssay 1912: 160). Houssay also caustically berated the North American practice of sterilization, which he described as a tool to perpetuate the social and economic unfairness suffered by the poor. Explicitly, the 'rich degenerates will slip out of the penalty of sterilization as they now slip out of all the others. It would be desirable, highly desirable', he continued, that these individuals 'should come under the grip of such a law, but in actual reality there is reason to fear that they would not' (Houssay 1912: 160). By stressing 'hygienic and moral' intervention as more eugenically efficient than any harsh 'penalty', Houssay interpreted eugenics as the synthesis of 'biological and moral principles' (Houssay 1912: 161). But eugenics was also fraught with problems: for example, how could eugenicists separate genetic from cultural and economic causes of social inferiority, given the limited knowledge of genetics at the time?

The Italian political economist Achille Loria agreed with Houssay. Loria warned that 'any theory which recognizes the existence of a relation, direct or indirect, between psycho-physical superiority and economic superiority leads fatally to a eugenic nihilism and destroys all practical action' (Loria 1912: 183). He recommended instead 'a minute and positive examination of individual characters, which must be directly ascertained and not inferred from the fantastic criterion of their economic position' (Loria 1912: 183).

Alfredo Niceforo, conversely, did not share Houssay's egalitarian interpretation of eugenic improvement or Loria's economic arguments. He made it clear that he did not 'deceive' himself when:

> asserting that the groups formed by individuals belonging to the lower *classes present, in comparison with subjects of higher classes, a lesser development of the figure, of the cranial circumference, of the sensibility, of the resistance to mental fatigue, a delay in the epoch when puberty manifests itself, a slowness in the growth, a larger number of anomalies and of cases of arrested development.* (Niceforo 1912: 192. Emphasis in the original)

Niceforo accepted the notion that poverty and social misfortune contributed to the degeneration of the 'lower classes'. These misfortunes masked the genetic superiority of some of the poor; conversely, some of the wealthy members of society had achieved their position not because of their genetic endowment but due to favourable environmental circumstance. Given this, eugenics as 'the study of the physical and mental amelioration of the race' should provide the much-needed "circulation" of the superiors who find themselves below, and of the inferiors who find themselves above, in order to group in the superior classes the greatest number of "superiors"' (Niceforo 1912: 194). Thus, eugenic engineering would serve to unite all members of the nation in the pursuit of social and biological improvement.

A similar interpretation, premised on the promotion of the biological capital of the nation, was proposed by another Italian statistician, Corrado Gini. Social efficiency, natalism and eugenics were – for Gini – intimately associated with demography and the fertility of social classes (Cassata 2006: 17–22). In his 1912 book *I fattori demografici dell'evoluzione delle nazioni* (*The Demographic Evolution of Nations*), he devoted a great deal of attention to the statistical investigation of what he called 'the cyclical evolution of nations'. Gini claimed that there was a 'circulation of elites in a population', with talented but oppressed groups recurrently rising to power and contributing their vitality and fecundity to the apex of the nation's bio-social structure, while the old elites degenerated, experiencing a decrease in prolificacy in each generation. Conveniently fitting his nationalist aspirations, Gini believed that the birth of the Italian nation in the mid-nineteenth century, the consequent mixing of long-separated peoples, and the emergence of the bourgeoisie from the ruins of the feudal past, were all evidence that Italy had entered a new 'cycle' of millennial growth. As a young, vigorous, and fertile nation, it was Italy's destiny to replace France, now exhausted and infertile, as the leading power on the European continent (Gini 1912a).

Gini developed 'a new statistical procedure – the mean difference – to study the variability of quantitative characteristics ("variability index") and qualitative ("mutability index")' (Cassata 2011: 34). This was to enable statistics to serve a mediating role between social and medical sciences. In his paper presented at the Congress entitled 'The Contribution of Demography to Eugenics', Gini employed similar statistical investigations of differential fertility to demonstrate the importance of social and biological improvement.

He outlined the following eugenic strategies:

> (1) Selecting the reproducers; (2) Placing the reproducers in the most favorable environment; (3) Regulating in the best way the circumstances in which the unions are consummated, both as regards the absolute and relative ages of the reproducers, and as regards the season in which the union take place, and the interval between successive conceptions; (4) Placing the offspring in the most favorable environment. (Gini 1912b: 296)

Gini's eugenics was cast in neo-Lamarckist terms, defined as the 'improvement of the environment in which the reproducers live and their offspring develop undoubtedly has beneficial effects upon the human race' (Gini 1912b: 296). Eugenics was not a programme of selective breeding as advocated by some American, German, and Scandinavian participants at the Congress. Like other Italian, French, and Belgian eugenicists, Gini claimed that the debate over the causes and variability of racial degeneration was as much social as it was biological.

While Gini and Niceforo emphasized the importance of statistical demography for Latin eugenics, the anthropologist Vicenzo Giuffrida-Ruggeri encouraged the application of Mendelian laws to the study of race. Yet for him, eugenics was more than just a collection of hereditarian facts; it conveyed a normative and prescriptive view of human history (Giuffrida-Ruggeri 1912: 38–47). In viewing eugenic knowledge in its diachronic perspective, Giuffrida-Ruggeri echoed the arguments put forward by psychiatrist Enrico Morselli, who highlighted racial specificity even when aiming to identify those desirable traits embedded by nature in all humanity. For Morselli:

> the general improvement of the species should not aim at the equalisation of men nor at that of races or nations. Each of these has its particular task in History, which is determined by its place in space, by its relation to the environment, by its experiences, by the products of its special mentality, by its ideals, and by its conception of life. (Morselli 1912: 62)

Accordingly, Morselli concluded, *universal* eugenics, one 'common to all civilized peoples', was to be accompanied by each of these peoples' *national* eugenics, so formulated as to 'keep in view the defence and the propagation of its proper physical type, always more differentiated, and of the proper mentality, always more characteristic' (Morselli 1912: 62). Morselli's invocation of humanity as the universal measure and model of eugenic rationality did not invalidate the understanding of eugenics as the science of improving the nation's social and biological health. In fact, eugenics reinstated the belief in the palingenetic myth of shared destiny, making it possible to connect notions of collective welfare with individual responsibility towards the nation.

The idea of eugenic 'responsibility' and its relation to social worth, fertility and pronatalism distinguished Lucien March's contribution to the Congress. March (see Figure 2.1) – described by one scholar as 'architect of the most important achievements in the field of early twentieth century econometrics and statistical economics' (Le Gall 2007: 164) – discussed France's low birth rate and 'the productiveness of families' (March 1912: 210).

Figure 2.1 Lucien March
Source: Collections École Polytechnique, Palaiseau, France.

Like Gini, March relied on statistics to demonstrate 'the influence on fertility of social status, social surroundings, and income' (March 1912: 218). Quantitative analysis of various factors contributing to population decline was especially important, March argued. To this effect, he brought statistics and eugenics together in providing a study of population in its economic and demographic aspects.

Italians and French eugenicists made a commendable contribution to the First Eugenics Congress, as can be seen easily from the published proceedings. Ignacio Valentí Vivó, a professor of legal medicine and toxicology at the University of Barcelona, was the only Spanish participant at the Congress. He presented the eugenic history of a family from Catalonia (Vivó 1912: 399). There was no paper from Latin America or Romania at the First International Congress of Eugenics, although the future founder of the Argentinian Eugenics Society, the physician Victor Delfino, did attend (Stepan 1991: 58).

In spite of the diversity of eugenic ideas proposed by the participants in the First International Eugenics Congress, all these authors had something in common: they aimed to contribute to their nation's population growth, along with restoring its racial strength. They stressed the need to observe both the population's eugenic *quality* and its numerical *quantity*. The First International Congress of Eugenics was thus a milestone in the development of the international eugenics movement, and spurred on the creation of national eugenic organizations in many countries. The Congress had a particularly profound impact on the Latin countries as well, not least in terms of encouraging French, Italian, and Belgian eugenicists to create their own national eugenic societies.

The first to be established was the French Eugenics Society (*Société Française d'Eugénique*) on 22 December 1912 at the Faculty of Medicine in Paris. The neo-Lamarckist biologist Edmond Perrier, the head of the Museum of Natural History, was elected president, together with vice-presidents Frédéric Houssay, Adolphe Pinard, and Charles Richet. Lucien March was appointed the Society's secretary. Other participants included the senator and future president of the French Republic Paul Doumer; the neurologist Louis Landouzy; the dermatologist François Henri Hallopeau; pediatricians Félix Jayle, Henri Méry, and Eugène Apert; the anthropologist Léonce Manouvrier; and the legal scholar Fernand Faure ('Fondation d'une Société Française d'Eugénique', *La Presse Médicale* 1913: 44).

In his opening address, Perrier insisted that the new society embraced an interpretation of eugenics that reflected the French realities. It was 'the country (*la patrie*) that defined the race', he concluded, not the opposite ('Fondation d'une Société Française d'Eugénique', *La Presse Médicale* 1913: 44). The papers presented at first meeting, then, appropriately focused on natalism, puériculture, physical education, and the fight against the *trois fléaux*: alcohol, tuberculosis, and syphilis. The Society's statutes adopted on the occasion, and published in the first issue of the journal *Eugénique*, stipulated:

> research and the application of useful knowledge about the reproduction, preservation and the improvement of the species, and the study of questions of heredity and selection in their application to the human species, and questions relating to the influence of environment, economic status, legislation and customs on the quality of successive generations and on their physical, intellectual and moral aptitudes. (Quoted in Léonard 1983: 145; Schneider 1990: 74–5)

Most eugenic societies existing elsewhere in Europe and North America at the time emphasized similar directions for future research and active involvement with the wider public. For example, when established in 1907, the Eugenics Education Society aspired, principally, to 'set forth the national importance of Eugenics, in order to modify public opinion and create a sense of responsibility, in the respect of bringing all matters pertaining to human parenthood under the domination of eugenic ideals'. The Society also endeavoured to 'spread knowledge of the laws of heredity so far as they are surely known, and so far as that knowledge might affect the improvement of the race'. Finally, attention would be devoted to 'further eugenic teaching at home, in the schools and elsewhere' (*Annual Report 1908*: 21).

The practical goals of the International Society for Racial Hygiene (*Internationale Gesellschaft für Rassenhygiene*), as outlined by Alfred Ploetz in 1910, were slightly more ambitious. In addition to an 'opposition to the two-child system' and the 'establishment of a counterbalance to the protection of the weak by means of isolation, marriage restriction, etc., designed to prevent the reproduction of the inferior', Ploetz pledged 'opposition to all germ-plasm poisons, especially syphilis, tuberculosis, and alcohol', the 'protection against inferior immigrants' and, finally, the 'extension of the reigning ideal of brotherly love by an ideal of modern chivalry, which combines the protection of the weak with the elevation of the moral and physical strength and fitness of the individual' (quoted in Weiss 1990: 23–4).

Read in the context of the early twentieth-century France, the programme of the French Eugenics Society embodied a social and medical vocabulary, which drew mostly on neo-Lamarckist paradigms and the country's social and demographic particularities. It shared with the Education Eugenics Society a preference for propaganda and public instruction, but was not predisposed towards negative eugenic measures such as those put forward by Ploetz and other eugenicists in Germany, the USA, and the Scandinavian countries (Schreiber 1913a, b, c, d, e, f, g). To this effect, when the British Royal Institute of Public Health held its second congress in Paris between 15 and 20 May 1913, the French Eugenics Society organized a section on 'Eugenics and Child Study', chaired by Lucien March.

Twenty-nine speakers (the French participants included Eugène Apert, Frédéric Houssay, Raphaël Raimondi, Georges Schreiber, Auguste Broca, and Raoul Dupuy; the neo-Lamarckist Caleeb Saleeby was one of the English participants) were grouped in two categories: 'eugenics, comprising those papers which have a direct bearing on the inborn qualities of future generations' and 'child study, including pre- and post-natal hygiene, etc.'. In his opening remarks, March defined 'the Eugenic ideal as the progressive development of race' and expressed the hope that 'the cruel methods of nature would not be re-introduced, but the attempt made to attain the same ends by more kindly means' ('Eugenics and Public Health', *The Eugenics Review* 1913: 157). In terms of the relationship between eugenics and puériculture, Apert made the necessary clarification:

> those engaged in excellent and much-needed social reform, concerning regulations of maternal health, facilities for infant feeding, school for mothers, and many other valuable institutions' are only indirectly involved with eugenics. Unless it was proved, he continued, that these 'institutions preserve a proportionally larger number of fit babies they are not eugenic, although they may reduce the death-rate. ('Eugenics and Public Health', *The Eugenics Review* 1913: 159)

Returning the courtesy and hospitality received a year earlier in London, the French eugenicists insisted on the necessity to synchronize English *eugenics* and French *puériculture*. It was clear, however, that the growing internationalisation of eugenics did not bring about the renunciation of specific national traditions of conceptualising national health, population growth, and the preoccupation with future generations. The preoccupation with demographic decline, for example, remained central to the French eugenic tradition, as illustrated by Lucien March's article on 'Depopulation and Eugenics'.

The article was translated from French and published in two parts in *The Eugenics Review* in late 1913 and early 1914. March focused extensively on the factors contributing to the demographic decline of the population, a process which, in turn, impacted negatively as much on France's political and economic power as on 'the value and vitality of the generations to come' (March 1913: 234). If in 1811, March further noted, the French represented 16 per cent of Europe's population, by 1911 this number dropped to less than 9 per cent.

By comparison to other European countries, France was losing population at an alarming rate (see Figure 2.2). This demographic decline eroded France's status as a major European and colonial power. In such circumstances, March wondered, will France 'remain strong enough in relation to the subject races or to possible rivals?' (March 1913: 236).

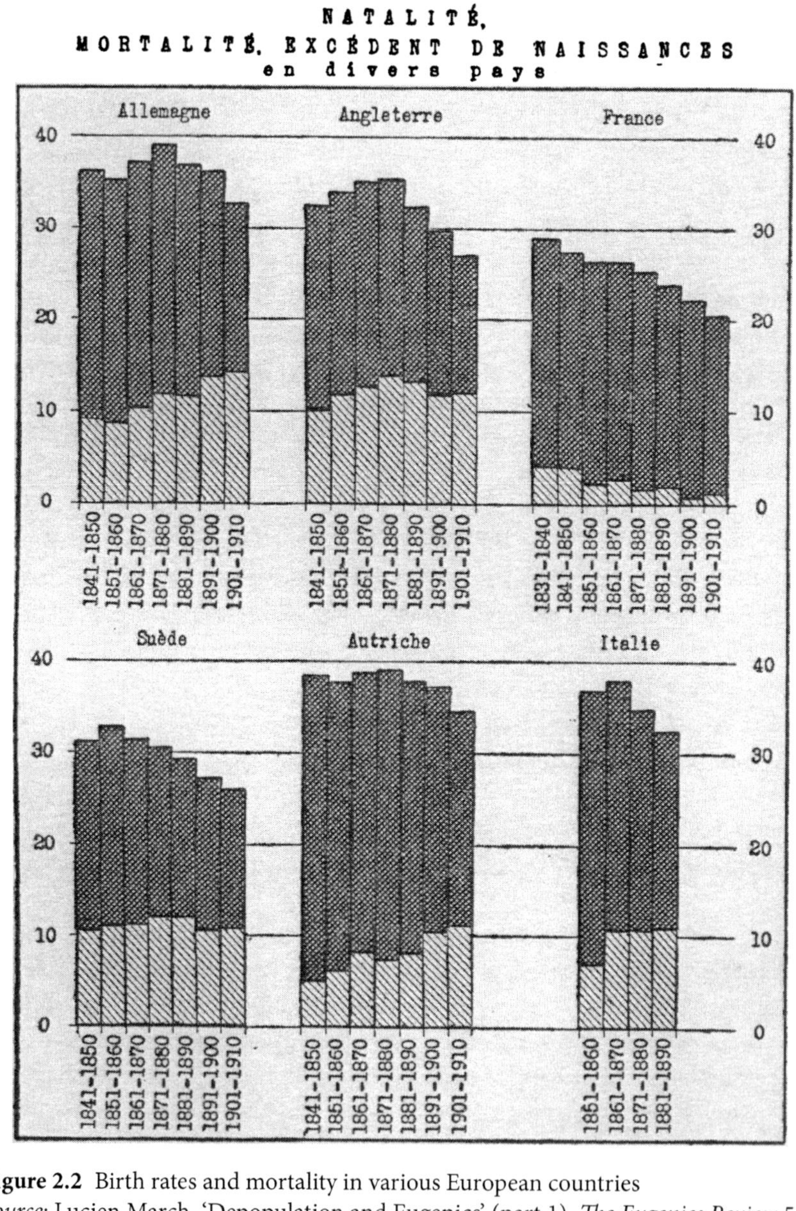

Figure 2.2 Birth rates and mortality in various European countries
Source: Lucien March, 'Depopulation and Eugenics' (part 1), *The Eugenics Review* 5, 3 (1913): 237. Courtesy of the Galton Institute, London.

Among the central factors contributing to France's protracted demographic growth, March enumerated high infant mortality, lack of financial resources required for a large family, living in urban areas, and feminism. The reproductive imperative, March grieved, was largely lost on modern Frenchmen and especially women. The priority accorded to fertility in previous centuries had been abandoned. At the beginning of the twentieth century, March argued,

> the decline of the birth-rate keeps step for step with regard for life and the progress of hygiene, with the improvement of the standard of living and the increase of wealth, with the tendency to live in towns, and the progress of education, with the protection of the weak, and the advance of individualism, especially of the individualism of women. (March 1913: 245)

What remedies did March propose? He endorsed a form of eugenic pastoralism, suggesting that 'the rejuvenation of society must be sought in [the] return to the land, and [in] the simple life of the country' (March 1914: 344). He praised the rural population for 'maintaining a higher degree of fecundity', while simultaneously building a defence against the degenerative effect of modernity. Another proposed measure was offering increased financial support for childcare and wider social assistance. In this sense, March referred to the recently adopted Law of 14 July 1913, which stipulated a grant of 7.50 francs per month for each child after the third. Economic incentives for large families were as important as encouraging them to reproduce more (March 1914: 347–48).

March concluded his analysis with the hope that the French Eugenics Society would take his observations seriously, and as a result 'collect further data dealing with the influence of heredity, of race, of environment and of social classes; so that it may contribute its share to the researches which in other countries are beginning to bring a little order and method into our ideas on the problems intimately connected with the question of population' (March 1914: 349). He also drew attention to the work conducted in the USA and Britain, in an attempt to mobilize other French eugenicists to engage with 'those questions' about human heredity 'upon which depends the life of the nation' (March 1914: 351). In this manner, March internationalized the French Eugenics Society's practical agenda, which could thus be conjoined with that of other eugenic societies, in Europe and elsewhere.

Italian eugenicists, too, had been emboldened by their participation in the First International Eugenics Congress (Ciceri 2009). Giuseppe Sergi and Alfredo Niceforo promoted the creation of a Committee of Eugenic Studies (*Comitato Italiano per gli studi di Eugenica*) within the Roman Society for Anthropology in March 1913; the first meeting of the new committee took place on 17 November

1913 ('Atti del comitato italiano per gli studi di eugenica', *Rivista di Antropologia* 1923: 543–4). Giuseppe Sergi was president and the psychiatrist Sante de Sanctis, vice-president. The psychologist Francesco Umberto Saffiotti was appointed secretary. Other founding members included the zoologist Cesare Artom, Corrado Gini, Antonio Marro, Alfredo Niceforo, and the gynaecologist Luigi Mangiagalli ('Italians take up Eugenics', *The Journal of Heredity* 1914: 138). The Committee focused on

> studying the factors that could determine the progress or the decadence of the race, both in terms of physical aspect and psychical aspect, carrying out, for example, research on the normal or pathological heredity of characteristics, on environmental influence and the life regime of parents on the characteristics of children, on the importance of the momentary conditions of the organism in the act of reproduction, or the environment in which the new organism develops. (Quoted in Cassata 2011: 40)

Although Sergi led the Committee of Eugenic Studies, it was clear from the Committee's proposed eugenic outline that he was very much in the minority with regard to his pro-Mendelian, anti-Catholic, and seemingly anti-Latin views. Although the Committee was short-lived (it dissolved in 1915), it succeeded in attracting a number of important Italian scientists among its members (for a full list, see Cassata 2011: 41). It supervised only one eugenic project: Gini's statistical investigation of the members of the Italian academia, 'in order to evaluate the relationship between birth, biological value of offspring, and prolificacy of families' (Cassata 2011: 41). The project also demonstrated Gini's broad engagement with a number of topics, including social worth, national efficiency, and differential fertility, all of which to receive further elaboration in the following decade.

Italian, French, and Belgian participants to the First International Eugenic Congress identified social worth, degenerative environments, family and children as their main eugenic concerns; to assume, however, that Latin eugenicists rejected more radical eugenic measures, such as the introduction of the premarital certificates and sterilization outright would be inaccurate. For instance, at the Fourth Scientific Congress held in Santiago (Chile) in 1908/1909, the prominent Argentinian physician Emilio R. Coni spoke favourably about birth control and eugenic sterilization, endorsing the need for eugenic legislation on procreation (quoted in Stepan 1991: 58). In 1908, the French journal *Archives d'anthropologie criminelle, de médicine légale et de psychologie normale et pathologique (Journal of Criminal Anthropology, Legal Medicine, and Pathological and Normal Psychology)* reported on attempts in

Romania to legislate marriage restriction for those 'suffering from epilepsy, tuberculosis and syphilis' ('Restriction du marriage', *Archives* 1908: 96). In Italy, Angelo Zuccarelli, the director of the Museum of Criminal Anthropology at the University of Naples, had been advocating 'the sterilization of extremely degenerate people' since the 1890s (Cassata 2011: 108–9).

The psychiatrist Miguel Bombarda articulated similar views in Portugal. In a series of articles published from 1904 to 1910 in the journal *A Medicina Contemporanea: Hebdomadario Portuguez de Sciencias Medicas* (*Contemporary Medicine: Portuguese Weekly of Medical Sciences*), Bombarda discussed the sterilization of 'degenerates' as well as eugenic marriage restriction (Henriques 2012: 64–5). Furthermore, in 1908 a proposal in the Portuguese parliament advocated legislative measures to prevent the 'marriage between syphilitics, those suffering from tuberculosis and leprosy, alcoholics, epileptics and those with heart problems'. It did not pass, however (Cleminson 2014: 38). As in Romania, a number of counter-arguments prevailed: that sterilization and marriage restriction were incompatible with Portugal's cultural and religious values; that it would not accelerate the country's eugenic improvement; that positive eugenics had greater value than negative eugenics; and that the populace would not accept eugenic sterilization. At the time, such proposals were however uncommon. Similar to the French neo-Lamarckian eugenicists, who 'were less obsessed by the lower social classes since they did not regard them as irreversibly and biologically inferior' (Lucassen 2010: 282–3), Romanian eugenicists worried more about the effect of disease and alcohol on the racial quality of the population.

If Romanian and Argentinian eugenicists were unable to implement their particular prescription for negative eugenics, their thinking was not – as we will see in the next chapters – without importance or influence. During the first decade of the twentieth century, the majority of Latin eugenicists articulated schemes of positive eugenics, reaffirming the critical importance of demography and social hygiene for furthering national improvement. In this, they were joined by social scientists, health reformers, and politicians. Together, these individuals were keen to diagnose the causes of the nation's declining health and fertility, propose solutions, and see them implemented.

The president of the Mexican Republic, Venustiano Carranza, provides one illustrative example. In 1917, Carranza proposed to the Constituent Congress the approval of the Law of Family Relations (*Ley sobre Relaciones Familiares*), which stipulated marriage restrictions for

> those suffering from incurable physical impotence, syphilis, tuberculosis or any other chronic and incurable conditions, which are also infectious or hereditary, as

> well as alcoholics, since anyone in the conditions cited can transmit pathological traits to their descendants, making them weaker and less able to work efficiently, both physically and intellectually, and in turn transmitting their weakness to future generations.

To allow these individuals to reproduce further, Carranza explained, was 'detrimental to our Nation, whose vigor depends on the strength of its children, and is likewise detrimental to the species itself, whose perfection requires a sane and judicious artificial selection added to natural selection so its rigors may be deflected and mitigated' (quoted in López-Guazo 2001: 145).

In Europe, however, most Latin eugenicists believed that the surest means to assist the emergence of a eugenically superior population was demographic growth, often through projects to assist mothers and children, both socially and financially. Yet this was not just another *natalist* project: social and biological technologies to improve the nation's health worked together with eugenic ideas for the long-lasting improvement of the race (Schreiber 1913). As discussed in the previous chapter, French demographers took up the issue of the dangers from de-population from the mid-1870s onwards. One of the most prominent demographers of the era, Paul Leroy-Beaulieu, noted that processes of modernization, such as urbanization and the desire for social and economic advancement, operated to make large families less desirable. Given France's declining birth rates, Leroy-Beaulieu was concerned that French culture was becoming 'feminized', with parents doting and lavishing attention on their disappointingly few children (quoted in Cole 1996: 666).

In 1896, the statistician Jacques Bertillon created the *L'alliance nationale pour l'accroissement de la population française* (*National Alliance for the Increase of the French Population*). It promoted propaganda regarding the extent of population stagnation, its causes, and possible solutions (Bertillon 1897: 538–9). These included government aid to pregnant women and the children of working women, and various benefits to large families, such as prizes, school benefits, reduction of military service, and a tax on bachelors (De Luca 2005: 11–35). The aim was 'to enshrine the principle of preferential treatment for *families nombreuses* in existing tax law and then to broaden continually the base of benefits until a whole system of family-based welfare was in place' (Quine 1996: 64). Excerpts from Bertillon's most important book, *De la dépopulation de la France* (*On the Depopulation of France*), published in 1896, would eventually be read by Benito Mussolini in the 1920s, and help him decide to launch a full-scale Latin eugenic programme in Italy (Ipsen 1996: 84).

To be sure, French eugenicists did not neglect issues surrounding birth control and voluntary abortion, both seen as factors responsible for the decline of birth rates. Together, eugenicists and political leaders suggested that the state's demographic needs had to come before personal reproductive decisions; thus the use of contraceptive methods was deemed 'unpatriotic'. The senator Paul Strauss, for instance, declared, 'France must not permit criminal abortion to decimate her natality' (quoted in Fuchs 1992: 197). Accordingly, some women had become so 'selfish', as to resort to abortions to avoid their maternal duties. Indicating the dire scope of the problem, even the French prime minister, Louis Barthou, commented on the issue in 1913. He estimated that there were 100,000 abortions in France each year. The deteriorating demographic situation for France became increasingly worrisome: there simply were not enough recruits to field an army that could match that of Germany, for example. To this effect, in that year, France lengthened its period of military service to three years for each conscript, in a rather desperate effort to keep up with its belligerent neighbour to the east.

Similar eugenic debates on population growth, family and child protection are to be found in Spain during the 1910s. Spanish eugenicists argued for wider popularization of ideas of racial improvement and of similar developments in other countries. For example, in 1910 Ignacio Valentí Vivó published his overview of eugenics as *La sanidad nacional: eugenesia y biometría* (*National Health: Eugenics and Biometry*), in which he promoted eugenics as a panacea to Spain's social and biological deterioration (Vivó 1910). The nation's growing health problems were filtered through a medicalized interpretation of society, which allowed for the emergence of eugenic programmes to control and ideally prevent the further spread of social pathologies. Thus, Nicolás Amador, another physician from Barcelona, who was also a member of the Education Eugenics Society, championed interventionist eugenic measures. He argued that 'genetic and eugenic investigation, which modern biology puts at our reach and disposal' could provide the solution to Spain's 'national problem (decadence)' (Amador 1914: 167–8). As a testimony to the engagement of these physicians with ideas of social and biological improvement, an *Institut Mèdic-Social de Catalunya (Medical-Social Institute of Catalonia)* was established in 1912, promoting broader ideas of social medicine and eugenics (Rodríguez Ocaña 1987: 33–6).

In Catalonia, in particular, eugenics was both a scientific and a political programme, reflecting anti-Spanish separatism (Caja 2009). Catalonian eugenicists viewed their community as a nation unwillingly under the control of the Spanish state. Thus, they tended to emphasize the need to keep the Catalonian people 'pure', and Catalonia free from large numbers of Spanish immigrants, who would inevitably weaken Catalonian 'national' identity and the prospect of eventual independence from Spain. The physician and leading Catalonian conservative eugenicist, Hermenegildo Puig i Sais, spoke of the 'fatal dangers of de-Catalanization and the need to increase the number of Catalans of pure stock' (quoted in Conversi 1997: 194).

Puig i Sais built his eugenic reputation on a book he published in 1915, *El problema de la natalitat a Catalunya* (*The Problem of Natality in Catalonia*). Similar to the demographic eugenics proposed by Corrado Gini and Lucien March, Puig i Sais claimed that population growth was essential to the nation's racial future (Sais 1915: 7). In terms of national efficiency, he also suggested that more workers would translate into higher commodity production and a flourishing economy. The nation would be more militarily powerful because there would be more soldiers. It would loom larger in the intellectual world, due to the birth of a greater number of 'superior intellects'. Conversely, any nation that suffered from a low population density would risk invasion and domination by peoples of higher population density, as had occurred to the Greeks and Romans and now threatened Catalonia. Above all, Puig i Sais worried about the low Catalonian birth rate relative to the rest of Spain, and the masses of Spanish immigrants attracted to Catalonia because of its relatively vigorous economy. He thus attributed the low birth rate to the effects of the 'exaggerated individualism' that had taken root among the rapidly modernizing Catalonian population. They increasingly ignored their obligations to the Church and the nation to produce large families, and were seduced into using birth control (Sais 1915: 66).

French and Spanish eugenic ideas of child protection, maternal care, and pronatalism, coupled with Italian demographic eugenics, found fertile soil in the Latin countries of South and North America. In puériculture, for instance, Argentina was most prominent, as it was in eugenics in general. A department for the 'Protection of Infants' (*Protección de la Primera Infancia*) was created in 1908 (Rodriquez 2006: 119), followed by the League for the Rights of Mothers and Children (*Liga para los Derechos de la Mujer y del Niño*) established in 1911 by Dr. Julieta Lantieri Renshaw. The League, in turn, organized the First National Congress of the Child (*Congreso Nacional del Niño*) held in Buenos Aires in 1913. Dr. Eliseo Cantón, the chair of Clinical Obstetrics at the University of Buenos Aires, was one of the contributors to the Congress. As Yolanda Eraso has noted, Cantón – considered to be the 'Father of Argentinian Obstetrics' – 'was a strong advocate for French puériculture'. His paper was 'imbued with French innovative ideas, especially those devoted to offer medical caring to destitute pregnant women' (Eraso 2013: 23–4).

The outbreak of the First World War in August 1914 dramatically changed eugenic rhetoric and practice. Eugenicists everywhere invoked themes of racial loss and disrupted birth rates, but more significantly, visualized these in terms of national virility and youthful sacrifice. Equally important for the topics discussed in this book, the war brought to the centre of national and international politics the dichotomous interpretation of history, positioning the Latin races' civilization versus the Teutonic races' barbarism.

The impact of the First World War

The ebb and flow of the war had a direct impact on attitudes in the Latin European nations regarding their own resilience and military potential. In particular, the Battle of the Marne, in the autumn of 1914, saw the French halt the German invasion and stabilize the Western front at a point beyond which the Germans would never effectively advance. Pan-Latinists were jubilant, portraying the victory as a 'miracle', and a vindication of the unfailing vitality of the 'Latin race'. Italy entered the war on the side of the Allies in 1915, the third Latin country to do so (after France and Belgium).

Previous collaboration between scientists from Latin countries – such as the *Union Médicale Franco-Ibéro-Américaine* (*Franco-Spanish-American Medical Alliance*), established in 1912 by two French doctors, Louis Dartigues and Gaullieur L'Hardy, and the Spaniard, Alberto Bandelac de Pariente – exercised this patriotic militancy in the sphere of science (Dartigues 1926). Described by the British medical journal *The Lancet* as 'a scientific alliance of Latin races' (see Figure 2.3), membership was 'open to all doctors throughout the world who speak Spanish or Portuguese' ('Scientific Alliance of the Latin Races', *The Lancet* 1913: 1081). Other organizations like the *Comité France–Italie* (French–Italian Committee), and its sister organization, *Italia–Francia*, dedicated to building bonds between Italy and France in the name of their common Latinity. Yet cultural and political unity did not inspire common activities among Latin eugenicists – as it did, for example, among German, Austrian, and Hungarian eugenicists. No Latin conferences on eugenics, population policy and race-protectionism took place during the war, comparable to the First German–Austrian–Hungarian Conference on Racial Hygiene and Population Policy (*I. deutsch-österreichisch-ungarischen Tagung für Rassenhygiene und*

Figure 2.3 French–Spanish–American Medical Alliance
Source: Dr. Dartigues, ed., *Livre d'or et Annuaire de L'Union Médicale Franco-Ibéro-Américaine* (Paris: Laboratoires Darrase, 1926), front page. Courtesy of Bibliothèque Interuniversitaire de Santé, Paris.

Bevölkerungspolitik), convened in Budapest in September 1918 or the Austro-Hungarian Conference on Child Rearing (*Österreichisch-Ungarische Tagung über die Fragen der Kinderaufzucht*), held in Vienna in October 1918 (Turda 2014: 223–4).

In a war that saw all sectors of society mobilized to serve the state in this bitter struggle, eugenicists were called upon to harden the nation's resolve to fight. This explosion of militarism was often accompanied by racial prejudice and stereotyping. French authors, for instance, consistently negated German claims of racial and cultural superiority, insisting – like the anthropologist George Poisson – that miscegenation and a succession of historical migrations had irremediably dissolved the 'pure' Aryan and Nordic nature of German racial type (Poisson 1916: 25–43).

With Latinity now fuelled by war propaganda, pan-Latinists proclaimed a new sense of 'Latin sisterhood' ('Vers la fédération: possibilité et conditions d'une federation latine', *Revue des Nations Latines* 1917: 60–4). In his *Les Pays méditerranéens et la guerre* (*The Mediterranean Countries and the War*), the French historian and novelist Louis Bertrand, for instance, proposed the creation of a Latin customs union. At the centre of Bertrand's programme was the argument that an inevitable racial fusion would occur as Italians and Spaniards settled in France, for example. As a result, the ancient Latin race would be resurrected and a union of 120 million revitalized Latins would be enough to counterbalance the equal number of ethnic Germans in the Central Alliance (Bertrand 1918).

Italian pan-Latinist proposals for union were quite similar to those of the French, with whom they were in frequent contact. The Italian jurist Pietro Bonfante's proposal was one of the best known. He argued that an Italian–French federation, of eighty million inhabitants, would likely be so prosperous that Spain and Portugal would eventually apply for membership (Bonfante 1915: 325–42). A young Benito Mussolini was among those Italian intellectuals sharing this vision. In October 1914, the Italian Socialist Party expelled Mussolini for his apparently sudden conversion to interventionism. He wished Italy to join the Allies, primarily because he felt that they offered Italy a better territorial deal.

Quite typically for socialist interventionists, Mussolini was also a pan-Latinist, although he interpreted Latinity as an aspect of Italian historical greatness. Certainly, he developed an obsession with Italy's Roman heritage, and dreamed of the creation of a new Roman Empire, and with Italian domination of the Latin peoples, if possible. In particular, during the war Mussolini attended meetings of an interventionist, Francophile circle among whom were the editors of the newspaper *Il secolo* and the *Revue des nations* (Luchaire 1965: 36). 'Many interventionists', as Cassata noted, 'conceived Italy's participation in the First World War as a decisive stage for the regeneration of the Italians through the test of war.' The war, in fact, 'became a factor of fusion between the myth of revolution and the myth of the nation' (Cassata 2011: 43).

Mussolini was convinced of the war's palingenetic potential. For instance, on 20 November 1916, Mussolini wrote the article 'Italia e Francia' ('Italy and France') for the newspaper *Il popolo d'Italia* (*The Italian People*) in which he explained that a Franco–Italian, Latin union was necessary for Italy to survive:

> Before the German and Austro-Hungarian powers, with seventy million inhabitants, neither the French nor the Italians would be able to have complete autonomy or security unless joined together [*se non congiunti*], in such a way as to counter the Germanic block with a Latin block of seventy million (and even eighty million, if one counts emigrant Italians). In the dynamic game of alliances, it wouldn't be wise for either of these nations to depend completely

> for its own national security on other ethnic groups, such as the Anglo-Saxons or the Slavs, since such a condition would create a state of dependence and inferiority. (Mussolini 1953: 199)

Such political ambitions did not obscure practical eugenic agendas, however. In France, for instance, the preoccupation with 'demographic warfare', remained central to the eugenic discourse during the war. Adolphe Pinard, for instance, argued that one of the most important offensive weapons in the hands of the French people was their ability to produce more children. The demographic consequences of the German invasion, which often led to the rape and impregnation of French women, provoked outrage and anxiety in Pinard and other eugenicists, as well as in the general public. For most of the war, ten French departments were under German occupation, and hundreds of instances of the German rape of French women were reported (Wishnia 1995: 37–8). As Elizabeth Vlossak has remarked, 'Despite the First World War's modernity, rape remained a widespread physical and psychological weapon used against women in an effort to humiliate the enemy and destroy morale, as well as blur national and racial borders' (Vlossak 2010: 144).

Similarly, parts of the Italian Veneto were under Austro-Hungarian and German occupation for a year, occasioning similar brutalities. Embracing nationalism, many eugenicists assumed that the children born of these unions would pollute the Latin race, and that there was even a deliberate attempt by the enemy to increase their population using Latin women as, in essence, forced surrogate mothers (Mantovani 2004: 152). Paul Hamonic, a French gynaecologist, indignantly denounced such 'low-bred criminals' who arrogantly imagined that they were honouring 'the Celto-Latins in violating their women and girls, under the pretext of creating, as supermen, choice human beings' (quoted in Harris 1993: 195). The idea of deliberate racial infection through rape was apparently so new to most Europeans that all sorts of possible horrors were imagined. Dr. Paul Tissier feared the possibility of 'physiological impregnation' of French rape victims. Imaginatively, Tissier believed that German sperm might be able to survive in the vaginal mucus, passing on their genes to the women's egg cells, and maybe even producing antibodies to destroy legitimate French sperm from successfully conceiving a true French child (Harris 1993: 196).

In Italy, the project of protecting women, as with all other eugenic schemes, involved mobilizing the medical profession and state resources. This is particularly well illustrated by the journal *La ginecologia moderna*'s (*Modern Gynaecology*), special 1917 issue on the 'protection of women and race' (Cassata

2011: 64–6). Its editor, Luigi Maria Bossi, problematized the protection of women in terms of its social and legal consequences:

> The defence, therefore, of women and race, in relation with neo-Malthusianism, criminal abortion and the right to abortion of women systematically violated by the Germans constitutes a large, complex problem that must be resolved through three indivisible relationships: social, juridical and medical. And it is above all pertinent to gynaecologists, because they are responsible, as is obvious, for the basal concept of conservation of the species, of the present life and health, of the mother, and subordinately, of the life and health that is the product of conception. The social and juridical sides must naturally be subordinate to the gynaecological side. (Quoted in Cassata 2011: 67)

The conflict between the role of women as racial guardians of the nation, and the physical abuse created by invading armies, became particularly evident in the eugenic propaganda directed at the future generations. Unborn children not only served as the ubiquitous focus of the nation's racial regeneration, but provided eugenicists with a mobilizing medical agenda.

The debate over wartime rape and racial pollution gives some idea of the psychological trauma and upheaval caused by the war. To win this struggle to the death, as the war came to be interpreted, all the resources of the nation were managed by central governments to an extent unimaginable before the conflict. Control of human resources was just as important as control over material assets. It was largely due to the demographic changes brought about by the war that political elites turned to eugenics as a means of promoting social and biological revivalism in the midst of the disillusioned post-war political environment. Eugenicists, in turn, alerted the government to the need for a rigorous health policy integrating hygienic and eugenic principles. What ensued was an intense debate not only about national health and societal protection but, ultimately, about national survival. Furthermore, eugenicists emphasized that society's purification with a view to its biological continuity depended upon the transmission of new racial codes and morality to the general public (Cassata 2011: 43–68).

Although some eugenicists were inclined to see the positive effects of the war in its early stages, especially with regards to fostering patriotism and national solidarity, by the end of 1916 the devastation brought about by this increasingly vicious military conflict convinced many eugenicists that war was dysgenic as it affected both the combatants and the civilian populations. In his 1917 *La guerra e la popolazione* (*War and Population*), the Italian demographer and eugenicist

Franco Savorgnan offered a convincing description of the qualitative and quantitative decline of 'the racial type' of Italian men, as well as their reproductive capacities, due to their participation in the war. Savorgnan further asserted that: 'The great majority will be undermined by privations, venereal diseases and tuberculosis, or, in the best hypothesis, will have brought home from the war a nervous system strongly prejudiced by the ceaseless fire of the artillery' (Savorgnan 1917: 93). Lucien March also feared that the war contributed to an increase of 'alcoholism, tuberculosis, and venereal diseases' among the civilian population, together with general mortality and a decline of birth rates. Such a catastrophe was even greater justification for eugenic legislation affecting the entire nation, March suggested (March 1919: 195–212).

Eugenic readings of war changed according to the author. The process of interpreting the dysgenic impact on the population was consequently not a process of assigning fixed meanings to war but, rather, that of reading the patterns of structured modification of the racial quality of the population. Corrado Gini was, illustratively, less inclined to wholeheartedly accept pessimistic visions of purported wartime racial destruction 'In practical eugenics', he maintained, 'the important thing is not to have a fixed eugenic ideal, which can seldom be realized, but to have adequate criteria for discriminating between favourable and unfavourable eugenic factors' (Gini 1923: 430).

The First World War had profound implications for eugenics in general. Issues such as low birth rates, infant mortality, reproductive health, and physical fitness became national concerns on an unprecedented scale. The material deprivation and suffering that intensified during the war also fuelled these concerns; virtually all measures of demographic health rapidly sank among the European combatants. France came through as the most ravaged Latin nation of the war: 1.5 million men were killed and 800,000 rendered infertile through battle wounds, with 400,000 more civilians dying than would have been normal for that period. This meant that about 2.5 million families would not form and produce children. Furthermore, with up to six million men mobilized for over four years, the birth rate of the country had plunged (Huss 1990: 39; Reggiani 1996: 731) Lucien March calculated that France, for instance, suffered a deficit of 1.4 million births during the war (March 1921: 399).

Demographic recovery after such loss of human life would be extremely difficult. The consequent reduction in the size of the next generation would naturally shrink the size of the army and the labour force. French political leaders understood the consequences of these dismal numbers even as they stood in the midst of victory in the world war. Compounding the problem, eugenicists noted that contagious diseases such as tuberculosis and venereal diseases had spread through the army and the civilian population. Alcoholism and nervous diseases also increased.

All European countries involved in the war emerged at the end of 1918 devastated by the spread of contagious diseases, economic ruin, and depopulation. The armistice with Germany was not yet signed when Latin eugenicists began planning the revival of their countries' national health. As one of the primary battlefields, France now devoted enormous attention to morally and physically rebuilding its body politic. To this effect, the introduction of the Act of 31 July 1920 – described by Angus McLaren as having 'endowed France with the most oppressive laws in Europe against contraception and birth-control' (McLaren 1983: 1) – clearly demonstrated the entrenchment of pronatalism as a means and a technique for managing the life of the French population (Accampo 2003: 235–62).

Considering the post-war importance bestowed on the nation's biological future, it is not surprising that post-war Latin eugenicists assumed social and biological improvement could not be effective without state intervention and support. Through financial subsidies and the creation of new institutions, the post-1918 European state prompted Latin eugenicists to see themselves as indispensable contributors to the project of national renewal. Eugenic conferences reflected this orientation, with crucial issues ultimately revolving around the tense relationship between the rights of the individuals and those of the state. As Europe recovered from the economic devastation caused by the war, Latin eugenicists sought to convince their governments to afford more importance to projects of public health, social hygiene, and puériculture, all with the aim to improve the general health of the population.

One telling example of this international collaboration is provided by the Interallied Congress of Social Hygiene (*Congrès Interallié d'Hygiène Sociale*), held at the Sorbonne between 22 and 26 April 1919. The French National Committee of Physical Education and Social Hygiene (*Comité National de l'Education Physique et de l'Hygiène Sociale)* – established 'for the safeguarding and regeneration of the French race' (Paté 1919: 42) – together with the French government and the American Red Cross were the main organizers. The Congress was attended by delegations from Belgium, Great Britain, Brazil, Greece, Italy, Japan, Poland, Portugal, Serbia, and Czechoslovakia.

As expected, French organizers determined the Congress' structure and scientific content. In his opening address, Henry Paté, president of the National Committee, saluted the 'representatives of the Allied nations who, once again, come together to strengthen and improve the race' (Paté 1919: 41). Sections were organized on diverse topics such as rural and urban hygiene, the protection of mothers and infants, puériculture, eugenics, school hygiene, physical education, sanitary prophylaxis, social assistance, and hygiene in educational and industrial institutions.

The papers presented at the Congress exuded optimism. Heralding a new international collaboration, Henri Doizy, president of the Commission for Public Health (*Commission de l'Hygiène Publique*), and the eugenicist Justin Sicard de Plauzoles proposed the creation of an Interallied Institute of Public Health (*Institut Interallié d'Hygiène Publique*), to be located in Paris and affiliated with the League of Nations. Four sections of the Institute were envisioned: the first dealing with statistics, economics and the aetiology of disease, along with associated research and legislation; the second comprising medical research, especially in pathology, immunology and therapy; and the third focusing on moral, sanitary, individual, public, social, national, and international prophylaxis. Finally, the fourth section brought together education and hygiene, with a special focus on eugenics, maternal hygiene, puériculture, pedagogy, school hygiene, physical education, and related topics (Doizy and De Plauzoles 1920: 9–11).

Eugenics and puériculture were well represented at the Congress ('Revue des Congrès', *Annales d'hygiène publique et de médicine légale* 1919: 50–7). Edmond Lévy-Solal, a gynaecologist and professor at the Faculty of Medicine in Paris, focused on puériculture. Following Adolphe Pinard, Lévy-Solal outlined the tripartite role played by puériculture in the 'expansion, improvement and preservation of race' (Lévy-Solal 1919). Conversely, eugenics figured prominently in the paper presented by the president of the Czech Eugenics Society, Ladislav Haškovec. He spoke at length on the benefits of requiring medical certificates before marriage (Haškovec 1920: 152–65). The introduction of this eugenic measure had preoccupied Haškovec since the early 1900s, and he promoted it at various international conferences during the 1920s and early 1930s. Considering the French preoccupation with the issue of premarital medical examinations, characteristic of the following decade, Haškovec could not have presented his arguments in a better location.

The fact that immediately after the war most European countries shared similar health concerns is revealing, suggesting that national traditions were less disconnected than previously assumed. But the overall impact of the peace treaties that followed the First World War should not be underestimated, particularly for the defeated nations: Germany and Hungary. However, victorious countries were no less pessimistic. France and Italy had particular reasons for bitterness: for

France, its erstwhile allies, Great Britain and the USA, lost interest in the issue of French security, and eventually refused to promise France protection from a revived Germany. In anticipation of such a promise, France had reluctantly let go of its most cherished project: an independent Rhineland buffer state.

With passion as great as their justification was weak, Italians became enraged as their promised territory along the Dalmatian coast was denied to them by the USA, driven by Wilsonian ideals of ethno-nationalism, and by the conspicuously weak support of France and Great Britain for Italy's wartime claims. To some extent, this disappointment would contribute to the emergence of fascism in Italy during the 1920s. Of the Latin nations, Romania benefitted the most from the victory of the Allied powers. Its pre-war area doubled with the addition of territories from the collapsed Russian Empire and defeated Austria-Hungary – an act meant more to punish these states than to reunite ethnic Romanians in a Wilsonian nation-state.

It seemed that the 'Latin alliance' was under threat after the war. Pan-Latinist Italians and French continued to hope that the Franco–Italian alliance would hold in the post-war world, and each partner would endorse the other's claims at the peace table. However, Otto von Bismarck's observation about Italy, that it had 'a large appetite but poor teeth' would prove all too true during the negotiations. The Italians not only demanded territories inhabited by a majority of ethnic Italians, which the Allies were perfectly winning to grant, but also demanded the fulfilment of vague wartime promises of distant lands where the Italian population was sparse, as in Dalmatia, or non-existent, as in Anatolia. The French, with their own dreams of a new Near Eastern empire, had no desire to compete with Italy for colonial dominion there, as they had experienced in the late nineteenth century in the western Mediterranean. Thus, pan-Latinist dreams that had seemed quite reasonable (if perhaps a bit overly idealistic) only one year earlier came crashing down in 1919 when faced with traditional nationalistic rivalries that quickly reasserted themselves after the pressure of Germanic domination had ended. Far from diminishing ethnic nationalism, the aftermath of the First World War had only increased it.

One can see the rapid disillusionment that spread throughout the pan-Latinist community in Mussolini's articles, published in the first half of 1919 in his newspaper *Il popolo d'Italia*. In mid-February of that year, when the French were deliberating over the fate of the small city of Fiume (Rijeka) that Italy now demanded, Mussolini wrote hopefully that:

> Many of our friends on the other side of the Alps [i.e. France] begin to realize that the alliance with Italy, a country in full development with forty million

> inhabitants, which will become sixty million within half a decade, is a lot more useful than the ignoble love of four wild Croats who until yesterday were the willing goons [sgharri] of Vienna.

He was prepared to grant the French their longed-for Rhine frontier, to help rein in the German menace, but Italy likewise had to expand, to face their common Germanic enemies on the Brenner Pass (i.e. the Tyrolean Alps). Thus arranged, the 'foolish revanchism' of the Germans would collapse, when faced with a 'compact Latin bloc of eighty millions' augmented by new territories and populations. The future Duce was adamant that there was no time to lose. 'We need to found this alliance between Italy and France with clear, solemn, precise treaties' – he insisted (Mussolini 1919a: 364).

Yet the French, whose territorial ambitions in the Near East now clashed with the Italians, refused to acquiesce to the enlargement of the Italian Empire in that region. Mussolini reacted immediately to what he perceived as French intransigence. On 24 May 1919, he wrote:

> Regarding France's pretension to a European hegemony: does it not seem to you a little ridiculous for a nation whose population is diminishing, which, in order to recover, will need Italian workmen, which Italy might always refuse! France is pursuing a stupidly and dangerously anti-Italian policy. Very well. Italy is free today to pursue an anti-French policy. (Mussolini 1919b: 104–6)

There would be no 'Latin Union' in the wake of the First World War. As one French commentator put it, the prospective bride and groom of the 'Latin Union' ended their engagement in a fury of recriminations. Mussolini also made clear in the declaration of the programme of the new National Fascist Party, in November 1921, that Italy would reaffirm 'Latin civilization in the Mediterranean' (quoted in De Felice 1966: 758). Once Mussolini gained control of the Italian government, he would promote his own pan-Latinist vision, this time under Italian, instead of French, domination, that had been the customary expectation a generation before. As will be discussed in the next chapters, this new, aggressive Fascist Italy also served as the main promoter of the interwar Latin eugenic movement.

3

Latin Eugenics in Interwar Europe

The main eugenic themes emerging before and during the First World War, such as national regeneration, demographic growth, the containment of contagious and venereal diseases, and maternal and child care, continued to dominate the research and public agenda of a number of health institutions created during the interwar period in the Latin countries. These included the Institute of Puériculture and Perinatology (*Institut de puériculture et de périnatalogie*) established in Paris in 1919, and the Ministry of Hygiene, Social Assistance and Prevention (*Ministère de l'Hygiène, de l'Assistance et la Prévoyance Sociale*) created in 1920. Immediately the Ministry set up a High Council of Natality and Child Care (*Conseil supérieur de la natalité et de la protection de l'enfance*), with the mandate to 'fight depopulation', by proposing means to raise the birth rate, expand programmes on puériculture, and assist large families. Similarly, in Spain the Institute of Social Medicine (*Instituto de Medicina Social*) was established in 1919, followed by the Italian Institute of Public Welfare and Assistance (*Istituto di Previdenza e Assistenza Sociale*) in 1922 ('Social Welfare in Italy', *Eugenical News* 1922: 62). The aim of these institutions was to 'renew or "regenerate" the health situation of the state, through the education of population and the guidance of authorities' (Rodríguez-Ocaña 2007: 50; also Peláez 1988a: 343–58).

But it was not only in the 'core' Latin countries that the eugenic conception of national improvement achieved prominence. The 1920s witnessed similar developments in Latin countries situated at the 'periphery', such as Belgium, Romania, Spain, and Portugal (Pimentel 1999: 477–508). Like their French and Italian counterparts, Belgian, Romanian, Spanish, and Portuguese eugenicists encouraged similar synergies between eugenics, social hygiene, and public health. They too placed preventive medicine, motherhood, and family at the centre of their eugenic agenda, one locating the nation within a renewable biological matrix. Yet, the active refashioning of various eugenic theories to accommodate local specificities produced interpretations of eugenics which were distinctively Belgian, Romanian, Spanish, and Portuguese, notwithstanding some commonalities with Anglo-American, German, and, more importantly, the Latin model advocated by French and Italian eugenicists.

Latin eugenics at the 'periphery'

The third eugenic society to be established in a Latin country was the Belgian Society of Eugenics (*Société Belge d'Eugénique*), in September 1919. The zoologist George Albert Boulenger was elected president, and Albert Govaerts its secretary. The Society was divided into several sections, each devoted to a particular field, including 'venereology', 'alcoholism', 'child welfare', 'morality', 'pedagogy', and 'criminality'. Other prominent members included the social hygienist and founder of the *Association Belge de Médecine Sociale* (Belgian Association of Social Medicine), René Sand and the criminologist Louis Vervaeck ('Société Belge d'Eugénique', *Eugenical News* 1920: 54; Nisot 1929a: 116). Prior to this, a eugenic 'working group' ('Cellule Eugénique') – created in 1912 by Emile Waxweiler – functioned within the *Institut de Sociologie Solvay* in Brussels. It included paediatrician Ovide Decroly, social hygienist Norbert Ensch, and the educationalist Louis Querton (Nisot 1929a: 117). The Belgian Society of Eugenics endeavoured to combine education and research with a practical programme, centred on the 'physiological, intellectual and moral amelioration of the human race and more especially [of] the Belgian nation' ('Belgian Society of Eugenics', *Eugenical News* 1920: 63; Nisot 1929a: 114–15). To this effect, it began publishing its own journal *Revue d'Eugénique* (*Eugenics Review*) in January 1921, and organizing public lectures on eugenics (Nisot 1929a: 117). As *Eugenical News* remarked in 1922: 'Eugenic education is the main business of the Eugenics Society of Belgium. The most striking result of this work is the recognition of eugenics by the Belgian government' ('Belgian Eugenics Society', *Eugenical News* 1922: 14).

As already mentioned in the previous chapter, Ensch and Querton participated in the First International Eugenics Congress in London, and then joined the Permanent International Eugenics Committee. In 1914, Albert Govaerts visited London and New York in order to familiarize himself with the work of the Eugenics Education Society and the Eugenics Record Office, respectively ('Revue d'Eugénique', *Eugenical News* 1921: 43; Nisot 1929a: 118). He returned to New York in 1921 as part of a cultural exchange programme between Belgium and the USA, entrusted by the Belgian government 'to make a careful study of the science of eugenics during his year in America with a view of the establishment of an Institute of Eugenics in Belgium' ('Foreign Notes', *Eugenical News* 1921: 72; 'Dr Albert Govaerts of Belgium', *Eugenical News* 1922: 64).

Indeed, a National Eugenics Office (*Office National d'Eugénique*) opened on 4 June 1922 in Brussels. The politician Emile Vandervelde was elected president, but Albert Govaerts and Willem Schraenen carried out the main responsibilities. The purpose of the new organization was educational, and focused on social hygiene, family, and puériculture (Nisot 1929a: 119–20). Specifically, the Office provided courses in eugenics to the State School of Social Service, which was an institution preparing 'students to undertake actual social service in connection with societies and institutions devoted to charity, the protection of children, and other welfare activities' ('The New Belgian Eugenics Office', *Eugenical News* 1922: 92; 'National Office of Eugenics in Belgium', *Eugenical News* 1922: 120–1).

A few months later, in early October 1922, the Permanent International Eugenics Committee met in Brussels. To draw attention to the meeting but also to increase its public visibility, the Belgian Society of Eugenics organized a series of public meetings and lectures under the name of 'international eugenic days' (7–11 October). For example, the French eugenicists Lucien March and Eugène Apert spoke on marital education and morbid heredity, respectively. Adolphe Pinard also attended, speaking in favour of 'rational sexual education'. These events also coincided with the establishment of the Belgian National League against Venereal Diseases (*Ligue Nationale Belge contre le Péril Vénérien*) on 8 October (Apert 1922a: 322–3).

While accepting the importance of heredity, the Belgian eugenic programme during the 1920s was deeply dedicated to social hygiene, preventive medicine, child protection, and Catholic morality. René Sand, for example, considered that the main reason for the practical application of eugenics in Belgium was the 'improvement of the sanitary condition of the population' (Nisot 1929a: 112–3). In his 1934 *L'économie humaine par la médicine sociale*, also translated into English as *Health and Human Progress*, Sand acknowledged that negative eugenic measures such as sterilization and segregation would eventually be effective in reducing 'the number of social derelicts'. But eugenics was not merely a question of preventing these individuals from reproducing; more importantly, Sand argued, '*eugenics may act on the population as a whole by educational measures and it may intervene by way of instruction and advice in individual cases*' (Sand 1935: 133. Emphasis in the original). Beyond the introduction of negative eugenic methods loomed something more meaningful. According to Sand, 'to spread knowledge relating to heredity; to enlighten parents, adolescents and young couples on these questions; to awaken the sense of responsibility in marriage and procreation' was 'to create a *eugenic conscience*' (Sand 1935: 133–4. Emphasis in the original).

Another prominent member of the Belgian Eugenics Society, the Jesuit Valère Fallon added that eugenic work in Belgium should be devoted to oppose neo-Malthusianism and to encourage the reproduction of large families. Fallon also emphasized the 'fight against the enemies of the family', such as 'the overworking of women and children, feminist abuse, infectious diseases and immorality'; while at the same time highlighting the eugenic importance of 'religious and moral education' for racial improvement (quoted in Nisot 1929a: 123).

When the Belgian Society for Preventive Medicine and Eugenics (*Société Belge de Médicine Préventive et d'Eugénique*) was established on 22 August 1929, it echoed many of the ambitions put forward by its predecessor in the early 1920s ('Society for Preventive Medicine and Eugenics', *Eugenical News* 1929: 173). By insisting on the importance of hygiene and health, the new society attempted to define eugenics within an explicitly social and medical realm. As in other Latin countries, Belgian eugenicists embraced an interpretative framework based on the protection of motherhood, child welfare, and the family, and promoted a range of pronatalist policies. This was also the eugenic programme that eugenicists and Catholic social hygienists exported to the colonies. As Nancy Rose Hunt convincingly argued 'The Belgian Congo's hygienic modality of colonial power' was 'distinctly maternalist, enumerative, and positive in its eugenics, unlike other imperial regimes whose sexual and bodily fixation were more obviously directed at eradication [...] or to halt signs of colonial degeneration such as miscegenation and venereal disease' (Hunt 1999: 10). 'The eugenic modality of Belgian colonial power' valued, first and foremost, 'sexuality, fertility, childbearing, and mothering', while at the same time accentuating the imperative of maintaining a healthy race (Hunt 1999: 10).

Not surprisingly, then, building on social hygiene, sanitation, and preventive medicine, Belgium provided a eugenic model that eugenicists in other Latin countries, such as Romania, found inspiring. The creation of 'Greater Romania' in 1918 prompted Romanian health officials, medical and social experts to engage in an unprecedented programme of institutionalization in the field of eugenics, social hygiene, and public health (Bucur 2002; Turda 2008). These experts believed that science offered the much-needed assistance to sustain the social and economic transformation of Romania into a modern country.

Accordingly, they planned the future of the nation through social policies built into the country's emerging public health and welfare system (Nisot 1929a: 433–4).

In 1919, the prominent eugenicist Iuliu Moldovan (see Figure 3.1) founded the Institute of Hygiene and Social Hygiene (*Institutul de Igienă şi Igienă Socială*) in Cluj, the capital of Transylvania. The Institute played a significant role in shaping the development of social hygiene, public health, biopolitics, and eugenics in Romania; it introduced courses on racial hygiene, eugenics ('igiena naţiunii'), and biopolitics in 1921 (Moldovan 1923: 77). In the same year, Moldovan proposed the creation of a Ministry of Health whose main focus would be the 'qualitative development of the entire population' (quoted in Bucur 2002: 191). Eventually, the Ministry of Public Health, Labour and Social Protection (*Ministerul Sănătăţii Publice, Muncii şi Ocrotirii Sociale*) was established in 1922.

Eugenics was regarded as a source of regenerative nationalism and scientific progress (Voina 1924a). In his 1925 *Igiena naţiunii: Eugenia* (*The Hygiene of the Nation: Eugenics*), Moldovan defined the nation as 'a biological reality, a human structure with its specific biology and pathology' (Moldovan 1925: 12). Moldovan then placed the nation at the centre of this theory of eugenics, envisioning hygienic and public health measures to protect it from both social and biological degeneration. 'The hygiene of the nation', he argued, dealt 'with factors and measures that determine the current and future biological prosperity of the nation' (Moldovan 1925: 37). However, Moldovan claimed that no medical prophylactics would be sufficient unless the Romanian population,

Figure 3.1 Iuliu Moldovan (centre) together with his collaborators
Source: Library of the Institute of Medicine, Institute of Public Health, Bucharest.

especially those affected by hereditary diseases, acquired 'a racial consciousness, a sentiment of biological responsibility' (Moldovan 1925: 46). What should unite the members of the national community was not merely a political ideology acting in the name of the nation, but a fusion of new nationalist and eugenic ideals into the Romanian biopolitical state (Moldovan 1926).

During the 1920s, Moldovan immersed himself in the difficult task of constructing a eugenic philosophy adapted to Romanian conditions. Like other Romanian and Latin eugenicists opposed to the concept of 'superior races', Moldovan rewrote the contents of the eugenic ideal to reflect his country's specific circumstances. Aiming to narrow the difference between the individual and the collective, Moldovan found in eugenics a theme intended to embrace all members of the Romanian nation. Moldovan tried to *nationalize* eugenics to better fit the Romanian nation-building project.

Dr. Ioan I. Manliu, Deputy-Director of the Social Insurance Central Bank in Bucharest, harboured a related eugenic vision suited to Romanian realities. In his 1921 *Crâmpeie de eugenie şi igienă socială* (*Short Commentaries on Eugenics and Social Hygiene*), Manliu proposed a number of eugenic initiatives – including the creation of a eugenics bureau; the establishment of chairs of racial hygiene at every university in Romania; a yearly eugenic congress; and the foundation of a museum of hygiene and eugenics – all with the purpose to promote a healthy Romanian nation and race (Manliu 1921: 39–42, 78–80). While the Romanian state was coming to terms with its enlarged size and ethnic composition after 1918, eugenic concepts that made the Romanian nation *stronger* asserted themselves.

Various Romanian politicians thus seized upon eugenics to demonstrate the necessity for a state-controlled programme of national rejuvenation based on 'therapeutic medicine, preventive hygiene and the hygiene of the nation' (Haţieganu and Voina 1925: 813). In 1925, the Romanian nationalist organization the Transylvanian Association for Romanian Literature and the Culture of the Romanian People (*Asociaţia Transilvană pentru Literatura Română şi Cultura Poporului Român*, or ASTRA), announced the establishment of a 'Biopolitical and Eugenics Section' (*secţia eugenică şi biopolitică*) within ASTRA's medical section, led by the physician Iuliu Haţieganu (Haţieganu and Voina 1925: 813–4).

In a note published in *Eugenical News* in July 1922, Iuliu Moldovan reported that a eugenic section was created within the Institute of Hygiene and Social Hygiene. In 1927, this section joined forces with the Biopolitical and Eugenics Section of the ASTRA Association (*Secţia de eugenie şi biopolitică a Asociaţiunii 'ASTRA*) to become Romania's first eugenic society, along with its journal, *Buletin eugenic și biopolitic (Eugenic and Biopolitical Bulletin)*. Further illustrating the growing impact of eugenic ideas in Romania, the prominent Romanian politician Alexandru Vaida-Voevod gave one of the first lectures to the newly established society. However, one commentator noted that other Romanian politicians were not 'accustomed' to eugenic theories of social and biological improvement (Bogdan-Duică 1924: 155–6).

French and Belgian eugenics served as an example for Romania. Pinard's ideas of puériculture, for instance, were considered one successful eugenic strategy of 'racial protection and improvement' (Voina 1924b: 266). For advocates of puériculture, such as paediatrician Gheorghe Popovici, the application of eugenics in Romania necessitated not only the adoption of a new biopolitical governing philosophy but also a national welfare programme centred on the protection of the family (Popovici 1928: 443–4). For others, such as Gheorghe Banu, social hygiene, rural biology protection, and child welfare were essential to the eugenic improvement of the Romanian nation. 'Rural biology is one of the greatest issues of national interest' – Banu declared in 1927. And furthermore,

> [t]hrough the study of rural biology we must understand social biology, *because social biology is the science dealing with all factors which condition the creation, selection and the defence of health.* Social biology has a sphere of action broader than social hygiene, which deals only with defending the health of the *subject*; social biology considers the individual not only as an independent factor, and a simple component of humanity, *but as a creator and generator of life and as a factor influenced by biological selection and perfection.* (Banu 1927: 87. Emphasis in the original)

There were three sets of factors influencing social biology: environmental, social, and medical; the latter encompassed 'hereditary diseases, epidemics, or social diseases' (Banu 1927: 87). Banu puzzled over whether the state and the public alike would respond positively to a shift from an environmentalist approach to diseases to an emphasis on heredity and eugenics. This highlighted the importance of social medicine in rural Romania, while prompting Banu to encourage the introduction of various biomedical measures overseen by the state and health officials. As such, it fed into his more general concern with Romania's national health, especially child welfare, as illustrated by his book *L'hygiène sociale de l'enfance* (*The Social Hygiene of Child Welfare*), published in 1930.

In this book, Banu offered an extensive discussion of the relationship between eugenics and puériculture, highlighting the intricate relationships linking family, children, and controversial reproductive measures such as marriage certificates and sterilization. He extended the definition of eugenics to include not only the importance of 'procreation' but also the study 'of the dysgenic factors that directly or indirectly influence the value of the race' (Banu 1930: 5). The protection of mothers and infants, the promotion of family morality, and the state's power to enact the corresponding eugenic policies became, for Banu, questions of Romania's national survival.

The Spanish pedagogue and eugenicist Luis Huerta expressed similar ideas (Lázaro 2009: 61–88). Before the First World War, Huerta studied at the International Faculty of Pedology in Brussels, becoming acquainted with Ovide Decroly's theories about the special pedagogy for abnormal children, alongside Belgian and French ideas of puériculture and child welfare. His 1918 *Eugénica, Maternología y Puericultura* (*Eugenics, Maternology and Puériculture*) linked eugenics to larger welfare concerns, providing a neo-Lamarckian justification for social assistance for mothers and children (Huerta 1918).

In 1922, Huerta became the editor of the journal *Eugenia* (*Eugenics*), founded a year earlier in Barcelona. Under his editorship the journal promoted an eclectic interpretation of human improvement, reflecting the following aspects of hygiene and eugenics: 'Racionalismo pedagógico. Naturismo. Eugenesia. Puericultura. Excursionismo. Vegetarianismo. Esperanto. Cooperativismo'. Appropriately, in 1927, the journal changed its name again to *Eugenia: Revista mensual de cultura ecléctica* (*Eugenics: Monthly Journal of Eclectic Culture*) (Lázaro 2009: 62).

Huerta was also the president of the Eugenics Section (*Sección de Eugénica*) of the highly respected *Gaceta Médica Española* (*Spanish Medical Journal*) and one of the main organizers of Spain's first public conference on eugenics. While Romanian eugenicists turned to the state for support of their vision of social and biological improvement, the situation was different in Spain. During the 1920s, the Roman Catholic Church, together with Miguel Primo de Rivera's conservative authoritarian regime, made official endorsement of eugenics difficult, if not impossible (Peláez 1988b: 77–95).

The fate of the public conference on eugenics, the 'Primer Curso Eugénico Español' planned for Madrid in the spring of 1928, is illustrative in this respect (Noguera and Huerta 1934). Organized at the *Gaceta Médica Española*'s initiative, the conference was sponsored by some of Spain's most prestigious biomedical organizations, such as the Society of Anthropology (*Sociedad Española de Antropología*) and the Gynaecological Society (*Sociedad Ginecológica Española*). Realizing the fragility of their undertaking, the conference organizers focused on puériculture and the neo-Lamarckist eugenic theme, 'The Defence of the Race in Children' ('La defensa de la raza en el niño'), and presented papers on infant health, *maternidad consciente* (conscious maternity), and the familiar panoply of Catholic eugenic issues (Ferrándiz and Lafuente 1999: 133–48; Rodríguez 2012: 133–45). However, midway into the conference the regime became perturbed with the public dissemination of its discussions, which it declared 'truly ruinous for the family and social foundations, and destructive of the sanctity of marriage and the dignity of woman' (quoted in Sinclair 2007: 55). The 'Primer Curso' was then unceremoniously cancelled (Nash 1985: 195–202; Cleminson 2000: 84–8; Barrachina 2004: 1005–6).

One of the lecturers at the 'Curso' was the endocrinologist Gregorio Marañón (see Figure 3.2). He was one of the founders of *Instituto de Medicina Social* in 1919, one of the most important Spanish eugenicists of the interwar period and, not least, Spain's most renowned endocrinologist. His work, as Michael Richards pointed out, 'was essentially concerned with the "life-curve" of the individual and its relation to evolution of the human species and the broad environment' (Richards 2004: 833). Marañón first outlined his neo-Lamarckian vision of medical eugenics, maternal education, and puériculture in a 1920 article, 'Biology and Feminism' ('Biología y feminismo'), published in *El Siglo Médico* (*Medical Century*). Several years later he wrote the influential *Tres ensayos sobre*

Figure 3.2 Gregorio Marañón
Source: Fundación Ortega-Marañón, Madrid.

la vida sexual (*Three Essays on Sexual Life*), linking sexual differentiation to women's maternal duties and reproductive mission (Glick 2005: 121–38; Bolea 2013: 1–9).

Working within this framework, Marañón extended his prophylactic agenda to include Spanish society at large. He believed Spaniards were 'racially vigorous', but suffering from a number of environmentally induced degenerative factors. Marañón's fusion of eugenics with *maternidad consciente* did not, however, encourage state intervention in family life or its control of reproduction. As he put it in 1929: 'The voluntary, systematic and arbitrary restriction of motherhood is an attack on society and for us, Catholics, a sin' (Marañón 1966, vol. 1: 476). Although he argued in favour of educating women about their sexual and reproductive roles, Marañón nevertheless placed motherhood at the centre of Spain's national and moral revival.

By the early 1930s, however, a growing number of Latin eugenicists dismissed the state's attempt to socialize health care as being too reliant on an outmoded utilitarian, liberal model incapable of biologically resuscitating society. Following developments elsewhere, Latin eugenicists advised the state to adopt eugenic policies to bolster health improvement and biological strength. This coincided with a new phase in the history of Latin eugenics, when the traditional components of eugenics in the Latin countries – namely a preoccupation with racial degeneration, puériculture and neo-Lamarckist interpretations of social and biological improvement – were augmented by new concerns and proposed solutions. As made clear by the Portuguese anthropologist and eugenicist António Mendes Correia in a lecture delivered in 1927 to the National Congress of Medicine (*Congresso Nacional de Medicina*) and entitled 'The Eugenic Problem in Portugal' ('O problema eugénico em Portugal'), 'the Portuguese population' was not yet 'doomed to perish', but the application and observance of eugenic principles could foster the country's much-needed racial regeneration (quoted in Henriques 2012: 73).

Other Portuguese eugenicists agreed, even though some – like the paediatrician António de Almeida Garrett – believed that it was puériculture that could best contribute to the improvement of the race. To this effect, Almeida Garrett established an Institute of Puériculture (*Instituto de Puericultura*) in Oporto in 1932, an endeavour he placed within a much broader Latin context, which – taking its inspiration from France and Italy – advocated the youth as the vehicle of national renewal (Garrett 1938). Moreover, following the example set by Fascist Italy, a law for the protection of large families was introduced in Portugal in 1934, followed by the creation in 1936 of the organization promoting family protectionism, Mothers for National Education (*Obra das Mães Para a Educação Nacional*), headed by Maria Guardiola (Pimentel 1999: 492–3).

As will be discussed in the next section, there was, however, another side to this Latin eugenic activism, especially during the 1920s. In conjunction with arguments for positive eugenics, puériculture and neo-Lamarckism, and within the much broader context of debates about social and biological degeneration, Latin eugenicists also engaged intensely with the issue of marital health certificates. It was argued that eugenic responsibility towards the nation required not only the protection of mothers and children, and social assistance for larger families; it also entailed sexual education and marital hygiene, coupled with a rational decision about whom to marry.

The debate on premarital medical examination

Eugenicists had long advocated the introduction of compulsory medical certificates before marriage, with continued eugenic supervision thereafter (Timm 2010). After the First World War, many Latin eugenicists assumed that, in the absence of more drastic measures such as sterilization, premarital health certificates were an efficient means of dealing with a number of social and medical problems associated with the war, including the spread of venereal diseases and the supposed increase in mental deficiency. Some proposals along these lines were applied in a number of European countries (Nisot 1927 and 1929a) and the USA (Hoffmann 1913; Mjöen 1913; Smith 1914) in the post-war period.

In Italy, Ferdinando de Napoli had already proposed the introduction of premarital certificates in January 1919 (Wanrooij 2001: 151; Mantovani 2004: 178; Cassata 2011: 91). This was seen, first and foremost, as a prophylactic method to prevent the spread of venereal diseases, especially syphilis, but its supporters also emphasized its eugenic value. Following a devastating world war, the eugenic importance of marriage was regularly discussed in Italy during the early 1920s, whether at political and scientific conferences – such as the Congress for Family Education (*Congresso per l'Educazione in Famiglia*), convened in Rome in 1923 by the National Council of Italian Women (*Consiglio Nazionale delle Donne Italiane*) or the Second National Meeting of the Italian Society for the Study of Sexual Questions (*Società Italiana per lo studio delle Questioni Sessuali*) organized in 1924 – or in popular newspapers such as *Il Resto del Carlino* and *Difesa sociale* in 1927 (Cassata 2011: 92–8).

Some Romanian eugenicists voiced similar opinions. For example, in 1921 the British medical journal *The Lancet* noted that 'largely due to the war, the number of registered cases of insanity' was 'steadily increasing year by year in all the Balkan states'. In the case of Romania, *The Lancet* referred to a medical report that concluded 'the science of eugenics must play a very important role in the prophylaxis of insanity, and should be carefully studied. When national eugenics become practical politics the problem of the prevention of insanity will have been largely solved' ('Insanity in the Balkan States', *The Lancet* 1921: 94).

French eugenicists, in particular, were concerned with the dysgenic effect of war on the civilian population. Many considered compulsory premarital eugenic certification as an efficient way of preventing the transmission of hereditary diseases to future generations. In a short article published in 1918, Eber Landau, a visiting professor at the School of Medicine in Paris at the time, advocated 'a law [in France] by which health certificates are made compulsory before marriage' (Landau 1918: 82). Landau's remarks about the need of premarital examination were phrased within the growing concerns – in France and elsewhere – with the dysgenic effects of the war on the civilian population. Adolphe Pinard, the foremost of the first generation of French eugenicists, even claimed that he had entered politics as a representative to the French Parliament's Chamber of Deputies, to promote a compulsory 'loi sur l'examen prénuptial', which he proposed as a law in 1920 (Gaudilliere 2006: 199).

That same year, the French Eugenics Society organized a series of public lectures in Paris devoted to examining the eugenic consequences of war. They brought together the main French eugenicists at the time: Eugène Apert, Lucien Cuénot, Frédéric Houssay, Lucien March, Georges Papillaut, Edmond Perrier, Charles Richet, and Georges Schreiber. The papers were then collected and published as *Eugénique et selection* (*Eugenics and Selection*) in 1922.

Most of the papers dwelt on the neo-Lamarckian effects of the environment on national health, pronatalism, social hygiene, and preventive medicine, all regarded as essential to the biological regeneration of the nation. The condemnation of eugenic sterilization was another important theme, notwithstanding contributions from Charles Richet and Georges Schreiber, who argued for the prevention of those with hereditary diseases from reproducing (Richet 1922: 33–57; Schreiber 1922: 159–89). Significantly, those speakers opposed to sterilization framed their arguments not in terms of scientific invalidity but in terms of 'our French mentality', 'generosity', and 'our moral traditions' – all deemed incompatible with negative eugenics (Houssay 1922:

21). In general, eugenicists who participated in the lecture series emphasized morality and familial duty. Many of the presenters also denied the existence of certain races (such as 'French', 'Germanic', or 'Slavic' races), in an effort to strengthen international solidarity (March 1922: 82).

The ease by which eugenicists could control the environment, instead of heredity, was another dominant theme of the papers. Lucien March admitted that eugenicists were 'powerless to change the innate qualities of the individual', highlighting instead the social and biological regenerative significance of the environment. Eugenics' main responsibility, therefore, was to 'encourage the health [and] educational work of promoters of social progress' (March 1922: 88). March applauded French eugenicists' efforts to connect social hygiene and public health in their quest for biological regeneration. As William H. Schneider noted, this series of lectures marked 'a new organizational strategy of the French Eugenics Society that aimed at reaching the intellectual, scientific, and political decision makers of France' (Schneider 1990: 77).

The conferences organized in the early 1920s confirmed the growing international reputation of French eugenicists, an honour already made clear during the first meeting of the Permanent International Eugenics Committee in 1913, and reiterated at the Second International Congress of Eugenics held in New York between 22 and 28 September 1921. For the plenary session of the Eugenics Congress, Lucien Cuénot was asked to give a lecture on 'Genetics and Adaptation' (Cuénot 1923: 29–58). Lucien March presented a paper on the 'Consequence of War and the Birth Rate in France' (March 1923: 243–65). Other participants from Latin countries included José Joaquín Izquierdo from Mexico City; the French racial anthropologist Georges Vacher de Lapouge; Vincenzo Giuffrida-Ruggeri, Fabio Frassetto, and Corrado Gini from Italy; Albert Govaerts from Belgium; and Domingo F. Ramos from Cuba.

Govaerts, who served as Secretary of the Belgian Eugenics Society, was appointed Secretary of the Permanent International Eugenics Committee at the Congress (renamed The International Federation of Eugenic Organizations in 1925). It was he who subsequently arranged for the next meeting of the Committee to be held in Brussels in October 1922 ('International Commission of Eugenics', *Eugenical News* 1922: 117; Nisot 1929a: 118). Govaerts increased the existing international visibility by other Latin eugenicists, such as Lucien March and Corrado Gini. Nevertheless, a sustained relationship among Latin eugenicists was slow to develop in the immediate post-war period.

Prior to the First World War, collaboration between Latin eugenicists was entirely cultural. Nothing similar to the International Society for Racial Hygiene

(*Internationale Gesellschaft für Rassenhygiene*), established by Alfred Ploetz in 1907 to foster collaboration between Nordic eugenicists, existed for the Latin eugenicists. At the time, their connection consisted mainly of translations of articles and reciprocal visits. In 1913, for example, the French eugenicist Rémy Perrier contributed with an article on eugenics and race improvement to the Spanish anarchist journal *Salud y Fuerza* (*Health and Strength*) (Cleminson 2000: 151–4). In Portugal, one of the earliest discussions on eugenics and modern genetics published in the journal of the *Portuguese Society of Anthropology and Ethnology* (*Sociedade Portuguesa de Antropologia e Etnologia*) was a translation of a text by the Italian anthropologist and eugenicist Vincenzo Giuffrida-Ruggeri.

Among Latin nations, links between French and Belgian eugenicists were closest. The French language, also spoken by many Belgians, for instance, certainly explains this particular affinity. Moreover, French and Belgian eugenicists sometimes attended sessions of each other's eugenics societies. For instance, Georges Schreiber attended the meeting of the Belgian eugenics society (established in 1919), held on 7 February 1926, when medical examination before marriage was discussed (Vervaeck 1929: 420; Nisot 1929a: 136–7).

Though their prevailing eugenic discourses and agendas shared many characteristics, the relationship between Italian and other Latin eugenicists, particularly the French, was more complicated. Nevertheless, when Corrado Gini established the international review of statistics *Metron* in 1920, Lucien March was invited to contribute, and also became a member of the editorial board. In 1926 March returned the support, proposing Gini as vice-president of the International Federation of Eugenic Organizations. Also when the international library of eugenics (*Bibliothèque internationale d'eugénique*) – a project initiated by the Italian Eugenics Society – commenced in 1924, it featured first the publication of a volume on Belgian eugenics, *Le problème eugénique en Belgique*, edited by Albert Govaerts ('Eugenics in Belgium', *Eugenical News* 1934: 148).

There was a general tendency after the First World War for Italian and French geopolitical differences to inhibit large-scale collaboration between them on eugenic projects. In the immediate post-war years, France's refusal to support Italian demands at Versailles echoed negatively during the 1920s. Differences grew more apparent after the Fascist takeover in Italy, in 1922. Italian eugenicists quickly fell into line with the regime's social and political programmes. However, France remained a democracy until 1940, and its eugenic community continued to show a much greater conceptual diversity. The agenda of the French and Italian eugenics programmes also differed: France attempted to arrest its decline in population and preserve its precarious international status and influence; Italy, conversely, was a status-hungry nation wanting to increase its population in the belief that this would enhance its geopolitical power (Pogliano 1984: 61–97).

Diplomatic tension between the two countries may explain why Italian eugenicists were not invited to the French Eugenics Society's 1920–1921 public lectures. Nor were the Italians invited to the conference organized by the same society at the Social Museum (*Musée Social*) in Paris in May and June 1926, which was devoted to a discussion of premarital medical certificates.[1] Papers presented on this occasion were then collected into a volume, which also included papers from Ladislav Haškovec and Henri Vignes that had been read at the Second Session of the International Institute of Anthropology, held in Prague in September 1924.

It may be suggested that the 1926 conference represented 'the beginning of the effort to pass the first eugenic legislation in French history' (Schneider 1990: 154). According to the editors of the conference proceedings – René Sand, Albert Govaerts, Lucien March, Lucien Apert, and Georges Schreiber – the purpose of these conferences was to inform 'the French public of the [eugenic] work and legislative measures taken in different countries' (Sand et al. 1927: 6). In accordance with this programme, Sand and Schreiber offered an overview of marriage restrictions and demands for prenuptial medical examinations in a number of European and South American countries (Sand 1927: 37–43; Schreiber 1927: 9–29).

However, the conference is most notable for its lively discussion of prenuptial medical certification. The debate occurred within the larger context that preoccupied Latin eugenicists of the late 1920s, namely the appropriate social application of eugenics. There was no disagreement at the conference about the need for eugenic reform and the biological improvement that the medical examination before marriage represented, but opinions differed on how such methods could be introduced in France.

Schreiber acknowledged that it was 'premature' to demand that existing matrimonial laws be rewritten according to eugenic principles, but he was adamant that if introduced, 'the medical examination before marriage could reduce the number of innocent victims and improve the future generations' (Schreiber 1927: 29). Govaerts took a middle position. He believed that there were alternatives to premarital eugenic certification that possibly had a similar effect. In his paper, Govaerts claimed that preventive medical examinations served a similar purpose, such as those practiced by the Clinique du Parc Léopold in Brussels, which registered the patient's complete hereditary history

(Govaerts 1927: 177–84). Henri Vignes took a stronger stance against prenuptial certification. He cited Alphonse Guérin, a French senator and physician, who considered marriage certificates not only 'unacceptable but also abhorrent'. Guérin insisted that the French, as a Latin people, resisted such radical eugenic measures since acquiescence meant 'surrender without reservation to the fiercest statism' and reduction 'to the level of breeding animals'. In short, compulsory premarital certification would cause the French 'to lose all human dignity' (Vignes 1927: 199).

Lucien March wrote the 'general conclusions' of the conference proceedings (March 1927: 239–46). He summarized the various opinions and suggestions put forward, ultimately deciding against the introduction of compulsory medical examination. To do otherwise would only serve to provoke the ire of moral conservatives and the Catholic Church, and thus threaten the French Eugenics Society's work to date. However, seeking to pacify the 'radicals' who advocated compulsory eugenic premarital certification, March encouraged a voluntarist approach towards the issue. He suggested that all future marital partners 'request medical consultation before marriage' (March 1927: 245), highlighting its preventive health benefits to prospective families.

An examination of the public conferences on prenuptial medical examination, such as those held in Paris and Brussels in 1926, also sheds light on the cultural interchange between French and Belgian eugenicists and, perhaps more importantly, on the growing importance of negative eugenics in France. The premarital certificate Pinard proposed in the Chamber of Deputies on 24 November 1926 illustrates this tendency. The proposal succinctly decreed that 'Every French citizen wishing to marry cannot be entered in the civil registry unless he provides a certificate, dated from the previous day, stating that he shows no noticeable symptoms of a contagious disease' (quoted in Arnould 1930: 114). The proposal's brevity may have been intended to assuage the pronatalists and the Catholics, who might have opposed a more invasive law. Although Pinard's proposed law enjoyed some support from both the French Eugenics Society and the general public, it did not receive the political support required to pass. Schneider has identified a number of factors militating against passage: the proposed law was too narrowly focused, as it required only the future groom be examined; the timing of the examination, just one day before the marriage, was too short; finally, its compulsory nature (Schreiber 1990: 158–9).

In January 1927 Gustave Guérin suggested an improved version of Pinard's proposal (Schneider 1990: 160). Not 'every French citizen', Guérin insisted, needed to provide a medical certificate, rather any 'person, French or not' who wanted to marry. He also recommended that penalties be applied to 'any physician who provided certificates' on false premises, or to a registrar who 'officiated a wedding without the required medical certificate' (quoted in Arnould 1930: 115). Guérin's proposal also failed, but Pinard's proposal continued to be discussed by various organizations, as well as by the French press, during the 1920s. Eventually, on 6 December 1927, Paul Nicollet, a deputy from Lyon, presented the Chamber's report on Pinard's proposal (Schneider 1990: 161–3). As a physician, Nicollet knew that the French medical establishment was convinced of the medical importance of premarital examinations, and his report highlighted this argument. Nicollet claimed that, after the introduction of medical examinations, 'marriages will be easier to arrange because they will no longer entail the same uncertainty for the young women, families and relatives about the health of the future spouse. With a healthy marriage, there will certainly be a greater confidence of having children' (quoted in Schneider 1990: 162). Not surprisingly, then, his overall conclusion regarding the bill was favourable, and he recommended the Chamber enact Pinard's proposal (Schneider 1990: 163).

In the discussion that followed, the deputies' opinions remained divided and a vote on the measure was postponed. However, the proposal and the ensuing parliamentary debate succeeded in engaging French politicians and the general public with eugenics. As Georges Schreiber noted in 1928:

> To be honest, we did not think that events would unfold with such a rapidity that has 'accelerated' public opinion to the point of considering the premarital examination as a necessary, beneficent measure of preventive and eugenic medicine. We have been brought to the point of 'restraining' the ardour of our deputies, who have suddenly become 'more eugenic than the eugenicists themselves'. (Quoted in Schneider 1990: 168)

In response to the growing public support for medical examination before marriage, the *Conseil Supérior de la Natalité* also discussed Pinard's proposal in its session of February 1928, though without adopting it (Arnould 1930: 115). However, such disappointments did not dissuade supporters from resuming the campaign for premarital examinations at the French Eugenic Society's May 1928 and May 1929 meetings. By then a broader version of Pinard's original proposal had emerged, and was adopted by the Society on 24 June 1930. It contained the following points:

> 1. The French Eugenics Society considers that the prenuptial medical examination is indispensable, and expresses the hope that a law make this examination mandatory.

> 2. It expresses the hope that from now on at the occasion of legal registration, notice be given to the interested parties emphasizing the fundamental importance of the prenuptial medical examination for future spouses and their descendants.
> 3. It believes that this examination should include the free choice of doctor.
> 4. It considers that the prenuptial medical examination represents a *health examination*, whose purpose is to inform about the state of health of those who requested.
> 5. It believes that this examination should result in the drawing up of a certificate establishing simply *which doctor, at which date*, examined M. X or Mlle. Z, and it was [this doctor] who declared that he [or she] had to be married at such a date. (Quoted in Schreiber 1930: 81–2 and Jourdan 1931: 64–5)

Although Pinard and his colleagues in the French Eugenics Society had little political success in their efforts to promote premarital medical examinations as a preventive health measure, they did succeed in introducing eugenics into the public discourse. The medical examinations proposal remained on the French eugenicists' agenda, reappearing regularly during the next decade, until eventually (as will be discussed in another chapter) it became law in 1942.

Similar public debates on the eugenic importance of premarital certificates also occurred in other Latin countries. For instance, in the 'founding text of eugenics in Portugal' (Cleminson 2014: 57) – namely, the above-mentioned 'The Eugenic Problem in Portugal' – Mendes Correia argued in favour of medical examinations before marriage, to complement what he described as 'healthy procreation'. It was supposedly essential that eugenic work be carried out in Portugal, in order to avoid a 'sad end for the race, a generation poor and incapable, of inept, evil, cowardly, dissipated, lazy, impotent, tainted people whose bodies and souls are ruined' (quoted in Cleminson 2014: 59).

At the same time the eugenic message was gaining support in Portugal, eugenicists in Belgium strengthened their efforts to improve popular knowledge about the importance of premarital examination. Thus, on 1 June 1929, the widely read *Revue Belge* (*Belgian Journal*) asked its readers the following question: 'Should premarital examination be made compulsory?' Pierre Goemaere, one of the editors of the journal, noted that what might have appeared to some as a straightforward individual decision was, in fact, 'a complex' medical and judicial issue. The practical difficulty its implementation entailed might have 'explained the absence of an official solution in Belgium' (Goemaere 1929: 408). Goemaere then offered an unrestricted debate on the topic by outlining a spectrum of possibilities, from voluntary to compulsory certification, for both men and women (Goemaere 1929: 409).

Dr. Louis Vervaeck, the director of the Penal Anthropology Service (*Service d'Anthropologie Pénitentiaire*), an institution created to coordinate all anthropological research in Belgian prisons, was the first contributor invited to discuss these proposals and their eugenic implications in the *Revue Belge*. Vervaeck – often described as the 'Belgian Lombroso' (Radzinowicz 1999: 56) – was an important Belgian anthropologist, eugenicist, and penal reformer. Vervaeck was also a practising Catholic and a neo-Lamarckist; because of these perspectives he promoted positive eugenics and social hygiene rather than negative eugenic measures, such as sterilization (De Bont 2001: 63–104).

These views were expressed in his lengthy article in the *Revue Belge*. From the outset, Vervaeck acknowledged that there was little disagreement in theory among physicians, criminologists, and sociologists with respect to the benefits of premarital certificates; rather, differences emerged about its practical application. Would it achieve the eugenic results desired, he asked, if the premarital certificate was imposed on the population? Conversely, would it not be better to first educate Belgians about 'this essential eugenic method', through a nationwide government and private advocacy propaganda campaign? (Vervaeck 1929: 410). This was a complex problem, Vervaeck continued, raising important concerns about individual and collective responsibility not yet completely and satisfactorily addressed by 'the science of eugenics'.

Vervaeck also offered a detailed definition of eugenics and an explanation of its aims. We can clearly see the affinity of Belgian and French eugenicists in Vervaeck's article. He acknowledged that eugenics originated in England, but then claimed that the French were first to give serious consideration of its practical application (as in premarital certification). As evidence for his claim, Vervaeck pointed to Henri Cazalis's proposal of such measures in 1902; Vervaeck also cited Charles Richet's 1919 book *La sélection humaine* (*Human Selection*) (Vervaeck 1929: 412). To adopt radical eugenic ideas of human selection, Vervaeck argued, would mean a return to 'pagan barbarism'. To care for 'the sick and the abnormals', he continued, did not mean ignoring 'the protection of the race'; in fact, there was no conflict between 'the human treatment of those [who were] infirm and defective' and a 'healthy social prophylaxis' (Vervaeck 1929: 413). He pointed out that countries such as Denmark and Turkey required premarital medical certificates, and people suffering from certain medical conditions, such as insanity or alcoholism, were banned from marriage. In Argentina, an obligatory premarital marriage law was proposed in 1926, to be applied only to those suffering from venereal diseases.

Vervaeck then explained that in France and Belgium, the campaign for the introduction of medical certificates before marriage centred mostly on voluntarism and education rather than obligation (Vervaeck 1929: 416–7). In this way, he claimed, the 'rights of the citizen and social interest' were reconciled (Vervaeck 1929: 418). Yet, he admitted that the eugenicists' efforts were rarely rewarded; public opinion in France and Belgium continued to remain 'in general indifferent, or otherwise hostile' to the introduction of compulsory medical examination before marriage (Vervaeck 1929: 420). Eugenicists had to explain better 'the great benefits for individuals, families and the Race' that such a measure would bring about, in the sense that the social and biological degeneration of the population could thus be prevented. Eugenicists also had to allay the moral, social, and medical objections raised against premarital medical certificates (Vervaeck 1929: 422–32).

Vervaeck suggested two reasons for public antipathy towards premarital examination: 'the imperfection and imprecision of eugenic science' and 'the small number of doctors familiar with eugenics' (Vervaeck 1929: 426). Clearly, successful promotion of eugenics would require specialized health centres to provide eugenic materials and information. In this respect, Vervaeck cited the Viennese Health Centre for Engaged Couples (*Gesundheitliche Beratungsstelle für Ehewerber*), opened in 1922, as an example.

Two main characteristics of Vervaeck's position should already be apparent. First, he stressed the overall importance of the 'eugenic reform' for the 'moral, social, medical and material' improvement of society; and second, eugenics could promote human regeneration – in 'our old exhausted societies, threatened by both the [declining] birth rate and the rising tide of degeneration' eugenics could provide the much-needed antidote (Vervaeck 1929: 441). In the end, Vervaeck intuited the nuanced synergy between private and public morality represented by the introduction of premarital medical examinations.

Other medical experts invited by the *Revue Belge* to discuss the issue were less accommodating. For instance, Manille Ide, professor of General Pathology at the University of Louvain, believed the eugenicists' aspiration to control human reproduction was 'an impossible dream' (Ide 1929: 50). Jules Destrée, a socialist Walloon politician, did not hesitate to portray eugenicists as 'terrible people'. He disapprovingly quoted the French eugenicist Georges Papillaut who claimed that it was 'wrong to protect the degenerates, the crowd of criminals, prostitutes and alcoholics'. According to Destrée, what followed next was a 'massacre of all those [whom] some doctors declared social waste' (Destrée 1929: 52–3).

The theologian Jacques Leclercq, a professor at the Facultés Universitaires Saint-Louis, was less caustic. He discussed the question of medical examination before marriage from a Catholic and ethical point of view, professing the 'profound respect for the integrity of human life' (Leclercq 1929: 53–5). In the same spirit, Valère Fallon reiterated the arguments he put forward in 1926 at the meeting of the Belgian Eugenics Society dedicated to the same subject of premarital examination. He questioned the eugenic definition of such imperatives as 'biological' and 'social responsibility'. Fallon positioned conjugal life and family above society and political community. In other words, 'marriage was not a public service' (Fallon 1929: 53). Yet Fallon did not completely dismiss the possibility of introducing premarital examinations; he only insisted that such examinations not be compulsory.

Willem Schraenen was given the final word in this exchange. Medical examination before marriage, Schraenen argued, was 'of real value' to society. No one could remain indifferent to the growing number of defective children being born (Schraenen 1929: 58–9). Therefore, he offered the following suggestions:

> 1. Medical examination before marriage must become a general routine;
> 2. We must reject coercive methods that never solved anything, [and] which are powerless to provide an effective and stable result and which present more disadvantages than advantages;
> 3. By contrast, persuasion is a powerful means of action [...]. It is therefore necessary to understand well the work of propaganda. This work will be carried out by physicians and organizations of preventive medicine, and where they are lacking they must be established. (Schraenen 1929: 61)

In the climate of post-war national reconstruction, such conferences and public debates clearly embodied Latin eugenicists' aspirations to play a greater role in national politics. Moreover, the specific debates on the introduction of premarital medical examinations reveal the Latin eugenicists' ambition to emulate strategies of biological improvement observed elsewhere, while simultaneously modifying these strategies in accordance with 'Latin' morality, individualism, and religion.

Fascism and eugenics

Nowhere in the Latin world was the synthesis of science and politics so blatant as in Fascist Italy. On 15 March 1919, eugenicists established the Italian Society of Genetics and Eugenics (*Società Italiana di Genetica ed Eugenica*). Ernesto Pestalozza, a professor of obstetrics and gynaecology at the University of Rome, was elected president, and Gini became vice-president. Cesare Artom served as secretary, and the statistician Marcello Boldrini as vice-secretary (Cassata 2011: 70). The Society began its operations with 100 members, eventually growing to 300 by 1925. This made the Italian Society of Genetics and Eugenics one of the largest in the world.

Benito Mussolini and his National Fascist Party gained control of Italy in 1922, and transformed the state into an authoritarian dictatorship in a little over two years. While Mussolini was notorious for his mercurial ideological shifts, he still abided by a set of stable principles. For one, the Duce knew that he should not unnecessarily antagonize the Church. Further, he intended Italy to become a great world power, expand its empire, and indisputably dominate the Mediterranean. To achieve this, Italians had to be both numerous and strong. As a neo-Lamarckist himself (Gillette 2002b: 33 and 53), Mussolini believed that Fascism could provide a transformative environment, remaking Italians in the image of their ancient Roman ancestors (or at least Mussolini's conception of these ancestors).

The First Italian Congress of Social Eugenics (*Primo Congresso Italiano di Eugenetica Sociale*) held between 20 and 23 September 1924 in Milan, revealed the diversity of Italian eugenics, alluding to many of the central themes developed by the Fascist rhetoric on national improvement during the 1920s. Organized simultaneously with the third meeting of the International Federation of Eugenic Organizations, the Italian eugenics congress demonstrated the conceptual versatility of Italian eugenics, 'with several interconnections between social hygiene and social medicine' (for a detailed discussion of the congress, see Cassata 2011: 147–59).

A number of prominent foreign eugenicists were present at the Italian congress. These included Leonard Darwin, president of the [British] Eugenics Education Society; Jon Alfred Mjøen, director of the Winderen Laboratory in Oslo; Nikolai K. Koltsov, president of the Russian Eugenics Society; and Søren Hansen, a Danish anthropologist. Other than the Italians, Latin eugenicists attending included Lucien March, secretary of the French Eugenics Society; and Victor Delfino, president of the Argentine Eugenics Society. As with other international eugenics conferences during the early 1920s, the German racial hygienists did not attend.

Corrado Gini offered the first paper given at the conference, entitled 'Le relazioni dell'Eugenica con le altere scienze biologiche e sociali' ('The Relationship of Eugenics with Other Social and Biological Sciences') (Gini 1927a: 3–26). From the outset, Gini indicated the need for increased interaction between eugenicists and the general public. He continued to view negative eugenics with apprehension:

> In Italy, as abroad, while Eugenics is alive and prospering as a discipline that interests experts in biological and social disciplines, some politicians, and some philanthropists, it is not however successful – it would be in vain to deny it – to capture the conscience of the masses, who consider it with persistent scepticism, if not with evident distrust. (Gini 1927a: 4)

In light of this state of affairs, Gini noted, the application of 'selective' and 'negative eugenics' was problematic. Therefore, he feared that 'selective action' could have 'radically erroneous and damaging consequences, rather than useful ones for the progress of the race' (Gini 1927a: 11). It must be conceded, Gini told the participants, that eugenics as a science of human improvement was still 'immature' and its numerous social applications had to await a wider consensus.

This perspective was shared by other Italian participants at the congress who, from their own areas of specialization, stressed the fact that the lack of scientific certainty about the nature of hereditary traits and the causes of physical and mental defects dictated a pressing need for further eugenic research; it was simply too early to apply eugenics to society, through sterilization or other means, in the attempt to control reproduction and hence the transmission of hereditary diseases. For instance, the gynaecologist Ernesto Pestalozza, in his paper 'Le operazioni operatorie in rapporto all'Eugenica' ('Surgery in relation to Eugenics'), questioned the efficacy of eugenic sterilization for the improvement of the race. There was no doubt, he argued, that if sterilization enabled eugenicists 'to cancel out, or at least to limit, the hereditary transmission of hereditary diseases that threatened the race, then its adoption would be justified' (Pestalozza 1927: 82). However, this point had not yet been demonstrated. Instead, Pestalozza argued for a synthetic approach to human improvement, combining preventive medicine, 'the new science of eugenics', and social hygiene, so that the 'benefits of hygiene that we are able to offer to the individual and society are extended to the race' (Pestalozza 1927: 85).

British and Scandinavian eugenicists attending the conference were skeptical of the Italian views on eugenics. Cora B. S. Hodson, General Secretary of the International Federation of Eugenic Organizations, questioned the pervasive neo-Lamarckism advocated by the Italian eugenicists. She rejected their demands that radical eugenic measures be eschewed for the foreseeable future. Moreover, she pointed out that control of heredity (i.e. Mendelism) was at the heart of eugenics, rather than environmental improvement (i.e. neo-Lamarckianism) (*Atti del Primo Congresso Italiano di Eugenetica Sociale* 1927: xxi–xxii).

In the end, however, the Italian congress not surprisingly adopted a resolution promoting scientific moderation and further research into human heredity, rather than radical negative eugenic measures (*Atti del Primo Congresso Italiano di Eugenetica Sociale* 1927: lxiii). Furthermore, by advocating 'social eugenics', the congress proposed a carefully styled version of Latin eugenics. Designed to reconcile nature and nurture, the congress ultimately favoured the latter, while not denying the importance of the former. Italian eugenicists, in particular, resisted the strict hereditarianism and social selectionism promoted by eugenicists in Britain, the USA, and Scandinavia; instead, they promoted a eugenic programme tailored to reflect their country's own cultural, social, and religious characteristics. When in the late 1920s, the Fascist regime launched its natalist population policy it reflected these values, seeking to rapidly modernize and homogenize the nation, as well as export the Italian version of Latin eugenics abroad, befitting the revived, vigorous, and expanding nation.

The First Italian Congress on Social Eugenics also effectively confirmed Corrado Gini's position as Italy's leading eugenicist (see Figure 3.3). By the mid-

Figure 3.3 Corrado Gini
Source: Photo Division, Ministry of Information & Broadcasting, Government of India.

1920s, Gini was recognized internationally as an important Italian demographer and eugenicist. After the short chairmanships of Pestalozza (1919–1921) and Achille Loria (1922–1923), Gini gained control of the Italian Society for Genetics and Eugenics in 1924, and remained at its helm until his death in 1965. During the same time, Gini became the first director of the Central Institute of Statistics, and consequently met with Mussolini quite often (Giorgi 2012: 620–6).

Gini's demographic theories emphasized natalism, national regeneration, neo-Lamarckism, European racial superiority, and Latinity. On 14 July 1926, in his inaugural speech as director of Central Institute of Statistics (*Istituto Nazionale di Statistica*, abbreviated as ISTAT), Gini reiterated his eugenic demographic ideas, while at the same criticizing 'other Latin nations' which did not take 'investigations of the population' seriously. Such investigations, he stressed, were 'decisive for the life of a Nation' (quoted in Cassata 2011: 137).

Gini also agreed with Mussolini's conception of the 'corporate state' (Gini 1927b: 99–115). Fascist corporatism subsumed the individual in the state. Membership in the national community, as represented by the state, supposedly brought greater personal fulfilment than did the selfish pursuit of personal goals. Harmony would be built through hierarchy, discipline, and obedience to state directives. The ultimate goal of corporatism was to increase national productivity and empower the state. Similarly, eugenics sought to elevate the good of society over the rights of individuals. Gini's version of eugenics saw the state as 'its own entity, much like an organism, having its own goals and needs' (Cassata 2006: 117). He was eager to employ eugenics to aid the Fascist state in maximizing the nation's human productivity. However, we should note the nagging tension between Fascist corporatism and Latin eugenic ideology. Latin eugenics supposedly emphasized the rights of the individual, humanism, and traditional Latin values, all of which the Italian Fascist state either threatened or denied. Blind adherence to the dictates of the state could destroy any coherence to Latin eugenics, as indeed would effectively happen in Italy in the last years of the Fascist regime.

With the rise of Latin eugenics during the 1920s, Nicola Pende's theories of biotypology and orthogenesis also flourished. Next to Gini, Pende (see Figure 3.4) was Italy's most influential eugenicist during the Fascist period. In 1922, while a professor of medicine at the University of Rome, Pende developed a biotypological theory based on constitutional medicine, pathology, and endocrinology, one that was also Latin, humanistic, and Catholic, treating the individual as a whole (Pende 1922). He claimed that every human would fit into one of six archetypical mind–body types. Each type had its own nutritional and

Figure 3.4 Nicola Pende
Source: Library of the Institute of Medicine, Institute of Public Health, Bucharest.

environmental needs, required its own distinct endocrine balance, and followed its own biological 'constitution'. Pende defined it thus:

> [T]he constitution is the morphological, physiological and psychological resultant (variable in each individual) of the properties of all the cellular and humoral elements of the body, and of the combination of these in a special cellular state having a balance and functional output of its own, a given capacity for adaptation and a mode of reaction to its environmental stimuli. (Pende 1928: 25)

Individual 'resultants' were inherited through a combination of Mendelian (primarily) and neo-Lamarckian mechanisms (secondarily).

In contrast to Anglo-Saxon and Nordic eugenicists who repeatedly emphasized the health of the collective and of the race, Pende described his theory of constitutional medicine as 'the prophylactic care of the individual'.

In other words, eugenicists ought to promote first and foremost a theory of individual improvement, as it applied 'to each individual, after a thorough preliminary study of his somatic and psychic personality' (Pende 1928: 236). The eugenicist, like 'the hygienist and champion of social medicine must never forget' – Pende argued – 'that he is first of all a physician, and the modern physician, like our good father Hippocrates, can only be an *individualistic and constitutional physician*' (Pende 1928: 237. Emphasis in the original). For Pende, methodical medical research led to familiarity with one's physical and psychological development. Once a 'complete *somatic and psychic constitutional formula* recorded from time to time in suitable *book of health*' would be accomplished for each individual, it will become possible to 'know the *biotypes* and to establish an individual *biotyprogramme*' (Pende 1928: 242. Emphasis in the original). To achieve collective improvement was ultimately to bring together all individual 'biotyprogrammes', to monitor them carefully and to achieve the ultimate synthesis between social policy, national efficiency, and health.

Pende believed that the endocrine glands were the specific sites for neo-Lamarckian action. The environment affected the endocrine glands, which then emitted hormones that, in turn, altered the sex cells. During the late 1930s, Pende celebrated his interpretation of endocrinology as 'the very modern [Latin] theory of emerging evolution, which is at odds with the [Nordic] mechanistic-physical chemical evolution' (Pende 1940: 569). Since individuals were best adjusted and most productive if they lived a lifestyle appropriate to their biotype, Pende complemented biotypology with orthogenesis. He determined a patient's biotype, then corrected any deviations from his or her biotypic norm with a life-long programme of exercise, nutrition, medicine, and behavioural adjustments (Pende 1933a).

Pende's emphasis on improving an individual through gradual environmental modification (or 'orthogenesis') appealed widely to neo-Lamarckian and Latin eugenicists. If implemented on a massive scale, his theories also promised to fulfil the dream of transforming a population by eugenically maximizing its biological health, sexual fertility, and economic productivity. To policymakers, such a doctrine offered low-cost, wide-scale preventive medicine, and held out the promise of an increase in national efficiency. Pende also described biotypology and orthogenesis as the utmost eugenic strategies able to provide the much-acclaimed biological rejuvenation of the nation. Rather than bemoan the terrible nutritional, sanitary, and healthcare deficiencies of Italy's growing population, one could dream about the transformation of Italy through a nation-wide biotypological healthcare programme. If biotypology could bring the Italian population to their full mental and physical capabilities, the country would reach its full potential as a world power. As fully outlined in his major work *Trattato di biotipologia umana individuale e sociale con applicazioni alla medicina preventiva, alla politica biologica, alla sociologia* (*Treatise of Individual and Social Biotypology and its Application to Preventive Medicine, Political Biology, and Sociology*), biotypology fused medicine and political biology into a new philosophy of national belonging, one aimed at transforming all Italians (Pende 1939b).

Pende agreed with the neo-organicism espoused by Gini and by Fascism in general. He welcomed state 'coordination' of the 'eugenic values' of the citizens for the good of the 'unitary cellular state'. If left to themselves, Pende worried that some individuals had the disturbing tendency to transform into 'malignant tumourous cells' that threatened the national body. However, by using biotypology, the state could determine the 'muscular, moral, and intellectual' capital of each citizen. Healthy individuals could then be 'rationally selected' for their place in society, as members of the ruling elite, an industrious working class, an aggressive military, and so on.

Biotypology could benefit the nation in other ways as well. Because certain psychological and physiological types were set as norms, those who deviated were, by definition, 'abnormal'. In this way, biotypology could help identify such individuals, and thus medically justify a legal decision to confine those deemed 'deviant' to a mental institution or a juvenile detention centre, as the situation dictated. Moreover, for nations experiencing large-scale immigration, biotypology could help assess the potential immigrants' mental and physical compatibility with the host-nation's dominant biotype(s), and so scientifically adjudicate their desirability. Borrowing a suggestion from Francis Galton, Pende incorporated into his orthogenetic programme the issuance of eugenic identification cards that would carry a variety of data on each person's physiological and psychological profile, thus aiding government efforts in controlling its citizens for the wellbeing of the nation.

Pende himself was very close to the Catholic Church and ensured that biotypology did not contradict the main tenets of Catholic ideology. Biotypology's comforting tendency to scientifically legitimize pre-existing stereotypes proved popular in Catholic settings. For example, biotypologists argued that children born of married parents were superior to illegitimate children – not out of inherent biological advantages, but simply due to the fact that the healthy family environment would confer substantial psychological benefits.

Pende's orthogenesis – and the Italian biotypological school in general – seemed tailored to the bio-social conditions and the medical concerns of eugenicists in the Latin countries (Aragon 1933: 136–8; Schreider 1933: 67–97; Marinescu 1934: 1–34). One clear example was the Society of Biotypology (*Société de Biotypologie*) established on 8 July 1932 in Paris, with the psychiatrist Édouard Toulouse as president (Traetta 2013: 149–58). The Society aimed to provide a 'scientific study of the human types by establishing correlations between various psychiatric, pathological, psychological, physiological and morphological traits' and then apply the accumulated data to 'various branches of human activity: eugenics, pathology, psychiatry, pedagogy', and so on ('L'activité de la Société de Biotypologie', *Biotypologie* 1932: 35–40). It consisted mostly of French physicians, biologists, and eugenicists (including Georges Papillault, Eugène Apert, Sicard de Plauzolles, and Henri Vignes), although foreign individuals, such as the Romanian physiologist Alexandru Rudeanu and the prominent Mexican physician Ignacio Chávez, were also accepted as full members. Pende himself honoured the Society with a visit and a lecture on 2 July 1933. On this occasion, he discussed the research conducted at his Institute of Biotypology in Genoa, on racial varieties to be found in the region of Liguria, in north-western Italy. It was, he declared, part of a larger study on the 'functional and metric biotypology of the white races' (Pende 1933b: 113–34).

Originating at a confluence of preventive medicine, constitutionalism, endocrinology, psychiatry, and demographic biology, the French Society of Biotypology and the work of its members transformed not only the eugenic discourse in France but also how social reform and the scientific study of labour was conceived and practised during the 1930s (Schneider 1991: 410–46). One of its international achievements was the creation of a Section of Differential Biometry and Biotypology at the International Congress of Anthropology and Ethnography held in London in July 1934 ('Notes et Informations', *Biotypologie* 1934: 140–1).

Equally important, Pende's theories, as developed particularly in his 1933 *Bonifica umana razionale e biologia politica*, also reflected the biological philosophy of Facism. Pende's enthusiasm for Fascism (he became a party member in 1924) and services to the Crown were rewarded with rapid promotion in the state medical hierarchy. By 1926, he received the funds necessary to establish an institute of biotypology and orthogenesis at the University of Genoa, which was meant to be the first in a national network of such clinics.

Gini, Pende, and other Italian eugenicists realized the importance of bringing their vision of social and biological improvement in alignment with Benito

Mussolini's fascist project of technological modernism and organic vitalism. If it facilitated his retention of power, Gini was often willing to modify those elements of his theories that clashed with the Duce's demands. For instance, though previously Gini estimated that the Italians were in the later stages of their millennial demographic cycle, and thus could expect to show signs of degeneration, he now rather conveniently decided that the advent of Fascism could allow Italy to reverse its demographic decline. After all, had not the Italians been reborn as a nation in the mid-nineteenth century? Was not Fascism a continuation of this youthful spirit? Could not demographic science combine forces with a powerful, authoritarian regime and thus contribute to the numerical growth of the Italian people through pronatalist policies (Gini 1930 and 1931)?

These ideas animated both Gini and Mussolini, particularly after the latter declared his hostility to Malthusianism in 1924 (Treves 2001: 128; Cassata 2011: 87), and then officially announced Italy's quantitative population policy in 1927 (Ipsen 1996: 84–5; Quine 2002: 35). David G. Horn has aptly argued that Italy's fascist demographic programme was a typically modern attempt by the state to manage the nation's demographic characteristics and medical needs, to the point that the relationship between the individual, society, and the state was dramatically altered (Horn 1994: 49). Furthermore, like other 'high-modernists' at the time, Mussolini combined his 'supreme self-confidence about linear progress', with a 'rational design of social order, the growing satisfaction of human needs, and, not least, an increasing control over nature (including human nature) commensurate with scientific understanding of natural laws'. As James C. Scott observed, Mussolini had 'a particularly sweeping vision of how the benefits of technical and scientific progress might be applied – usually through the state – in every field of human activity' (Scott 1998: 89–90).

We must also address the impact of Mussolini's pan-Latinism on the Italian variety of Latin eugenics. Mussolini had been a pan-Latinist during the First World War, and had absorbed some of the demographic and pan-Latinist rhetoric that attempted to promote the creation of a Latin confederation. These ideas, combined with Mussolini's neo-Lamarckism, his desire for public reconciliation with the Catholic Church, and the impact of Italian nationalist theory, contributed to Italy's complex eugenics programme during the fascist era. Furthermore, Mussolini sought to export Italian eugenics to Latin America, as one component of a scientific and cultural policy in accord with Italy's new status as a world power.

Regarding fascist family and population policy, Mussolini claimed that population growth would enhance Italy's economic and geopolitical power.

He boldly imagined a future Italy, swollen by an enormous population, and with it a Europe divided between Italy (which would lord over the Latins), Germany (leader of the Germanic peoples), and Russia (patron of the Slavs). For this scenario to become reality, a strong state was essential. Fascist Italy was to become such a state, one capable of 'continually physically transforming the people' (Mussolini 1929: 90).

In the 1920s Mussolini enacted a series of neo-Lamarckian-inspired measures to increase the health and fertility of the Italian population, virtually all of which were based on long-standing staples of Latin eugenics. Mussolini eventually realized that reviving the nation's 'racial vitality' necessitated a massive process of social engineering that presupposed new state institutions and the commitment of vast resources. The first such eugenic institution set up by the Fascist state, in October 1925, was christened the National Agency for the Protection of Maternity and Infancy (*L'opera nazionale per la protezione della maternità e infanzia*, abbreviated as ONMI). The purpose of ONMI was 'the protection of mothers and infants', not for their benefit per se, but to further the nation's moral, economic, and cultural progress through the political application of eugenics, hygiene, and social assistance (Somogyi 1934: 11; Glass 1936: 108; Monacelli 1937: 386–90). Maternal health and childcare, two of the central tenets of puériculture, became central to fascist population welfare and its corollary, national rejuvenation.

The state now vigourously promoted the protection of motherhood, social hygiene, puériculture, and the general health of children and women. According to Victoria de Grazia, 'In the interests of promoting the race, the welfare of the mother was subordinated to that of the infant' (De Grazia 1992: 60). Yet the reasons for such solicitousness were not based on concern for the common weal. As David G. Horn noted, 'Rather than purification, the goals of fascist demographic politics were social defence and multiplication; rather than selective breeding and sterilization, its means were improved hygiene, diet, and education' (Horn 1994: 60). To this effect, human fertility and reproduction were now deemed state priorities. The reconfiguration of the traditional private sphere, and of individual, gender, and religious rights, were important consequences of this transformation. Essentially, the boundary between private and public spheres was eroded by the fascist conception of 'public responsibility for the nation'. As a result, it became possible to connect notions of collective welfare with individual responsibility towards the nation.

In the famous 'Ascension Day' speech of 26 May 1927 (the day Catholics marked the Ascension of Mary into Heaven), Mussolini announced the second stage of his broad eugenic programme with appropriate bombast. It is worth reproducing some fragments at length:

> For five years, we have continued to assert that the population of Italy was like a river overflowing its banks. That is not true. The Italian nation is not growing but diminishing in size [...] The Italian nation still has an excess of births amounting to half a million per year. But this excess is no longer even as high as it was during the War. [...] To count for something in the world, Italy must have a population of at least 60 millions when she reaches the threshold of the second half of this century [...] It is a fact that the fate of nations is bound up with their demographic power [...] Let us be frank with ourselves: what are 40 million Italians compared with 90 million Germans and 200 million Slavs? Let us look at our Western neighbours: what are 40 million Italians compared with the 40 millions of France and the 90 millions in her colonies, or with the 46 millions and the 450 million inhabitants of her colonial possessions? [...] With a declining population a country does not create an empire, but becomes a colony. (Quoted in Glass 1936: 106–7. See also Somogyi 1934: 5–6)

According to Stefano Somogyi, a Hungarian-born statistician at ISTAT, the quantitative population policy was only one – albeit central – component of fascist demographic policy. The impression was, he remarked, that fascists 'sometimes neglected eugenics' and that they were 'hostile to qualitative methods of demographic policy adopted by other nations' (Somogyi 1934: 11). This was simply incorrect, Somogyi maintained: fascist demographic policy was both *quanitative* and *qualitative*; in short, it was 'totalitarian'. However, he acknowledged that the fascist regime, 'did not pursue a qualitative demographic policy inspired by the Anglo-American school of birth-control' and that 'the results obtained so far in the domain of eugenics' might be less impressive than in other countries. But fascist demographic policy was not 'one-sided', as some believed: it drew sustenance from the 'political goals of the nation' in order to ensure the 'the physical integrity of the race' (Somogyi 1934: 11). In fact, Somogyi asked, was ONMI not the most poignant example of the fascist regime's pursuit of a qualitative demographic policy?

Mussolini denounced the widespread practice of birth control, which he claimed falsely promised greater wealth and happiness if families limited the number of their children. Mussolini argued that nations had historically become wealthier as they became more populous. He grandiloquently predicted that if Italy could increase its population to sixty million by the end of the twentieth century, 'Italians will feel the weight of their mass and their force in the history of the world' (Mussolini 1928: 7–23). Such a fertile and dynamic people would then be able to 'have a right to Empire' and 'propagate their race over the face of the earth' (Mussolini 1928: 7–23).

The nation's health and physical fitness was central to the fascist revolution. The general social hygiene of the population was to be improved through the provision of popular health courses and the encouragement of adult physical education. Medical experts were charged with combating the dysgenic scourges of malaria, tuberculosis, and endemic cretinism. The state now provided insurance for the sick and disabled. Eugenicists hoped that alcoholism would decrease after the government mandated the closure of some taverns. Fascist youth groups were created, and subjected to regular physical training, gymnastics, and outdoor activities. The aim was to create a New Italian, reminiscent of his ancestors' physical prowess but modern in his commitment to the state and the national community (see Figure 3.5). The regime appropriated the 'the cult of

Figure 3.5 Silvio Canevari, Il Littore, Stadio dei Marmi, Rome
Source: Aaron Gillette's personal collection.

classical beauty because this cult was functional for the fascist plan to make Italians virile, by means of special attention given to physical robustness and to eugenics' (Gori 2000: 39).

No one embraced the cult of virility and masculinity with more enthusiasm than Mussolini. Like their Duce, Italian eugenicists believed that cities were physically and morally unhealthy environments, promoting sedentary lifestyles, exposure to industrial toxins, and dangerous 'modern' ideas such as feminism and limitation of family size. As we pointed out in the introduction, such imperatives seem to contradict the eugenic drive for national prosperity and empowerment, which necessitated modernization and urbanization. The Catholic Church enthusiastically supported the fascist regime's strong encouragement of traditional family life, and its punitive measures against birth control. Father Agostino Gemelli, who was increasingly falling under Mussolini's spell, was particularly delighted with the new policies (Gemelli 1928: 647–8).

Italian scientific research was also reoriented towards supporting the regime's demographic policies and social eugenic programmes. Mussolini mandated that Italy's National Research Council (*Consiglio Nazionale delle Ricerche*) promote 'social and political medicine' suitable for a prolific people (Canali 2001: 143–67). As part of this concerted effort, an Italian Committee for the Study of Population (*Comitato italiano per lo studio dei problemi della popolazione*), led by Corrado Gini, was created in 1929. By claiming that the social and biological study of population was intrinsic to fascist ideology, Gini placed demography and eugenics at the centre of Fascist Italy's much-publicized programme of national revival. This appeared to be a promising model to emulate, and eugenicists in other countries were impressed with Italian accomplishments.

If French eugenics, in its various permutations (puériculture, preventive medicine, social hygiene, and natalism), had attracted Latin eugenicists from other countries before the mid-1920s, towards the end of the decade Italian eugenics had assumed a central role within the international Latin eugenic movement, due largely to the fascist state's ability to concentrate resources in the hands of a dictator determined to secure for Italy a leading position in international eugenics. Recognition of the Duce's success in this regard came from admiring foreign eugenicists, who attended the Eighth Meeting of the International Federation of Eugenic Organizations held in Rome in 1929. One of them, the German eugenicist Eugen Fischer, for instance, praised Mussolini as the 'great statesman who, in the Eternal City, shows more than any other leader today, both in deed and work, how much he has taken the eugenic problems of his people to heart' ('The Meeting of the International Federation of Eugenic Organizations', *Eugenical News* 1929: 154).

By adopting Mussolini's vision of a new, healthy, and numerically strong national community, Italian eugenicists succeeded in significantly broadening the scope and outreach of their agenda, and acquiring access to state power and resources that eugenicists elsewhere could only admire from afar. Moreover, with the emergence of fascism as an international ideology, some Italian eugenicists, such as Corrado Gini, Nicola Pende, and Agostino Gemelli, advocated a new version of Latin eugenics, symbolized not only by Italy's preeminence within the international eugenic movement, but also by an image of social and biological improvement manifestly opposed to Anglo-Saxon and Nordic eugenics as practised in Britain, Germany, the USA, and Scandinavia. As we will see in the next chapter, debates on sterilization, racism, and the role of religion would intensify these trends, as Latin eugenicists organized themselves on the international level in order to thwart the greater threat of Nazi German racial hygiene.

4

Latin Eugenics, Sterilization, and Catholicism

Sterilization is arguably one of the most contentious of all eugenic measures. First legalized in the state of Indiana, USA, in 1907, sterilization had fascinated many eugenicists as the most 'efficient' instrument of negative eugenic control. The disruptions caused by the First World War and the Great Depression attracted a wider audience in Europe and Latin America to the eugenic discussion. Hitherto restricted to medical specialists, eugenics now increasingly attracted other professionals, especially legal scholars, sociologists, and statisticians, some of whom were open to the eugenic use of sterilization. By 1937, eugenicists had succeeded in enshrining compulsory eugenic sterilization laws in a dozen countries in Western and Northern Europe.[1]

A number of Latin eugenicists, admiring the 'progress' of eugenics in these countries, became advocates of restricted eugenic sterilization in their own countries. Positive eugenics mediated through social medicine and public health – a dominant theme in Latin eugenics – seemed to them to be time-consuming, in a period when immediate renewal of nation and race seemed essential. However, as eugenically inspired legislation to prevent the physically and mentally 'unfit' from reproducing proliferated across the world, biological theories of human improvement gradually came into open conflict with the teological dogmas advocated by Christianity. Essentially, the eugenicists demanded control over reproduction and marriage, postulating the primacy of the state's biological aims. As Sharon M. Leon noted, the 'fundamental dispute' was 'whether the state has a right to sacrifice an individual for the common good or whether the state has the obligation to care for all individuals, as an integral part of the common good' (Leon 2013: 4).

Any inventory of the Latin eugenics during the interwar period must therefore take into account two constitutive elements, in particular: the debate on eugenic sterilization and the relationship between religion and eugenics. As some of the most committed critics of negative eugenics were Catholics, and considering the dominant position of the Church (Catholic and Orthodox) in local and state politics in Latin countries, it is important that the role of religion in shaping the contents of Latin eugenics is properly recognized.

Latin eugenics and sterilization

Some Latin eugenicists discussed eugenic sterilization laws as early as the turn of the twentieth century. Such discussions seemed most common in France and Romania, for example, perhaps because the Catholic Church had less political influence there than in other Latin countries. In Romania, the physician Eraclie Sterian first broached the subject of sterilization in a 1910 article published in the journal *Medicul Poporului* (*People's Physician*). Sterian sternly criticized the North American sterilization laws implemented several years before; nor did he spare Lombrosian criminal anthropology for stigmatizing the alleged 'born criminal'. Even though he approached the subject from a hereditarian perspective, Sterian claimed that sterilization would not bring to an end criminality, alcoholism, or insanity (Sterian 1910: 113–4). Albeit contentious, eugenic sterilization did not lack early supporters in Romania, such as the gynaecologist Constantin I. Andronescu. Already in 1914, at the National Medical Congress (*Congresul național de medicină*), Andronescu affirmed the need for the introduction of premarital health certificates and eugenic sterilization of the feeble-minded and the mentally ill in Romania (Andronescu 1943: 15 and 44).

During the economic crises and political instabilities characterizing the 1920s, eugenic sterilization began to attract considerable attention from both the medical profession and health reformers. In his 1921 *Crâmpeie de eugenie și igienă socială*, the physician Ioan I. Manliu recommended that:

> 1) Every degenerate individual should be sterilized and, if possible, returned to society. 2) Every degenerate and sterilized individual should be kept in isolation in asylums and colonies until he/she can be returned to society as a useful member. 3) Only those individuals who still pose a danger to society after their sterilization should be isolated for life, while they should sustain themselves and society through work in gardens, workshops, and so on. (Manliu 1921: 21)

That American-inspired sterilization had an eager following in Romania may seem at odds with the country's broad agenda of positive eugenic improvement. Romanian medical and scientific communities learned, however, from a variety of sources, not least the influential journal *Buletin Eugenic și Biopolitic* (*Eugenics and Biopolitics Bulletin*). When Romania's most important eugenic journal began its publication in 1927 it featured a commissioned article by Harry Laughlin, deputy director of the American Eugenics Record Office, and one of the leading advocates of eugenic sterilization (Laughlin 1927: 253–7).

Yet these views did not go unchallenged. Like other Latin eugenicists, most Romanian eugenicists agreed that, if introduced, eugenic sterilization would contribute decisively to reducing the number of those considered mentally and socially degenerate; they also shared the conviction that the nation was in need of biological regeneration – yet only a few advocated the introduction of a sterilization law in Romania.

Another important factor to be considered here is Romania's social composition. At the time, the Romanian peasantry was valued as an untainted source of the nation's racial vitality. To this effect, many social hygienists and health reformers promoted a form of eugenic pastoralism, reinventing the Romanian peasants as degenerate and 'backward' eugenic subjects in need of medical care, modern hygiene, and social welfare, rather than sterilization. The physician Iosif Leonida was one of many who phrased the discussion of eugenic sterilization in Romania in these terms. As a neo-Lamarckist, Leonida minimized the importance of heredity, claiming that poor environmental conditions more likely caused immediate biological damage to the individuals concerned. Much like some French and Italian eugenicists, Leonida also highlighted the incompatibility of sterilization with Romanians' 'Latin mentality', in contrast to the 'Anglo-Saxon mentality of countries where sterilization had been introduced' (Leonida 1935: 367).

The psychiatrist and neurosurgeon Dimitrie Bagdasar also spoke in favour of the sort of public medicine-focused, positive eugenics so common in the Latin countries. He doubted that eugenic sterilization could achieve the goal of human betterment. 'Repeated sterilizations', he contended, 'could only lead to the numerical reduction of the race, but would not guarantee its future improvement, not even a relative one' (Bagdasar 1923: 1). Bagdasar then argued that improvement of the Romanian nation's health should be based on moderate population policies and qualitative eugenics.

Grigore Odobescu, a psychiatrist at the Central Hospital in Bucharest, further articulated scepticism about the introduction of eugenic sterilization in Romania. In *Eugenie pentru neamul românesc* (*Eugenics for the Romanian Nation*), Odobescu argued that in Romania 'neither the voluntary sterilization practised in Switzerland, nor the social prophylactic sterilization practised in the USA will be received favourably' (Odobescu 1936: 12). Employing the same neo-Lamarckist vocabulary used by Leonida and Bagdasar, Odobescu argued that 'degenerates' in Romania were largely the result of ruinous economic, sanitary, and hygienic conditions. He identified poor nutrition, lack of hygiene, and rampant contagious diseases (but not, significantly, hereditary diseases) as leading causes of racial degeneration. Without such obstacles, numerous examples proved the 'wonderful quality of the biological substance' intrinsic to the Romanian nation. For Odobescu, 'the [eugenic] education of the masses' was the solution to Romania's social and biological improvement. Such an educational programme would increase economic and social standards; more essentially, it would 'improve the biological condition' of the population. In effect, he regarded this as the 'eugenic policy most suited to our country and nation' (Odobescu 1936: 15).

The neurologist Dumitru Enăchescu also believed that 'there is no need for eugenic sterilization to protect our race from degeneration' (Enăchescu 1936: 279). This category of Romanian eugenicists was indeed careful to distinguish between the excesses of negative eugenics and positive methods of biological improvement. In return, supporters of eugenic sterilization, such as the racial hygienist Iordache Făcăoaru, were unimpressed with the fusion between social medicine and hereditarian eugenics. He declared that it was the responsibility of 'the spiritual leaders of the nation' to adopt negative eugenics. The neo-Lamarckist version of national improvement was welcomed, but it was not 'eugenics'; rather, Făcăoaru claimed neo-Lamarckist eugenics was 'curative and preventive medicine, demography, as well as the hygiene of the individual and the social education of the nation' (Făcăoaru 1936: 131).

Negative eugenic measures such as prenuptial marital certificates and sterilization lacked, however, the necessary political support in Romania. For instance, no mention of sterilization was made in the country's 1930 sanitary law. On the contrary, the law reflected the conception of positive eugenics, social hygiene, and public health held by its architect, Iuliu Moldovan, at the time sub-secretary of state in the Ministry of Work, Health and Social Protection.

Latin eugenicists in other countries similarly emphasized their distinct cultural and religious environment, when discussing the topic of eugenic sterilization. Like Leonida, the Italian military psychiatrist Placido Consiglio dismissed compulsory sterilization as unsuitable for the Latin world, 'where the traditional humanitarian instincts rebel against it' (Consiglio 1914: 461). The neurologist Augusto Carelli likewise appealed to the virtues of Latin eugenics as opposed to the inhumanity of 'Nordic' eugenics of the United States, a country he described as being driven by 'a mechanical brain and a mechanical heart'. In contrast, 'Latin gentleness, a consequence of the race's ancient genius, intuitively rebuffs this useless brutality. It feels that the remedy for human ills is not to be found in barbaric crudities, but in divine piety, in solidarity, and in faith in ideals as opposed to blind materialism' (Carelli 1928: 341–5).

Prominent members of the Italian Society for Genetics and Eugenics such as Cesare Artom also denounced negative eugenic ideas based on Mendelian genetics, which 'fortunately' had not found 'a climate hospitable for their expression in our country' (Artom 1926: 48). His colleague Marcello Boldrini wrote a stinging rebuke of American eugenics and, in particular, the radical negative eugenics campaign orchestrated by Harry H. Laughlin. Boldrini noted caustically that 'no country in the world is as full of blind people, deaf-mutes, the insane, the deficient, criminals, epileptics, sufferers from venereal diseases, and drug addicts as is the United States'. Italy, however, would achieve the improvement of its population through much more humane eugenic means. Not Protestant individualism, but 'entirely Roman and Catholic' principles would be employed. In accordance with the fascist demographic policy, Italy's eugenic programme would focus on *increasing* the population not *decreasing* it (Boldrini 1928: 51–4).

The issue of sterilization also caused notable tensions among French eugenicists. In 1925 the French Eugenics Society rather weakly endorsed eugenic sterilization 'if it can be shown that the descendants of the abnormal [person] constitute an overwhelming burden for family and society'. With this proviso, 'it would be wise to envisage the possibility of sterilizing undesirable procreators, after a minutely detailed study of each particular case' (quoted in Nisot 1927: 436). Even this endorsement would not find resonance among most French eugenicists. In 1929, the Belgian eugenicist Marie-Thérèse Nisot mourned the fact that 'sterilization of the abnormals', which so strongly preoccupied English, German, Swiss, and American eugenicists, was less discussed in France (Nisot 1929b: 595–603). The fact that the only papers dealing with eugenic sterilization at the 15th International Congress of Anthropology and Prehistoric Archeology (*XVe Congrès International d'Anthropologie et d'Archéologie Préhistorique*), held in Paris in 1931, were given not by the French but by the British eugenicists, C. P. Blacker and Cora B. S. Hobson, seems to substantiate Nisot's statement.

As the 1930s unfolded, the main French eugenicists did, however, engage with the topic of eugenic sterilization. Tellingly, in 1932 Henri Vignes, general secretary of the French Eugenics Society, sent out a questionnaire to twenty prominent French eugenicists regarding sterilization of the 'socially inadequate', basing the term on Charles Davenport's 1912 study, *Proper Care for the Socially Inadequate* (Vignes 1932: 228). Ten of them replied, including Eugène Apert,

Henri Briand, Georges Papillaut, Georges Schreiber, and the geneticist Raymond Turpin (Vignes 1932: 230–42). While some contributors, such as Apert and Schreiber, were in favour of limiting the fertility of those deemed 'inadequate', others – Papillaut and Turpin, for example – remained unconvinced. Accepted measures for controlling the reproduction of 'socially inadequate individuals' were institutionalization and isolation. Equally important, it was 'positive eugenics' (*eugénique positive*) that was described as the most adequate strategy of human improvement rather than 'negative eugenics' (*eugénique negative*). Sterilization was rejected, and 'public education' and the introduction of 'the prenuptial certificate' proposed instead (Vignes 1932: 243–4).

Similar preoccupations and disputes about eugenic sterilization existed during the 1930s in Portugal and Spain, where those dedicated to Latin and Catholic cultural traditions joined in this rejection of radical eugenic measures. In an article published in 1931, the Portuguese priest Domingos Maurício Gomes dos Santos (who published under the pseudonym Riba Leça) criticized negative eugenics, arguing that to promote it in Portugal was no less than a 'crime against morality and nationhood'. Like other Catholic eugenicists, Leça identified the centrality of Christian morality to 'the preservation and perfection of the species' (quoted in Cleminson 2014).

Leça also criticized the Brazilian eugenicist Renato Kehl for his support of eugenic sterilization (discussed in the next chapter). On 24 October 1932, Kehl visited the University of Oporto and gave a lecture to the Portuguese Society of Anthropology and Ethnology. The fact that Kehl had endorsed eugenic sterilization during his lecture, prompted Leça to declare 'this conference to be extremely infelicitous in respect of its nature, intent and content' (quoted in Cleminson 2014: 79). This was more than just a disagreement between two eugenicists. As Richard Cleminson noted, 'the rejection of sterilization – but not of "eugenism" – from a renowned Catholic commentator [Leça] in a prestigious scientific and cultural review of the faith set the parameters for the development of eugenics in Portugal' (Cleminson 2014: 80). This can be seen in the comments made by the psychiatrist and biotypyologist Barahona Fernandes, who although embracing a strong hereditarian interpretation of eugenics did not hesitate to describe the Nazi sterilization law in 1934 as 'a product of desires and ambitions for a eugenic culture and for racial perfection in Germans and coeval with the megalomaniacal theories of superiority' (quoted in Cleminson 2014: 148). During the same period, other Portuguese eugenicists such as António Mendes Correia also contemplated more interventionist eugenic measures, though not endorsing compulsory sterilization. Not surprisingly, then, when on a visit to Brazil in

1935, Mendes Correia attempted to understand Kehl's eugenic radicalism. Upon meeting Kehl, they agreed that 'there must be prudence in the application of practical measures' (quoted in Henriques 2012: 67).

In Spain, too, proposed eugenic methods of population control were set against Catholic traditions (Peláez 1988b: 77–95). Take, for example, the book *Eugenesia y Matrimonio* (*Eugenics and Marriage*) written by physician Francisco Haro in 1932. Based on extensive knowledge and sources, Haro requested the introduction of marriage certificates in Spain. But, as noted by his mentor Gregorio Marañón in the preface, Haro professed a moral eugenics that operated in more dimensions than the one defined by the opposition between *negative* and *positive* methods of human improvement (Haro 1932).

A national version of eugenics was thus proposed in tune with international developments but also responsive to the new social and political context in Spain after the end of Primo de Rivera's dictatorship. As such, another conference on eugenics *(Primeras Jornadas Eugénicas Españolas)* took place in Madrid from 21 April to 10 May 1933 (Cleminson 2000: 97–107; Barrachina 2004: 1011–26). More than fifty intellectuals attended the conference, including doctors, lawyers, teachers, politicians, and theologians. Topics included eugenics and divorce, 'conscious maternity' (*maternidad consciente*), criminality, prostitution, birth control, and abortion (Noguera and Huerta 1934). As Luis Huerta explained in his introductory lecture, eugenics was to be discussed from the vantage point of its four component disciplines: 'biology, medicine, social and pedagogy' (Huerta 1934, vol. 1: 9). In Spain, Huerta continued, 'eugenics has a legitimate right to be practised', not least for historical reasons due to the 'monstrous and terrible forms of human improvement' promoted by officials of the Inquisition such as Pedro de Arbués and Tomás de Torquemada, which lowered Spain in the eyes of the world, ultimately leading to her 'unbearable and shameful decline' (Huerta 1934, vol. 1: 11).

The conference promoted 'a variety of different theories and techniques geared to population improvement' (Cleminson 2000: 104). With respect to sterilization, the gynaecologist José María Otaola, in his lengthy discussion of the eugenic importance of reproduction, offered the moderate view that curtailing the 'unlimited reproduction' of those deemed 'unfit' was justified in some cases, but remained problematic in general (Otaola 1934, vol. 1: 273–309). By contrast, Federico Castejón, professor of law at the University of Seville and a long-standing supporter of sterilization, maintained that negative eugenic measures were not only justified (medically and legally) but also needed in order to abate the growing anxiety about national degeneration (Castejón 1934, vol. 2: 181–201).

Interestingly, the Papal Encyclical *Casti Connubii* (discussed below) did not elicit any commentary from the conference participants, including Jaime Torrubiano, professor of canonical law, who discussed the symbiotic relationship between Catholicism and eugenics. Like other Catholic eugenicists, Torrubiano was interested in reforming and improving society in coalition with science not against it. After all, he maintained, Christianity was the 'eugenic science's best assistant' (Torrubiano 1934, vol 1: 60–84).

Yet in Spain – as elsewhere in the Latin world – the significance of Catholicism to the nation's spiritual and biological renewal could not be ignored, even when all religious barriers have been eliminated, as it happened in Catalonia during the Civil War (1936–1939). Prior to 1936, eugenicists in Catalonia promoted pronatalism and the strengthening of national identity. For instance, in 1934, the manifesto for 'the protection of the Catalan race', was published by Josep A. Vandellós, the philologist Pompeu Fabra, the jurist Francesciso Maspons, and physicians Santiago Pi Sunyer and Hermenegildo Puig i Sais. Following this public display of racial awareness regarding the biological future of the Catalans, the Catalonian Eugenics Society (*Societat Catalana d'Eugenèsia*) was established in 1935, with Puig i Sais as president, and Vandellós as secretary general (Domingo 2012: 18).

Vandellós broadly outlined the aims of the Catalonian Eugenics Society in his 1935 book, *Catalunya, poble decadent* (*Catalonia, Decadent Nation*). Vandellós feared that the Catalonian people had reached the latter stages of their population cycle, which would account for the fact that they had the second lowest birth rate in Europe, next to France. Typical for a Latin eugenicist, Vandellós denounced the causes of population decline: feminism, birth control, and anti-familial, individualistic egotism. Not surprisingly, he also condemned sterilization. The remedies he proposed to safeguard the 'Catalan race' reflected as much his Catalan nationalism as his commitment to natalist eugenics (Vandellós 1935: 223–32).

Such eugenic views were, however, promptly revised after 1936 by the Catalan government. Most notably, on 25 December 1936, Dr. Félix Martí Ibáñez, Director General of Health and Social Assistance (*Sanitat i Assistència Social*) of the *Generalitat de Catalunya*, issued the 'Eugenic Reform of Abortion' ('Reforma eugénica del aborto'), and thus effectively legalizing abortion in Catalonia (Cleminson 2000: 236–42). The decree, containing the guidelines under which abortion was permitted, was then published on 9 January 1937 ('Decret', *Diari Oficial de la Generalitat de Catalunya* 1937: 114–5).

The fact that many liberal, leftist, and anarchist eugenicists in Spain were anti-clerical only accentuates the importance of religion to eugenic programmes of social and biological improvement, especially to conservative Latin eugenicists. In the late 1930s, one of them, in particular, exalted Catholic values and the importance of traditional Spanish family to propose a form of national eugenics that would revitalize the race: the psychiatrist Antonio Vallejo-Nágera (see Figure 4.1).

In his 1937 *Eugenesia de la Hispanidad y la regeneración de la raza* (Hispanic Eugenics and the Regeneration of the Race), Vallejo-Nágera outlined a eugenic programme that aimed to restore Spain to her former historical magnificence through the purification of the body politic. While glorifying the 'raza hispánica' (Hispanic race), Vallejo-Nágera did not propound negative eugenic measures to protect it from degeneration but a form of Catholic eugenics, one legitimized by religion and anti-communist nationalism. 'The regeneration of the race', he thus argued, 'must necessarily be based on the regeneration of the family institution, because the family constituted in accordance with the traditional principles of Christian morality is the nursery of social virtues, a shield against a corrupt social environment, a sacred reservoir of traditions' (Vallejo-Nágera 1937: 118).

Figure 4.1 Antonio Vallejo-Nágera
Source: Archivo General de la Administración, Alcalá de Henares, Spain.

The eugenic ideas that supported Vallejo-Nágera's vision of a racially revived Spain were therefore based on positive theories of human improvement, as well as on constitutionalism and biotypology. 'The regeneration of the race', he argued, was based not only on the 'selection of biotypes', but on the 'improvement of phenotypes through constant action to improve each individual, physically and morally' (Vallejo-Nágera 1937: 77). Through positive eugenics a new Spanish biological elite, a 'eugenic aristocracy' ('aristocracia eugenésica') would eventually be formed, symbolizing Spain's racial vitality and global hegemony (Vallejo-Nágera 1937: 122).

Catholicism was central to Vallejo-Nágera's eugenics, and this in turn created a comfortable symbiosis with the official nationalism of the regime during the late 1930s. The vehemence that fuelled Vallejo-Nágera's racial vision of the new Spanish state will be discussed in another chapter; here we retain only his concern with the imperfections of the Spanish physical body and the eugenic strategies he envisioned as a result. Given the role played by Catholicism in shaping the political and national traditions during the interwar period, it is important to examine this essential component of Latin eugenics in a wider context.

Eugenics and religion

The different attitudes of Protestants and Catholics towards eugenics only reinforced the growing divide in the worldwide eugenics community. In general, Protestants of that era did not object to state involvement in reproductive control; indeed, some of the Protestant Churches had originated as extensions of state power in the first place. Many Protestants embraced eugenics as part of a broader trend towards acceptance of the secularization of modern societies. In Christine Rosen's inspired description, 'preachers had become enamoured by the possibilities science presented; in eugenics they found a science whose message moved effortlessly from laboratory to church' (Rosen 2004: 4). Many Protestant theologians were outspoken eugenicists, and some even supported eugenic sterilization: in Britain, William Ralph Inge, the dean of St. Paul's and Ernest William Barnes, the bishop of Birmingham; in Germany, Hans Harmsen and the *Innere Mission* (National Missionary Work); in Romania, Alfred Csallner and the Saxon Evangelical Church. Such direct involvement with eugenics demonstrates that a religiously sanctioned programme of human improvement was possible (Mayer 1930: 43–51; Dietrich 1992: 575–601; Richter 2001).

Conversely, the Orthodox Church expressed a reserved attitude towards social and scientific movements in general. In Romania, however, religious journals often published commentaries about eugenics, and the eugenicists often appealed to the Church to contribute to the 'biological rejuvenation' of the Romanian nation. Ioan Manliu, for instance, called on the Orthodox Church 'to use its overwhelming moral authority, declare itself in favour of biological purification, and act accordingly'. In a country where Orthodox Christians made up 72.6 per cent of the population, the Orthodox Church played an important role in all aspects of life. 'The moment has come', Manliu further believed, 'for [the Orthodox Church] to take part without delay in this [eugenic] movement, in order to ensure scientifically and biologically the happiness of its believers. If the Church firmly popularized eugenic ideas and collaborated devotedly in their realization, it could provide an invaluable service in our struggle against the degeneration and Asiatization of our race' (Manliu 1931: 382–3).

The Roman Catholic Church showed a much more proactive attitude against the increasingly aggressive secularization of national reproductive control. The Church's role in society, as well as its theological authority, was obviously under threat. Some Catholic eugenicists feared that the encroaching powers of the state even threatened the survival of human individuality. Thus, it is hardly surprising that the Church failed to support the bold demands of some eugenicists for the power to eugenically control marriage and reproduction.

Catholic eugenics did exist, and was tolerated by the Church, but its premises had to steer clear of any doubts regarding Church theology or authority. All Catholics were expected to endorse the sanctity of human life and its reproductive faculty, both divinely endowed by God. The goals of Catholic eugenics – the reformation and improvement of the nation and society according to modern ideas of medicine and public health – were consistent with the Church's views on modern medicine and education, and were thus tolerated. Henri Le Floch, the rector of the French Seminary (*Collège Français*) in Rome and an advisor to the Holy Office, thus remarked categorically in 1931 that only the Catholic Church was 'the true promoter of eugenics' ('vera fautrix *Eugenicae*') (quoted in Betta 2008: 142).

Some Catholic eugenicists accepted the arguments put forward by supporters of premarital medical examinations, particularly in France and Belgium in the late 1920s. However, the majority of Catholic eugenicists adhered to neo-Lamarckism, with its focus on nurture and environmental conditions, and to puériculture. As the historian Édouard Jordan, editor of the journal *Pour la vie* (*For Life*) and member of the Association of Christian Marriage (*Association du*

Marriage Chrétien) put it in his 1931 book *Eugénisme et Morale* (*Eugenics and Morality*), eugenics in France 'would continue its legitimate warnings against unfortunate births, but it would concentrate again more on the improvement of the milieu, on the progress of medicine, on general hygiene, urban planning, and why not say it, morality' (Jordan 1931: 173). It was not sufficient, Jordan warned, to expose the 'immoral excesses' of eugenics; one also needed to aspire to a new epistemology of human improvement based on Christian morality. 'Eugenics was wrong', Jordan concluded, 'not when it addressed the serious question of racial degeneration, but when it pretended to solve it without considering the other great interests of humanity, and when it acted as the only Law inspired by Science' (Jordan 1931: 174).

Such was the prevailing attitude at the 8th National Congress of the Association of Christian Marriage (*VIIIe Congrès National de l'Association du Mariage Chrétien*), held in Marseilles between 25 and 27 April 1930, under the chairmanship of Monsignors Maurice-Louis Dubourg, archbishop of Marseilles, and Emmanuel-Anatole Chaptal, assistant to the archbishop of Paris. The main theme of the Congress was 'Church and Eugenics'; its goal, however, was to 'denounce and demonstrate the immorality and the danger represented by the regrettable alliance between eugenics and neo-Malthusianism' (Jordan 1930: v).

Édouard Hawthorn, a physician, set the tone of the Congress, by criticizing the 'neo-Malthusian movement and doctrine'. Henri Brenier, president of the Federation of Large Families (*Fédération des Familles Nombreuses*) in Provence, pointed out the nation's economic problems caused by the control of fertility and family limitation (Brenier 1930: 11–30; Hawthorn 1930: 3–18). Both papers identified 'Anglo-Saxon' eugenics, as displayed at the First World Population Conference held in Geneva in 1927, as reprehensive international endeavours to recast marriage and reproduction according to neo-Malthusian principles. However, it was generally assumed that the adoption of family limitation, together with birth-control and sterilization, were cautiously advocated by the French eugenicists. For instance, the physician Jean Piéri put it simply: 'The idea [of sterilization], I say it here, is not French' (Piéri 1930: 72). Piéri also quoted from a letter sent to him by Eugène Apert, president of the French Eugenics Society. 'In the Latin countries', Apert wrote, 'we believe that human improvement can be achieved equally through the encouragement of reproduction' as through negative eugenic measures. However, Apert could not simply ignore the official 1925 pronouncement of the French Eugenics Society, which sanctioned sterilization in some cases. Rather, Apert attempted to reassure the Catholic Church that this declaration was not an endorsement of a eugenically interventionist state. Rather, sterilization was to be applied only 'with the greatest consideration and be left to the judgment and common sense of the physicians to use it when needed (similar to therapeutic abortion)' (Piéri 1930: 72–3). Piéri interpreted Apert's letter as an endorsement of voluntarism and medical responsibility, reflecting deeply entrenched ideas about the individual and the family in France as elsewhere in the Latin world. Piéri used other legal and religious arguments to reject sterilization, but it was ultimately 'respect for life' that had to be the foundation underlying 'racial selection' (Piéri 1930: 77–88).

The other papers at the Congress were variations on these themes. Robert de Vernejoul signalled out Christian morality in his discussion of family limitation (De Vernejoul 1930: 161–73). Jean Dermine, professor of the Theological Seminar of Bonne-Espérance (in Belgium), claimed that 'Christian eugenics' was a form of 'integral eugenics' ('l'eugénisme intégral'), based on morality, prenuptial chastity, and the respect for the family (Dermine 1930: 174–190). Most authoritatively, Maurice-Louis Duborg, the archbishop of Marseille, explained the connection between Catholicism and eugenics. Speaking of 'True Eugenics' ('le veritable eugénisme'), Duborg recognized the necessity for biological awareness in preserving the race as long as it did not contradict Christian morality and reasoning (Duborg 1930: 224–7). Duborg thus articulated a scientific vision of eugenics aimed at national improvement, but still in accordance with the main tenets of Catholicism.

The Church's fears regarding eugenics were unequivocally expressed in Pope Pius XI's encyclical *Casti Connubii* promulgated on 31 December 1930. This was the Roman Catholic Church's belated, but powerful, response to Galton's vision of eugenics as 'the religion of the future' (Wattiaux 1981: 801–17; De Raes 1989: 457–60; Lepicard 1998: 527–44). The encyclical chastised those who 'put eugenics before the aims of a higher order, and by public authority wish to prevent from marrying all those whom, even though naturally fit for marriage, through some hereditary transmission, bring forth defective offspring'. Equally, it rejected the use of such notions as 'unworthy' life. Abortion and sterilization were both seen as expressions of excessive secularization and of the state's attempt to control the 'private sphere' of individual and family:

> But another very grave crime is to be noted, Venerable Brethren, which regards the taking of the life of the offspring hidden in the mother's womb. Some wish it to be allowed and left to the will of the father or the mother; others say it is unlawful unless there are weighty reasons which they call by the name of medical, social, or eugenic 'indication'. Because this matter falls under the penal

> laws of the state by which the destruction of the offspring begotten but unborn is forbidden, these people demand that the 'indication', which in one form or another they defend, be recognized as such by the public law and in no way penalized. (Casti Connubii 2007: 81–2)

The Church could not condone the termination of life sought by some eugenic programmes: 'to wish to put forward reasons based upon them for the killing of the innocent is unthinkable and contrary to the divine precept promulgated in the words of the Apostle: Evil is not to be done that good may come of it' (Casti Connubii 2007: 81–2). By condemning contraception and sterilization as immoral, the encyclical reinforced Catholic notions of marriage as essential to a well-functioning, modern society. Yet Catholicism's traditional approaches to contraception, marriage, and family life were not necessarily in conflict with eugenic teachings: '[w]hat is asserted in favour of the social and eugenic "indication" may and must be accepted, provided lawful and upright methods are employed within the proper limits' (Casti Connubii 2007: 81–2).

Reaction to the papal encyclical by the international eugenic community ranged from indignation to agreement. In its April 1931 issue, the British journal *The Eugenics Review*, for instance, described the *Casti Connubii* as 'a defiant return to mediaevalism'. This unflattering remark indicated the Eugenics Society's frustration with the Pope's pronouncements on eugenics. It became clear the encyclical 'will undoubtedly drive many from the fold and deter many more, while it will regiment the faithful into a compacter body of ardent and well-disciplined militants' ('On Catholicism', *The Eugenics Review* 1931: 41). A chasm thus opened between the eugenicists and the 'crusaders of Rome', *The Eugenics Review* noted defiantly ('On Catholicism', *The Eugenics Review* 1931: 44), while at the same time suggesting a vigorous defence of the idea of intellectual and scientific freedom.

Set against the backdrop of an assertive Roman Catholic Church, eugenics was redefined by its Catholic supporters after the promulgation of *Casti Connubii*. The Italian fascist state lent tacit support to this trend. The new penal code, introduced in 1931, officially condemned eugenic sterilization in accordance with the 'indefatigable principles of Catholic morals'. So argued Francesco Leoncini, professor of legal medicine at the University of Florence and one of the participants at the Second Congress of Catholic Physicians, held in Florence in October 1932 (see below). Such attitude illustrated powerfully, he believed, the merits of the new civilization promoted by Italian fascism, one 'shaped by Latin genius and the spirit of Christianity' (quoted in Cassata 2011: 138).

Catholic physicians sympathetic to eugenics endorsed the correlation between Latinity and Christianity (Betta 2008: 133–44). In doing so, they also fundamentally revised the traditional Christian model of humanity's corporeality, in which the body was devalued as fallen and ultimately dismissed as insignificant. Eugenics, they argued, offered a new epistemology of the human body, emphasizing its central importance to the life of the individual and the national community. Thus René Brouillard, a French Jesuit theologian, spoke in favour of a 'Christian eugenics' promoting life and religious morality as opposed to the 'eugenics of death' advocated in the USA, Britain, and Germany. In his synthesis of puériculture, public health, preventive medicine, and Catholicism, Brouillard delineated the objectives of 'a moral, pro-family, social and Christian eugenics':

> Sanctification of marriage and the duties of spouses; attention to morality and health at the time of conception and birth; action by the state, associations and the church against public immorality, social diseases, alcoholism, slums, etc. to develop the economic well-being, general hygiene, puériculture, healthy dwellings, the prosperity of families. (Quoted in Schneider 1990: 197)

A similar vocabulary would be used throughout the 1930s by Latin eugenicists in their common effort to rescue eugenics from the confines of racial hygiene (López 1935: 6–7). Brouillard was, however, not the highest-ranking Catholic theologian to suggest a compromise between eugenics and religion; or even the most representative Catholic figure of Latin eugenics in Europe. This distinction belongs to the Italian Franciscan theologian Agostino Gemelli (see Figure 4.2).

Today Gemelli is most remembered as the founder and rector of the Catholic University of the Sacred Heart in Milan (Bocci 2003). However, in the interwar period, Gemelli was the most outspoken advocate of Catholic eugenics in the Latin world (Ottaviani 2010: 69–92). Gemelli obtained his medical degree in 1902 from the University of Pavia and began his scientific career as a psychologist. He soon joined the Franciscan order, and was ordained a priest in 1906 (Cosmacini 1985: 53; Colombo 2003: 338). Nevertheless, Gemelli remained interested in experimental psychology and psychiatry throughout his life (Foschi, Giannone, and Giuliani 2013: 130–44).

Gemelli began publishing on eugenics in 1915, promoting a Catholic version of eugenics that exalted Christian morality and family values. Gemelli's emphasis on the effect of environment on human behaviour accorded well with neo-Lamarckism (Gemelli 1915). To some extent, he also shared Corrado Gini's populationist views, believing, however, that the decline in population was due

Figure 4.2 Agostino Gemelli
Source: Archivio generale per la storia dell'Università Cattolica del Sacro Cuore, Milan.

primarily to the negative effects of secularization, de-Christianization, and socialist ideas on family life (Cassata 2011: 53–4).

Gemelli was one of the participants at the First Italian Congress of Social Eugenics held in Milan in 1924, and assumed a leading role in the Italian Eugenics Society in the period that followed. In his paper to the Congress, he discussed the relationship between eugenics and religion, with a particular focus on its social dimension (Gemelli 1927: 53–66). It is also on this occasion that Gemelli reiterated his conviction that science and religion were compatible, as long as the guiding principles of reconciliation were based on morality. He recognized eugenics' appeal to modern society. As a Catholic and eugenicist, Gemelli described his work as 'positive' in terms of its moral purpose, and as 'practical', presupposing an active engagement with 'eugenic propaganda to save our race' and 'to facilitate her improvement'. This two-pronged approach to eugenics enabled Gemelli to bridge the gap between the scientific planning of human life and its moral, ethical, and religious purpose. Catholicism, he asserted, could provide 'valuable assistance' to eugenics, hastening its practical application to society (Gemelli 1927: 54–5). In this respect, Gemelli referred to the work of theologians in England, Belgium, and Germany, including Dean Inge, Valère Fallon, and Hermann Muckermann, and their description of eugenics as both a science and a religion (Gemelli 1927: 56). But it was not only eugenicists who tried to appropriate religion for their purpose; ecclesiastical authorities also realized the distinctive potential of eugenics, one that was not merely biological but also religious and historical. In this respect, Gemelli mentioned the favourable attitude of the Diocesan Board of Milan towards eugenics expressed in May 1924, encouraging Catholics to pursue 'the true and healthy eugenics' and to make use of 'the means that social and biological sciences offered in this field' (Gemelli 1927: 57).

There remained, of course, a fundamental disagreement between the Catholic Church and the eugenicists over sterilization (Gemelli 1927: 61). No eugenic intervention in the family and reproduction was justified, Gemelli insisted. Marriage existed primarily to provide conditions appropriate for procreation, which was under God's control. Thus, Catholicism vehemently opposed 'artificial' means of birth control as a sinful thwarting of God's will. An unhealthy person should either live in chastity or accept a marriage without sexual relations (Gemelli 1927: 63). Gemelli further stressed the need to guide eugenic regeneration through moral improvement, rather than corporeal human evolution through the manipulation of reproduction and sterilization. The entwinement of eugenics and Catholicism was thus to be encouraged as the much-needed alliance 'in the battle against immorality' and 'for the improvement of the race' (Gemelli 1927: 66).

After the release of the encyclical *Casti Connubii*, Gemelli remained committed to Catholic eugenics, both in Italy and abroad (Gemelli 1931: 606–14. Also Cassata 2011: 137–8; Pozzi 2012: 161–74). For example, he was one of the main participants at the Congress of Catholic Physicians (*Congresso dei Medici Cattolici*) held in October 1932 in Florence and at the Second International Congress of Catholic Physicians (*II. Internationalen Kongress katholischer Ärzte*) convened in May 1936 in Vienna. In promoting eugenics within the confines of Catholic morality, the participants at the Italian Congress agreed on the following course of action:

> 1) To invite Catholic physicians to keep abreast of scientific progress in genetics and invite Catholic scholars to cooperate with such studies and promote the good and healthy applications of this young and already greatly progressed science;

> 2) To ask that the civil public authority prevent the diffusion in Italy of foreign propaganda of those eugenic methods that represent a violation of moral laws;
> 3) To vote that Catholic physicians explain to the profane how the moral and physical improvement of humanity cannot be obtained with the hurried and unjustified application of genetics to the human race, and neither with the propagation of those eugenic norms that contradict divine laws and are contrary to human dignity, but rather through the moral laws taught for centuries by the Catholic Church, norms that also govern the real progress of social hygiene and genetics. (Quoted in Cassata 2011: 139)

In this resolution, the community of Italian Catholic physicians articulated its own version of Christian eugenics in accordance with the teachings of the Church on family, reproduction, and the sanctity of human life.

The introduction in Germany of the Nazi Law for the Prevention of Hereditarily Diseased Offspring (*Gesetz zur Verhütung erbkranken Nachwuchses*, GezVeN), on 14 July 1933, added additional pressure on the Catholic Church to denounce eugenic sterilization in no uncertain terms. Thus, when the Pope addressed the Fourth International Hospital Congress in Rome in May 1935, he warned against other countries emulating the sterilization law adopted by Nazi Germany: 'We must express the conviction that if such practices were accepted by people, states, governments, if they should enter into the practices of life, if, in a word, should by adopted, then Our duty suggests to Us that as Supreme Pastor, We shall have to use every means to protest' (quoted in Leon 2013: 107). Relations between the Vatican and Berlin continued to deteriorate, eventually leading to Pope Pius XI's public rebuke of the Nazi government in the papal encyclical *Mit brennender Sorge* (On the Church and the German Reich), read in German churches on Palm Sunday, 21 March 1937.

Catholic physicians meeting in Vienna in 1936 for their Second International Congress were united in their unfailing endorsement of Pius XI's encyclical and his condemnation of eugenic sterilization. Presided over by Cardinal Theodor Innitzer, archbishop of Vienna, the Congress brought together Catholic physicians from all over Europe (Löscher 2009: 128–32). Agostino Gemelli read a message from the Pope, reiterating the Catholic Church's official position against eugenic sterilization and emphasizing *Castii Connubii*'s statements regarding 'modern biology in general and medicine in particular' ('Kongressbericht', *St. Lukas* 1934: 74–6).

Octave Pasteau, a French doctor and founder of the Central Secretariat of the National Associations of Catholic Physicians (*Secrétariat Central des Sociétés Nationales de Médecins Catholiques*), explained that eugenic sterilization was as much a religious as a social issue. State-imposed sterilization presupposed that man was only a biological creature and reproduced accordingly. It subsumed individual interests to those of the state. 'Christian eugenics', by contrast, 'encompassed the individual as a whole, body and spirit; it examined the individual's full physical and spiritual value.' Pasteau also criticized the state's 'materialist-egotistical interests' in guiding social and biological selection. The introduction of the German Sterilization Law confirmed the ascendancy of the biopolitical state, which proclaimed as official policy the 'prevention of worthless life' (*Verhütung unwerten Leben*), by means of 'birth control, sterilization and abortion' and ultimately, 'euthanasia' (Pasteau 1936: 79).

Gemelli's paper was perhaps the most elaborate attempt at the Congress to combine the new science of genetics and religion into a 'Christian eugenics'. Gemelli sought to prove that '[t]rue science was not against religion'. Like Catholicism, science 'should be the champion of altruism and charity'. This had to be the guiding principle of the science of eugenics as well. As a scientist, Gemelli recognized the value of rational social planning and the need to protect future generations from further degeneration; as a Catholic, however, he believed that eugenic methods such as compulsory sterilization contravened the basic premise of Christianity: compassion and respect for life in all its forms. If it assisted the state in its right to introduce prophylactic sterilization then it failed its mission and became an instrument of abhorrent barbarism (Gemelli 1936: 96). A type of eugenics consistent with Christian morality should rely on positive methods of human improvement, not 'anti-humanist' racial hygiene. The papers of other participants, most notably the Austrian physician Albert Niedermeyer; the Belgian anthropologist Louis Vervaeck; Paweł Gantkowski, professor of pastoral medicine in Poznań, Poland; and the young Spanish priest Pedro Arrupe, reiterated similar themes.

The Congress concluded by issuing a number of statements on eugenics, 'influenced not by any preconceived notions of biological and medical research' but in accordance with Christian morality. The first was to reject 'sterilization as a means by which to prevent hereditary diseases'. Not only was sterilization deemed 'inappropriate' but it was also described as dangerous for the 'individual, the society and the nation'. Health authorities were warned that 'state regulated birth-control, abortion, and legal sterilization' would ultimately lead to 'euthanasia' (a forecast that unfortunately proved only too accurate). Castration for eugenic reasons or as judicial punishment was also rejected.

Yet not all eugenic measures were discarded. The Congress suggested that the Catholic fraternal society 'Catholic Action' should promote an active, positive eugenics, which would include the 'establishment and promotion of Catholic marriage counselling centres'; provide more opportunities for Catholic physicians to study social hygiene and eugenics; and assist state and local officials in their health campaigns against social and biological degeneration. Finally, the Congress proposed that all Catholic medical associations collaborate with each other when dealing with 'questions of eugenics, genetic research and especially sterilization', and that studies pertaining to 'all aspects of sterilization' be published and collected to form the core of a new medical library devoted to eugenics in the Vatican ('Ergebnisse und Beschlüsse', *St. Lukas* 1936: 173–5). Although the Congress continued to support traditional Latin eugenics free of reproductive control, the tone was somewhat more skeptical and cautious than at the previous meeting in Florence.

The Second International Congress of Catholic Physicians represents an important moment in the history of Latin eugenics. The Catholic Church was losing ground to the state over control of reproductive rights, and the state's efforts to engineer society. In response, the Catholic Church endorsed Latin eugenics insofar as it emphasized positive eugenics, individualism, Christian morality, and family. It regarded Latin eugenics as a cultural mission as well as a means of biological improvement. In this sense, the Catholic Church also looked to Latin eugenicists to cooperate in their efforts to combat the growing appeal of Nazi eugenics.

The impact of the Nazi Sterilization Law of 1933

The growing gulf between Latin eugenicists, who opposed eugenic sterilization, and Nordic eugenicists, who supported it, was further widened with the Nazi takeover of the German Reich in early 1933, and their promulgation of the Eugenic Sterilization Law on 14 July 1933. The Swiss-German racial psychiatrist Ernst Rüdin had been appointed some months earlier as president of the International Federation of Eugenic Organizations (IFEO), at its tenth meeting, held in New York during the Third International Congress of Eugenics (21–23 August 1932). Rüdin still held that post after the Nazis came to power in Germany, and was one of the racial hygienists directly involved in writing the Nazi Sterilization Law. Thus, Rüdin was in a position to act as a conduit for Nazi eugenic activism in the most important international eugenic organization.

The French eugenicist Georges Schreiber realized exactly what this meant for eugenics. He remarked that supporters of Hitler in Germany, Rüdin among them, considered 'racist policies to be essentially eugenic', while other countries saw them as 'deeply anti-eugenic' and discriminatory against parts of the population (Schreiber 1933: 388). Schreiber was forced to admit that, as 'each nation has the right to adopt its own eugenic policy', any 'official protest' against German racial eugenics, and by extension racist and immigration legislation adopted in the USA against members of the 'Latin or Mediterranean race', had 'no standing' (Schreiber 1933: 388).

Schreiber expressed his anti-Nazi views again at the 11th meeting of the IFEO, held between 18 and 21 July 1934 in Zürich, Switzerland. The meeting attracted some of the most important Nordic eugenicists such as: Jon Alfred Mjöen, Alfred Ploetz, Ernst Rüdin, and Otmar von Verschuer. Eugenicists from Great Britain, the Netherlands, Denmark, Poland, Hungary, Austria, and Czechoslovakia also participated (*Bericht über die 11. Versammlung der Internationalen Föderation Eugenischer Organisationen* 1935). No Italian or Belgian eugenicist attended. Schreiber was astonished to see that after Hitler's rise to power, 'an international scientific organization', such as the IFEO, could continue to be chaired by 'a German'. No German scientist, Schreiber believed, could 'write and say what he really thought. He must write and speak "Nazi", which is essentially anti-scientific' (Schreiber 1935a: 79). Falk Ruttke's presentation was illustrative in this sense. Ruttke was a member of the Reich Committee for Public Health Policy, as well as a member of the Advisory Board for Population and Racial Policy at the Reich Interior Ministry. Ruttke told the participants that after Hitler's accession to power, the 'knowledge of genetic laws was invoked towards the creation of a healthy race, and regulations were brought into conformity with scientific knowledge and a commonsense application thereof'. The Sterilization Law was one of the main methods of applying these principles of social and biological selection, and Ruttke anticipated that it would 'serve as a model for the world' (*Bericht* 1935: 67).

Schreiber knew all too well that most German eugenicists were academics employed by the state. After 1933, in the scientific hierarchy, the course of one's career came to be determined much more by signs of devotion to the Führer than by the quality or quantity of one's scientific work. It did not take long for German eugenicists to discover that cooperation with the regime, and coordination of one's research with Nazi racial ideology, could be rewarded with funding and other resources. Yet Schreiber – like many other eugenicists at the time – was impressed with the German eugenic programme. When comparing the number of sterilizations carried out in the USA from 1900 to 1928 with those performed in Germany within a year since the introduction of the Sterilization

Law – 16,000 cases according to Schreiber – he wrote approvingly of the need for state intervention to ensure the race's biological improvement (Schreiber 1935b: 82–4 and 1935c: 84–91). He knew, however, that qualitative eugenic measures to eliminate the 'unworthy and the unfit' were not endorsed by French general public and political elites. From his vantage point as a French eugenicist, Schreiber considered his country as 'a nation whose sentiment resents the power of the state to interfere with the individual'; as such it would 'not be converted to obligatory sterilization by reasoning of any kind'. Sterilization, he predicted, would 'probably never become firmly established in France' (Schreiber 1936: 105).

This was also the position adopted by the Italian politicians and eugenicists. Mussolini set the tone by denouncing the German sterilization programme in September 1934. The paediatrician Ivo Nasso similarly condemned 'German barbarism'. According to Nasso, 'Germany does not hesitate to engage in the most cruel and inhumane acts, such as those which deny the most sacrosanct individual rights' (Nasso 1935: 721–22). The legal scholar Giuseppe Bettiol also saw the differing attitudes towards sterilization as a reflection of the contrast in Latin and German racial psychologies. Bettiol suggested that the different attitudes to sterilization between the two races was 'based largely on the great respect of individuality rooted in the Latin soul, and in second place the circumspection and moderation with which Latins discuss and [show a reluctance to] accept scientific hypotheses that still lack strong experimental evidence' (Bettiol 1934: 754–5).

Eugenicists in other Latin countries displayed a range of reactions to the German Sterilization Law (De Raes 1989: 452–7). The Belgian eugenicist Réne Sand, for instance, accepted that sterilization could reduce the number of 'social wastes' ('déchets sociaux'), but questioned whether the 'general hereditary patrimony' of a nation could be improved by 'changing the relative fertility of different social classes' (Sand 1935: 145–6). In 1937, the Portuguese journal *Acção Médica* (*Medical Action*), however, defined the German Sterilization Law as the product of 'delirious nationalism' and predicted that it 'will quickly be modified in order to purify the race more effectively and to banish all that are not blond and Nordic' (quoted in Cleminson 2014: 136).

Spanish eugenicists, like Antonio Vallejo-Nágera, had a more nuanced reaction (Vallejo-Nágera 1934). Similarly, in his commentary accompanying the Romanian translation of the German Sterilization Law, Gheorghe Banu expressed his concerns over the 'authoritarian' nature of the law, though he conceded that its 'main principles, intended to protect and develop the biological qualities of the race, entirely correspond to the ideal of protecting the highest biological values' (Banu 1933: 554). Another Romanian eugenicist, Iordache Făcăoaru, praised the law even less reservedly. Făcăoaru agreed with the German racial hygienists who considering '[human] biological capital as the supreme treasure of the nation' and the law for 'assuring the priority of family and the ethnic body over the individual' (Făcăoaru 1934: 235). In fact, he even believed the law was *overly* restrictive. He had expected that a much wider variety of 'degenerates' should have been subject to sterilization, including 'moral degenerates, sexual offenders, internees of houses of correction, drug-addicts, prostitutes and vagabonds' (Făcăoaru 1934: 236).

As the German Sterilization Law continued to demand the attention of eugenicists around the world, the Nazi government sent a questionnaire to various countries hoping to collect evidence for the widespread appeal of eugenic sterilization in Europe. The copy sent to Romanian eugenicists, for instance, included the following questions:

> Do laws or legal decisions exist [in Romania] with respect to the prevention of hereditary diseased offspring, to the encouragement of those hereditarily healthy, and especially of those hereditarily healthy with many children? [...] What are the reasons for sterilization? Are they eugenic, medical, social? On what type of decision is sterilization based: judicial, sanitary policy, voluntary? Is sterilization performed itinerantly [by mobile stations]? What methods are used? Are those sterilized kept under observation after their release? Do card indexes about sterilization exist? When was sterilization introduced, and how many individuals were sterilized by the end of 1934? (Marinescu 1936: 70; Marinescu Personal Archive)

The Romanian response, prepared by Gheorghe Marinescu and Gheorghe Banu, was, 'evasive, because, in reality, in Romania systematic and coordinated measures to encourage healthy elements and prevent the development of unhealthy ones, anti-socials, etc., had not been introduced' (Marinescu 1936: 71).

German eugenicists sought international sanction of their negative eugenic policies amidst growing criticism of the Nazi programme of racial engineering. As Stefan Kühl has remarked, 'the German race scientists were able to use the international context that they had built up over decades to form close connections and to be official sources of information between the National Socialist race politicians and the often critical outside world' (Kühl 2013: 94). The 11th International Penal and Penitentiary Congress (*11. Internationaler Kongreß für Strafrecht und Gefängniswesen*) held in August 1935 in Berlin is one such example.

The legality and applicability of eugenic sterilization was discussed at the Congress' third section (20–23 August), which focused on crime prevention (and was chaired by Ernst Delaquis, the renowned Swiss legal scholar). The representatives were asked to consider: 'In what cases and according to what rules should sterilization be applied in the modern penal system, whether by castration or by vasectomy or salpingectomy?' (Van der Aa 1937: 293). Participants from Latin countries (Belgium, Brazil, Chile, Cuba, France, Italy, Portugal, Romania, and Spain) were all against eugenic sterilization. Delfin Camporredondo Fernández, director of Women's Prison in Madrid, rejected 'both sterilization and castration'. In a more nuanced statement, Jules Simon, professor of penal law at the University of Ghent, considered that 'castration of sexual offenders' could be permitted, but he too 'rejected sterilization on eugenic grounds'. Georges Paul-Boncour, one of the vice-presidents of the French Eugenics Society, argued that 'our knowledge of heredity [is] not sufficiently reliable for the practice of sterilization on eugenic grounds, even with the consent of those concerned'. Silvio Longhi, attorney general at the Court of Cassation in Rome, was slightly less definitive. He approved 'laws for eugenic sterilization with the consent of those concerned' but opposed 'compulsory sterilization or castration on any grounds whatsoever' (Van der Aa 1937: 295).

Eugenicists from the United States and Germany asserted a starkly different position. Paul Popenoe, secretary of the Human Betterment Foundation, Pasadena (California), recommended the eugenic sterilization of certain feeble-minded individuals; Arthur Gütt, the German Justice Minister, believed that 'eugenic sterilization should be employed by all States as a means of preventing the transmission of hereditary taints' (Van der Aa 1937: 296–97). The next day, Gütt offered a broader presentation of the ostensible 'medical and legal importance' of the Nazi sterilization law (Van der Aa 1937: 309–12). He then recommended that the Congress proclaim sterilization admissible '*by the law in all the States if it seem[ed] necessary for reasons of health or eugenic nature*' and that 'compulsory castration and sterilization may be coordinated with other measures of security provided by the existing law'. Furthermore, the 'principles governing the sterilization of criminals should not differ from those which admit the sterilization – for reasons of *health* and eugenic nature – of other persons' (Van der Aa 1937: 313. Emphasis in the original).

In response, Nicolae Iorgulescu, a medical advisor to the Romanian Ministry of Justice, worried that acceptance of sterilization as an 'appropriate method to reduce criminality' was actually promoting 'the application of sterilization as a dangerous anti-conceptional [contraceptive] weapon having as a consequence the diminution of the natural power of procreation and therefore the diminution of the population of a country' (Van der Aa 1937: 313–14). Iorgulescu emphasized the detrimental medical and psychological effects of sterilization and castration on the individual, arguing that '[f]rom the scientific point of view, sterilization, no matter in what way it is performed, constitutes a mutilation; castration in particular produces a definite and profound degradation in all the biological, psychological and social functions and manifestations of the human being. From the medical and social standpoint such methods should be rejected'. Not surprisingly, Iorgulescu recommended that the Congress adopt a very different resolution than that proposed by Gütt. Rather, Iorgulescu wanted the Congress to stipulate that sterilization 'must not be applied in a general manner as a prophylactic weapon against criminality'. Attempting to save some degree of collaboration between the two factions, Iorgulescu was willing to accept sterilization in asylums and hospitals if the subjects had been 'under observation for no less than two years', and provided that medical officers 'observe[d] the necessary legal guarantees, without incurring the risk of becoming accused of mutilation in the sense of corporal injury' (Van der Aa 1937: 315–6).

In the discussions that followed, the Spanish criminologist and sociologist Quintiliano Saldaña suggested a more accommodating compromise, namely that 'Sterilization can and even should be admitted as a preventive measure in certain cases'; yet he opposed the use of sterilization in the case of 'normal individuals for the only reason that they are recidivists' (Van der Aa 1937: 324). Most participants from Latin countries, while retaining their eugenic interpretation of criminality, advised against the sterilization and castration of those stigmatized as inherently criminal. Eventually, however, Gütt's proposal triumphed, receiving the majority of votes from the participants (Van der Aa 1937: 329).

During the concluding discussion, Iorgulescu tried again, albeit unsuccessfully, to persuade the other attendees to revise Gütt's proposal (Van der Aa 1937: 332–5). The final resolution, adopted during the third section of the Congress, thus reflected the German views on eugenic sterilization. With respect to castration, the resolution specified that 'all States ought to amend or supplement their respective laws, so as to facilitate the performance of such operations upon demand or with the consent of the person concerned'; the same condition applied 'to sterilization for reasons of health or [of a] eugenic nature, provided the person to be operated upon consents to the operation'. The resolution also recommended compulsory sterilization as a 'measure of prevention, as it will reduce in the future the number of abnormal persons from among whom criminals are recruited to a great extent'. It further recommended that 'principles governing the sterilization of criminals should not differ from those which admit the sterilization – for reasons of health or eugenic nature – of other persons'. Finally, to ensure that such procedures were applied rigorously, the resolution requested that national legislatures 'guarantee, from all points of view, that compulsory castration and sterilization is undertaken with the greatest precaution only, and in proper proceedings which provide for a thorough investigation of the case by a committee of jurists and medical men' (Van der Aa 1937: 340–1).

This resolution on castration and sterilization, presented to the Congress's General Assembly on 24 August, provoked strong reactions from some participants who had not attended the deliberations of the third section during the previous days. One of them, the Brazilian criminologist Candido Mendes De Almeida, informed the Congress that his 'entire nation [was] against a measure [such as castration] which would mean a mutilation of the human body'. He opposed the resolution and demanded that 'the vote should be taken by roll-call' of those countries represented at the Congress (Van der Aa 1937: 477). Vespasian V. Pella, a Romanian jurist serving as the chair of the first section (on penal legislation), endorsed de Almeida's proposal, but sought to accommodate its opponents' views to some extent. He proposed that those Congress participants not accepting the resolution, or having doubts about it, should come forward. Twenty-one participants thus recorded their opposition to the resolution on castration and sterilization adopted, including: Mendes de Alemeida (Brazil); Léon Cornil, R. Rubbrecht (Belgium); V. V. Pella, C. Rătescu, H. Aznavorian, I. Ionescu-Dolj, N. Iorgulescu; G. Solomonescu (Romania); F. Nielsen Reyes (Bolivia); Louis V. de Porto-Seguro (Chile); E. Labougle (Argentina); Aurelio F. Concheso (Cuba); Asenjo Garcia (Nicaragua); Alberto Benavides Canseco (Peru); J. Beleza de Santos (Portugal); and Quintilian Saldaña (Spain) (Van der Aa 1937: 486). There was no opposition from the French and Italian delegations.

These two international congresses – one for Catholic physicians and theologians; the other for criminologists and legal experts – represent the complexity and variety of positions on eugenic sterilization in most Latin countries. While participants from these countries tolerated different versions of eugenics, they also offered organized opposition to Nazi racial hygiene and its corollary, sterilization. In both congresses it became clear that Latin eugenicists shared more than just religious, cultural, and linguistic ties; they also thought of themselves as representatives of a distinct type of Latin, 'humanitarian' civilization, reflected in their interpretation of eugenics.

5

Eugenics in Interwar Latin America

Not long after the turn of the twentieth century, many Latin American countries began promoting eugenic ideas within their progressivist reforms of working conditions, education, hygiene, and public health. The Peruvian eugenicist Carlos Enrique Paz Soldán claimed that this synthesis of social reform and eugenics would 'enlighten the public in regard to racial and eugenic duties as a new and exalted form of patriotism' (Soldán 1919: 95–6 and 'The Eugenization of America', *Eugenical News* 1920: 14). In pursuit of these goals, Latin American eugenicists formed their own national organizations, and held periodic, international social-medical conferences that were partially or entirely devoted to eugenics. They drew upon eugenic models from Latin Europe (beginning with France, then Italy, and finally with Spain as the chief sources of inspiration); from the United States; and from their own developing cultures. In the process they established their own unique eugenic identities, while remaining within the broader spectrum of Latin eugenics.

Puériculture and biotypology in Argentina

In many ways, Argentina was the country most closely associated with Latin eugenics in South America. The country's institutionalized eugenics traces its origins to Victor Delfino, the editor of the Argentinian medical journal *La Semana Médica* (*Medical Week*). Delfino was a well-known researcher on the destructive physical and social effects of alcohol consumption (Delfino 1907). These concerns led him to attend the First International Congress of Eugenics in 1912; thereafter he devoted himself to establishing eugenics in Argentina. In 1914, Delfino established an Argentinian Eugenics Committee (*Comité Eugenésico Argentino*), followed by the Argentinian Eugenics Society (*Sociedad Eugenica Argentina*) in 1918. Other founding members included

Ubaldo Fernández, professor of obstetrics at the Faculty of Medicine; Gregorio Aráoz Alfaro, president of the National Department of Hygiene; and Antonio Vidal, director of the Bureau of Hygiene (Delfino 1925: 123–4; Nisot 1929b: 69; Cecchetto 2008: 42). In 1921, the Society became a member of the International Federation of Eugenic Organizations, with Delfino as its official representative.

As in the other Latin countries, the first Argentinian Eugenics Society showed special interest in puériculture and neo-Lamarckist eugenics. The same neo-Lamarckist emphasis on family and child welfare also characterized the discussion of eugenics at the Second National Congress of Medicine (*Secundo Congreso Nacional de Medicina*) held in October 1922 in Buenos Aires. In their reports to the Congress, Delfino and Fernandez explained that puériculture, sexual education, child welfare morality, and modern ideas of hygiene and public health were essential for eugenic improvement (Delfino 1925: 124; Nisot 1929b: 69 and 73–4).

In the early 1920s Delfino strengthened his connections with other eugenicists in Latin Europe. He participated in the First Italian Congress of Social Eugenics held in 1924 in Milan, where he presented a general outline of his eugenic programme for Argentina. It included the promotion of extensive research into (a) the mechanisms of heredity, and their application 'to the conservation and improvement of the human species', and (b) 'the causes of human degeneration'. Eugenic propaganda and education were also advocated, especially with respect to 'pauperism, social diseases, morbidity and infant mortality, population growth and immigration'. Delfino also recommended the establishment of chairs in eugenics at the Faculties of Medicine and Institutes of Higher Education across the country (Delfino 1927: 447–8). This was an ambitious programme that embraced the regenerative power of eugenics and applied it to the specific conditions in Argentina.

Delfino continued to promote the eugenic achievements of his country abroad. As a member of the International Federation of Eugenic Organizations, he attended the 5th Meeting held in Paris in 1926. In his report, and much like his French and Italian eugenicist colleagues, Delfino claimed that consanguineous marriages led to racial degeneration. Some of these ideas influenced the Argentinian premarital sexual hygiene law of 1924, followed shortly thereafter by a law on 'racial protection' (Zimmermann 1992: 44–5;Vallejo and Miranda 2004: 429–31). This 'Ley de la Raza' gave the state the responsibility to 'fight any disease or social customs leading to the degeneration of the race and to adopt the means necessary to improve and invigorate national health'. To accomplish this, men wishing to marry were required to first obtain an 'official certificate of venereal health issued by the appropriate hygiene authority' (Lavrin 1995: 169). Delfino's ultimate international recognition came in 1932, when he was one of the vice-presidents of the Third International Congress of Eugenics held in New York.

The possibility of creating a modern Argentinian nation through population management and the control of society more generally appealed to large segments of the political and scientific establishments. As argued by Andrés Reggiani and Hernán Bollo, biotypology endowed Argentinian eugenics 'with a holistic and technocratic perspective', enabling the 'health management' of the country's 'human biology' according to scientific norms and regulations (Reggiani and Bollo 2007: 39). Eugenicists promised the state that biotypology would allow more efficient use of the nation's human resources, and thus accelerate national development. It also encouraged the fusion of eugenics with demographics and statistics with the ultimate aim of purging the nation of unwanted social and biological factors affecting its development.

At first, Argentinian eugenicists borrowed principally from French models of human improvement. For example, the Argentinian League of Social Prophylaxis (*Liga Argentina de Profilaxis Social*), established in 1921 by Alfredo Fernández Verano, derived its conception of eugenics largely from Adolphe Pinard's theories of puériculture (Verano 1929). Beginning in the late 1920s, however, as part of Mussolini's effort to promote Italy's 'World Power' status, Italian eugenicists replaced their French colleagues as the decisive foreign influence on Argentinian and other Latin American eugenic programmes (Stepan 1992: 749–56). The Argentinian Institute of Italian Culture (*L'Instituto Argentino de Cultura Itálica*) was established in 1924, serving as a cultural outpost for fascist science (Reggiani 2010: 15). During the next decade a number of prominent fascist scientists, including Filippo Bottazi, Carlo Foà, and Nicola Pende, were invited to Buenos Aires by the Institute to promote Latin medicine and science.

Pende, for example, visited Argentina in November 1930 (Vallejo and Miranda 2004: 433; Reggiani 2010: 17), as the most important stop on a eugenic mission to several countries in the region (including Uruguay and Brazil). While in Argentina, Pende gave a short lecture course to medical doctors and students, held a series of scientific conferences, and toured medical institutions in Buenos Aires. On this occasion, he also announced that his Institute of Biotypology and Orthogenesis in Genoa would henceforth provide fellowships for Argentinian doctors to study there: the first such fellowship was awarded to the physician Arturo R. Rossi.

Pende returned to Argentina in 1932, promoting, among other projects, the idea of creating a 'Latin Atheneo' in Buenos Aires, dedicated to forming a 'spiritual web' between Italy and Argentina (Reggiani and Bollo 2007: 38–9). The 'Atheneo' was designed to sponsor such events as a grandiose scientific 'Latin Congress'. In such a way, Italy would advance towards its goal (Reggiani 2010: 20–1). It is within this context of a close relationship between Argentina and Fascist Italy that eugenicists in both countries promoted common intellectual agendas.

One of them, Arturo R. Rossi (see Figure 5.1) became the leading Argentinian eugenicist of the 1930s, and a particularly enthusiastic proponent of Latin eugenics (Reggiani and Bollo 2007: 39). After studying at Pende's Institute, Rossi returned to Argentina determined to offer his newly acquired scientific expertise to the country's transformation according to eugenic principles. To this effect, Rossi worked with Victor Delfino and Bernaldo de Quirós – another of Pende's disciples – to create the Argentinian Association of Biotypology, Eugenics and Social Medicine (*Asociación Argentina de Biotipología, Eugenesia y Medicina Social*) in 1932. The Association drew together Argentina's most notable obstetricians, paediatricians, demographers, criminologists, and medical legal theorists. Pende was elected an honorary member.

Figure 5.1 Arturo R. Rossi
Source: *Anales de Biotipología, Eugenesia y Medicina Social*, vol. 3 no. 42, 15 April 1935, p. 2. Courtesy of the National Library of Argentina.

In the next decade, biotypology came to dominate Argentinian medicine (Vallejo and Miranda 2011: 57–75). In institutional terms, the Association established a School of Biotypology (*Escuela de Biotipología*), within the Institute of Biotypology in 1935 ('Ha sido inaugurado nuestro Instituto de Biotipología', *Anales de Biotipología, Eugenesia y Medicina Social* 1935: 2 and 12–6). A Department of Eugenics, Maternity and Infancy (*División de Eugenesia, Maternidad y Primera Infancia*) was set up in 1938 by Josué A. Beruti (Biernat and Ramacciotti 2008: 337; Eraso 2013: 68), followed in 1939 by the creation of the National Institute of Biotypology and Medicine (*Instituto Nacional de Biotipología y Medicina del Trabajo*) (Rossi 1940: 20–4; Vallejo 2004: 219–44).

Rossi edited the Association's journal, *Anales de Biotipología, Eugenesia y Medicina Social* (*Annals of Biotypology, Eugenics and Social Medicine*), which ran from 1933 to 1941. The *Anales* focused on the classic issues of Latin eugenics: demography, fertility, natality, immigration, neo-Lamarckian heredity, gender, and reproduction. The pronatalist, pro-family, and anti-abortion agenda of the journal, aimed at increasing the fertility and productivity of the population, showed obvious similarities to the dominant eugenic issues in Fascist Italy (Reggiani 2010: 7–26; Haidar 2011: 317–32).

Rossi also initiated the mass biotypological analysis of the Argentinian body with his biotypological identification card (*ficha biotipológica escolar*). The card, based on Pende's models, was assigned in 1933 to each student in the provincial school district of Buenos Aires. It recorded biotypological eugenic information such as the child's health history, education, and employment (Palma and di Vincenzo 2009: 1–21; Reggiani 2010: 19–20).

As in France, Italy, and Spain, eugenicists in Argentina placed the health of the mother and the family at the centre of their programme of national social and biological improvement. Inspired by Nicola Pende and Gregorio Marañón, some Argentinian eugenicists displayed a growing attraction to biotypology and endocrine studies. Octavio V. López, the vice-president of the Argentinian Association of Biotypology, Eugenics and Social Medicine, emphasized this connection. In a lecture on 'the new biological foundations of eugenics', delivered to the Academy of Medicine in Barcelona on 25 April 1932, López highlighted the importance of the Italian and Spanish schools of endocrinology and constitutional medicine to the development of eugenics and social medicine in Argentina. He placed eugenics on a 'biological tripod', consisting of 'inheritance of acquired characteristics, adaption and natural selection'. The nation's 'biological personality' could be discerned through the cumulative eugenic culture of each individual type (López 1932: 143). To strengthen these cultural and scientific connections, and in preparation of the next International Congress of Latin Culture, the Association of Biotypology, Eugenics and Social Medicine created a 'Spanish Section' in 1935 ('Primera reunion de la Sección Española de la Asociación Argentina de Biotipología, Eugenesia y Medicina Social' *Anales de Biotipología, Eugenesia y Medicina Social* 1935: 3).

The growing acceptance of the biotypological management of the country's national body, and the prescribed orthogenetic cures, was paralleled by successful attempts in Argentina to reform marriage legislation. Already in 1919, Emilio Coni, a public health reformer, proposed that the National Hygiene Department (*Departamento Nacional de Hyigiene*) introduce prenuptial medical certification (Lavrin 1995: 168). Argentinian eugenicists were eventually able to overcome the objection of liberals and Catholics, and convince the government to legislate a prenuptial certificate as part of its 1936 *Ley de Profilaxia Social* (Law of Social Prophylaxis). The certificate stipulated that all men contracting marriage were required to have a health certificate no earlier than fifteen days before the wedding ceremony (Lavrin 1995: 172). The National Directorate of Maternity and Infancy (*Dirección Nacional de Maternidad e Infancia*) was also established in 1936, inspired by the Italian National Maternity and Child Health Centres (*Opera Nazionale Maternità e Infanzia*). Next to combating infant mortality and the protection of mothers and infants, eugenics figured prominently among the goals of the new Directorate (Eraso 2013: 69).

Argentinian eugenicists also became interested in the work of the Spanish eugenicist Antonio Vallejo-Nágera. As a Catholic, Vallejo-Nágera was very aware of the Church's opposition to interference in marriage and reproduction. His solution was to encourage the development of voluntary eugenic prenuptial counselling. Counsellors would study a prospective couple's genealogies, determine their 'biosocial diagnostics' and access their 'psychobiograms'. Eugenic counselling would also include instruction on the eugenic principles important to prospective parents. Argentinian eugenicists popularized Vallejo-Nágera's marital guidelines and neo-Lamarckism through the public school eugenic education programme, by providing lessons to teach children the proper eugenic criteria for choosing a spouse (Nari 1999: 356–67).

These nationwide programmes demonstrate the commitment of Argentinian physicians, health reformers, and politicians to eugenics. In recognition of the growing prestige and influence of Argentinian eugenics, the Second Pan-American Congress on Eugenics and Homiculture (*Segunda Conferencia Panamericana de Eugenesia y Homicultura de las Repúblicas Americanas*) was held in Buenos Aires from 23 to 25 November 1934. Representatives from twenty Latin American countries and the USA participated. The Second Congress was presided over by the renowned Argentinian paediatrician Raúl Cibils Aguirre, with prominent Argentinian eugenicists such as Victor Delfino, Josué Beruti, and Alberto Peralta Ramos in attendance. Topics included 'Latin American eugenics and its prospects'; 'race in the Andes'; 'immigration'; 'prenuptial medical examination'; 'eugenics and maternity'; 'eugenics and infancy'; and 'eugenic sterilization'. Such a robust variety of subjects demonstrated the growing influence of eugenics in Latin America, and also offered Argentinian eugenicists an opportunity to reassess their vision of human improvement in comparison to the success of eugenicists in formulating similar policies in the USA, Nazi Germany, and Fascist Italy (Reggiani 2010: 14).

In his opening address to the Congress, the Argentinian Foreign Minister, Carlos Saavedra Lamas, agreed with many participants in defining eugenics as 'medical science, social hygiene and preventive medicine'. He placed the health of the nation under the eugenicists' 'fatherly protection' (Lamas 1934: 20). Raúl Cibils Aguirre, the Congress's President, agreed. Positive eugenics, puériculture, biotypology, and homiculture – all provided promising foundations for a strong and healthy nation, and thus informed the appropriate eugenic methodologies envisioned for each country, Aguirre believed. Argentina's methodology, in the words of the political philosopher Juan Batista Alberti: 'To govern is to populate' (Aguirre 1934: 23). To this effect, Aguirre reassured his audience that Argentinian eugenics promoted both the 'quality and the quantity' of the population. For Josué A. Beruti this meant, first and foremost, puériculture and maternal care. 'Maternity', he remarked, was at the centre of 'most of the eugenic problems requiring urgent solution, both in our country and in the majority of American nations' (Beruti 1934: 164).

Accordingly, eugenics was in complete harmony with the medical and hygienic values prevalent at the time in Latin America. The eugenic–medical–social welfare nexus, which characterized Latin eugenics, was also recognized by Carlos Enrique Paz Soldán, honorary president of the Pan-American Health Organization (Soldán 1934a: 24–31 and 1934b: 198–208). Uruguayan attendees pointed out that their country's 'Children's Code' (*Código del Niño*), based on the law with the same name and introduced on 6 April 1934, was an excellent example of the harmonious synthesis of positive eugenics with medical and social welfare measures.

Not surprisingly, mandatory prenuptial medical certificates and compulsory sterilization were once again rejected by the Congress's participants, notwithstanding the renewed efforts of the Cuban eugenicist Domingo F. Ramos and his North American allies. The main tenets of Latin eugenics were reaffirmed by the final resolutions adopted the Congress, which emphasized the eugenic importance of education, child protection, puériculture, maternal welfare, the centrality of family, and the relationship between eugenics and homiculture ('Mociones y resolutiones', *Actas de la Segunda Conferencia Panamericana* 1934: 275–84). The Congress also elected new officers of the Pan-American Office of Eugenics and Homiculture: Raúl Cibils Aguirre and Gregorio Aráoz Alfaro of Argentina were elected as honorary presidents, and Rafael Lorié of Cuba was chosen as president ('Constitution y Nomina de la Oficina Central Panamericana de Eugenesia y Homicultura', *Actas de la Segunda Conferencia Panamericana* 1934: 284–5).

During the 1930s and 1940s, as in Latin Europe, some Argentinian eugenicists rose to positions of power throughout the country's national healthcare and medical system. These eugenicists accepted the transformative role of the state, which in turn subsidized many eugenic projects. Positive eugenics, puériculture, and biotypology came to dominate the state health system and shape the nation's political demography. By these means, Argentinian eugenicists hoped to improve 'the nation's biological balance, ridding it as much as possible of the mediocre and the unproductive, of precocious invalids, of the weak and the morally and the intellectually mediocre' but without coercive eugenic methods.

The primacy of neo-Lamarckism, biotypology, and orthogenesis within biopolitical discourses in Argentina during the 1940s exemplifies the transformation of eugenics into a national science devoted to the protection of the nation's health and pronatalism. This transformation was reaffirmed at the First Congress on Population (*Primer Congreso de Población*) held in Buenos Aires between 26 and 31 October 1940 (Miranda 2005: 195–6); and by the integration of the National Institute of Biotypology and Related Subjects (*Instituto Nacional de Biotipología y Materias Afines*) into the National Ministry of Public Health's (*Secretaría de Salud Pública de la Nación*) Directorate of Health Policy and Culture (*Dirección de Política y Cultura Sanitaria*) in 1943 (Haidar 2011: 318). By the time Juan Domingo Perón came to power in 1946, the regime could benefit from a discursive culture of eugenic improvement, with the ultimate goal of increasing the population and enhancing the Argentinian nation's 'racial qualities' (Vallejo and Miranda 2004: 440).

Race and eugenics in Mexico and Brazil

Eugenic developments in other Latin American countries, particularly Mexico and Brazil, were often quite different. While Argentina sought to reinforce its

Latin heritage in the face of its population's growing heterogeneity as a result of 'undesirable' non-Latin immigration, Mexico and Brazil endorsed racial nativism, glorifying the country's diverse races and its indigenous traditions, together with racial mixing (Hedrick 2003). North American eugenicists, in particular, were explicit in their disapproval of such practices. For instance, in a lengthy study published in *Eugenical News* in 1922, the biologist Reginald G. Harris – at the time an assistant at the Cold Spring Harbor and Charles Davenport's son-in-law – discussed in detail the dysgenic effects of racial interbreeding in South America (Harris 1922: 17–42). 'Eugenically', Harris noted, 'the crossing of widely different human races, viz., Indians, Negroes, and whites, in South America has not been successful, and its continuance is undesirable'. As a remedy, he recommended that 'the hybrids [...] be replaced, and the general stock of Europeans renewed by abundant selective immigration'. In this context, 'the knowledge and practice of eugenics' provided the rationale for national renewal in 'those countries of South America where inferior races and hybrid stock [were] present in large numbers' (Harris 1922: 42).

Mexican authors, however, such as José Vasconcelos, rejected claims that Latin Americans were degenerate hybrids. Instead, the people of Mexico were stronger and healthier due to their mixed genetic heritage. The *mestizo* was not 'the epitome of degenerate humanity', but the 'virile and vigorous hybrid of the European and the Indian' – a new racial ideal (Stern 1999a: 2). Alfredo M. Saavedra, one of Mexico's leading eugenicists, further claimed that miscegenation allowed the population to be better adapted to its environment, and swamp out regressive, degenerative traits, such as illness, insanity, and sexual perversion. Neo-Lamarckian principles were useful in supporting these arguments (López-Guazo 2005: 101–7), and 'mestizophilia' – according to Alexandra Minna Stern – became 'a doctrine of state-building and nationalism' (Stern 1999a: 2). However, eugenicists in Mexico and Brazil still availed themselves of most elements of traditional Latin eugenics, such as neo-Lamarckian theory, biotypology, emphasis on biological assimilation, and so on. Some of the leading Mexican eugenicists endorsed the new racial paradigm; others opposed it entirely.

In many respects, Mexico became a leader in the changing Latin American views on race in the early twentieth century (Knight 1990: 71–113). The radical transformations the country underwent as a result of its political revolution had a profound impact on the country's racial ideology and the politics of eugenics. The Mexican Revolution, beginning in 1910 and spanning the entire decade, ushered in one of the first modern Latin American governments: anti-clerical, leftist in orientation, and fiercely nationalist (Bethell 1998: 209). But the country was also devastated from the destruction and dislocation of the Revolution. From 1910 to 1921, Mexico lost over 14 per cent of its population; the nation's sanitation systems were ruined; malnutrition and infectious diseases were widespread.

The duty of Mexican eugenicists was self-evident: to help the nation recover stability as rapidly as possible, and tackle the glaring public health problems that marred the country's racial future. The respected physiologist José Joaquín Izquierdo, for instance, associated this programme of national revival with renewed interest in family history and local patriotism. He thus recommended demographic and genealogical projects into prestigious Spanish families in Mexico, based on the type of eugenic investigation he himself presented in 1921 at the Second International Eugenics Congress (Izquierdo 1923: 348–73).

Given their country's high infant-mortality rate, Mexican eugenicists also took a great interest in improving child and maternal care (Cházaro 2005: 65). These concerns were evident at the First Mexican Congress of the Child (*Primer Congreso Mexicano del Niño*), held in Mexico City in 1921. One section of the Congress was devoted to eugenics; its discussions revolved around the danger to the nation's genetic patrimony caused by poor environmental influences and disease. Most papers in this section showed the continued indebtedness of Mexican eugenics to the Latin model, and the dominance of neo-Lamarckian concepts of heredity (Albarrán 2008: 42–3). By addressing the danger posed by alcohol, syphilis, and unhealthy work conditions, the Congress advocated ways to combat child criminality, proposed legislation regarding child labour, and requested the establishment of clinics and social services for mothers and children (*Memoria del Primer Congreso Mexicano del Niño* 1921: 264–6).

Not all of the Congress's participants were content with positive eugenic reforms. Antonio F. Alonso, a member of the Mexican National Academy of Medicine, advocated the castration of 'certain degenerates'. Most saw this recommendation as too extreme; nonetheless, Ángel Brioso Vasconcelos and Isaac Ochoterena also promoted the benefits of eugenic sterilization (*Memoria del Primer Congreso Mexicano del Niño* 1921: 34–5 and 42), a view the latter revised, however, in 1923 at the Second Mexican Congress of the Child (*Primer Congreso Mexicano del Niño*). On this occasion, Vasconcelos – while insisting on the 'importance and practical value of eugenics for society and state' – emphasized the importance of education in shaping the morality and physical education of the Mexican family (Nisot 1929a: 355–6). With the creation of the Infant Hygiene Service (*Servicio de Higiene Infantil*) in 1929 under the directorship of child health advocate and eugenicist Isidor Espinosa de los Reyes, puériculture and eugenics 'translated into a commitment to preserving children's health to safeguard and preserve the Mexican race' (Blum 2009: 159).

Mexican eugenics during the 1920s and 1930s generally retained its neo-Lamarckian orientation, and therefore focused more on social prophylaxis, puériculture, population growth, and positive eugenics (Nisot 1929a: 355–56). As Alexandra Stern noted, 'Desirous of recrafting the nation in accordance with their understanding of the emergent disciplines of psychology, genetics and bacteriology, Mexican eugenicists gravitated toward three key elements of reproduction and socialization: motherhood, sexuality, and children' (Stern 1999b: 370).

Like their counterparts in the Latin countries, Mexican eugenicists in the 1920s adopted French ideas of puériculture and pronatalism, and promoted the notion of 'conscious maternity' (*maternidad consciente*) and the centrality of the family. This was reflected in the mission of the School of Health and Hygiene (*Escuela de Salubridad e Higiene*) established in the Department of Public Health (*Departamento de Salubridad*) in 1928. The School certified doctors and nurses in child hygiene and puériculture, paralleling the broader efforts initiated by eugenicists to ensure the healthy reproduction of the nation.

As a modern science, eugenics appealed to Mexico's secular, post-revolutionary regimes. The eugenicists, in turn, promoted a materialist vision of a healthy body politic that echoed the state's nationalist ethos. Quite exceptionally for a Catholic country, some Mexican eugenicists were able to institute negative eugenic measures that were practically impossible in other Latin countries. As early as 1914, the Public Health Department decreed that medical certification before marriage was required for individuals suffering from 'habitual alcoholism, impotence, syphilis, insanity or any other potentially contagious or hereditary chronic and incurable disease' (quoted in Stern 1999b: 378). A law with similar goals was enacted in 1928, but essentially ignored (López-Guazo 2005: 193–4).

Among the negative eugenic measures attempted in Mexico of this period, the 1932 eugenic sterilization law was the most extreme. The law was conceived by the governor of Veracruz, Adalberto Tejeda, a 'true believer' in the application of science to society. As part of a wider programme of social and biological engineering, in 1932 Tejeda established a Section of Eugenics and Mental Hygiene (*Sección de Eugenesia e Higiene Mental*) in the state's General Department of Public Health (*Dirección General de Salubridad del Estado*) ('Ley quecrea la Sección de Eugenesia e Higiene Mental', quoted in López-Guazo 2005: 266–7). A mandatory eugenic sterilization law was passed by the state legislature in December that year. According to Article 6 of the Law, those judged by the Section of Eugenics and Mental Hygiene to be 'insane, idiots, degenerates or demented' were targeted for sterilization ('Reglamento de eugenesia e hygiene mental' quoted in López-Guazo 2005: 269–71; see also Stepan 1991: 131–3).

Whether or not eugenic sterilizations were carried out remained unclear, although there is evidence that state physicians performed some 'operations' under this law. Although the Section continued its work until 1934, it did not focus on sterilization, but on decreasing prostitution and venereal disease, and expanding maternal and infant hygiene (Stern 2011: 440–2). These, of course, had long been standard goals in Latin eugenics. Yet, this short-lived experiment with legal eugenic sterilization in Veracruz is important to the history of Latin eugenics; as Alexandra Stern pointed out, 'Tajeda's dream of eugenic social engineering was similar to the neo-organicism put forth by some Italian scientists, and his concerns about sexual hygiene were commonly voiced among leading Spanish eugenicists' (Stern 2011: 443).

As shown above, Mexican eugenicists established a number of eugenic institutions to intensely scrutinize 'the nation's future and flesh' (Stern 1999b: 370). The Society of Puériculture (*Sociedad Mexicana de Puericultura*), founded in 1929, was among the most important. It included a section devoted to eugenics (*Seccíon de Eugenesia*), led by Rafael Carrillo, a paediatrician and professor at the National School of Medicine. Likewise, the Society of Puériculture was instrumental in creating the Mexican Eugenics Society for the Betterment of the Race (*Sociedad Mexicana de Eugenesia para el Mejoramiento de la Raza*) two years later ('Eugenics in Mexico', *Eugenical News* 1934: 144–5). Founding members of the Society included many of the nation's foremost health authorities and physicians, such as Rafael Carillo, Salvador Bermúdez, Fernando Ocaranza, José Rulfo, Adrián Correa, and Alfredo M. Saavedra, its first president (López-Guazo 2005: 113–4).

The Mexican Eugenics Society's mission was 'the improvement of the species', 'individual health', 'the creation of Clinics for Hereditary Health and Eugenics (*Consultorios de Salud Hereditaria o Eugénica*)', and the spread of knowledge and support for eugenics among the general public. However, it also reiterated its support for prenuptial certification, eugenic sterilization, and the eugenic screening of immigrants ('Declaración de principios de la Sociedad Mexicana de Eugenesia para le Mejoramiento de la Raza' quoted in López-Guazo 2005: 260–1). With respect to the latter goal, Rafael Carrillo offered the following explanation:

> We naturally do not think of selecting superior eugenic individuals; we do not seek a Marañón, a Shaw, a Mussolini, a Hindenburg nor an Edison, yet neither

> do we accept epileptics, alcoholics, feeble minded or syphilitics. We only wish to inject eugenically selected individuals into Mexican *mestizos* who, according to Galton's scale of values, do not ostensibly deviate from the mean. (Quoted in López-Guazo 2001: 149)

Yet emphasis on the biological rejuvenation of the Mexican nation through hygiene and sanitary measures remained central to the programme of the Mexican Eugenics Society, as illustrated by the public conferences organized in Mexico City between 17 and 21 August 1936, under the auspices of the National Athenaeum of Sciences and Arts ('Eugenics in Mexico', *Eugenical News* 1936: 114).

Equally durable was the influence of Italian eugenics and biotypology (Laugier 1933: 145–9). Corrado Gini, for instance, attended one of the first meetings of the Mexican Eugenics Society, and became an honorary member. Between August and September 1933, he and two of his colleagues from the University of Rome, Giuseppe Genna and Dino Camavitto, travelled to Mexico for a demographic expedition, with the purpose of investigating those groups deemed 'racially' pure among the Indian populations, such as the Seri tribe. The investigation included anthropological, biological, medical, and demographic research ('An Investigation of Some Indian Tribes in Mexico', *Eugenical News* 1934: 114–5; Camavitto 1937: 40–59).

Mexican delegations were also sent to Italy to study the Italian fascist government. The most promising young Mexican eugenicists, such as Gilberto Loyo, the country's first professionally trained demographer, were awarded fellowships to study Italian eugenics and statistical demography. Upon his return, Loyo established the Mexican Committee for the Study of Population Problems (*Comité Mexicano para el Estudio de los Problemas de la Población*). While in Italy, Loyo wrote his most important work, *La política demográfica de México* (*The Political Demographics of Mexico*). Loyo's text was pronatalist, and emphasized the value of puériculture and public health programmes. Eugenic improvement of Mexicans would be furthered through prenuptial certificates and the selective admission of 'assimilable' foreigners. The governing National Revolutionary Party (*Partido Nacional Revolucionario*) published Loyo's work in 1935, and a year later the General Population Law (*Ley General de Población*) was introduced, reflecting Loyo's ideas (Stern 1999a: 7–8).

In 1939, at Loyo's initiative, the Mexican Committee for the Study of Population Problems and the Mexican Eugenics Society agreed to collaborate on a common demographic survey of the Mexican population. As in Argentina, Mexican eugenicists of the 1940s attempted to synthesize eugenics and population management. Showing Gini's influence, Mexican eugenicists and demographers embraced organicism and the vision of the corporatist state – a model they hoped reflected Mexican realities. The state was thus reconceptualized along eugenic lines to reflect the growing need to protect its biological and human capital.

The growing acceptance of population management also led to a redefinition of eugenics and its practical importance to the Mexican nation. According to Alexandra Stern, 'the underlying premises of Mexican eugenics were being transformed' while at the same time 'in place of neo-Lamarckism' Mexican eugenicists 'began, often grudgingly, to accept Mendelism' (Stern 1999a: 12–3). For example, in his 1940 editorial to the journal *Eugenesia* (Eugenics), Saavedra spoke of 'a rational concept of hereditary' based on Mendelian genetics. A year later, Saavedra restated the importance of hereditary studies of the Mexican population, while at the same time recommending more efficient control of its biological quality. The country's 'Sanitary Authorities', he advised, 'should concern themselves with the problems of Racial Hygiene, establishing an institution for genealogical research of the Mexican family, in order to study the application of socially beneficial measures' (quoted in Stern 1999a: 13 and 16).

Biotypology at this time enjoyed a new influence in Mexico, since it appeared to be a 'third way' between neo-Lamarckism and Mendelism. Similar to their counterparts in Argentina, Italy, and France, Mexican eugenicists, 'found in biotypology a vision of the individual and society that they could embrace' (Stern 1999a: 14–5). To some extent, however, the resurrection of biotypology during the 1940s and 1950s also reflected the dominant position occupied by the ideology of Mexican *mestizo* nationalism. In this respect, biotypology was fully compatible with the 'recoding [of] racial categories in a neologistic, seemingly more neutral lexicon based on understandings of the uniqueness of individual hereditary difference rather than around "pure" races' (Stern 1999a: 21). Given developments in medical genetics and social anthropology (Suárez-Diaz and Barahona 2013: 101–12), as well as the deepening influence of racial science in Europe and the USA, eugenicists and biotypologists in Mexico came to reappraise their views on racial degeneration and the sort of state intervention required to reverse or prevent it.

As in Argentina, biotypology in Mexico became an element in a much larger political and cultural programme, combining social and biological research into the nation's diverse ethnic communities with the institutionalization of public welfare. In broader terms, then, the specific intellectual and cultural terrain on which the public health policies of the Mexican government were articulated during the 1950s drew inspiration from the eugenic narratives about a healthy and racially strong nation first developed in the early twentieth century.

***

As elsewhere in Latin America and Latin Europe, the institutionalization of eugenics in Brazil must be understood in the context of debates on national and racial identity, and the future of the state (Skidmore 1990: 7–36; Dávila 2003; Hochman, Lima and Maio 2010: 494–9). Equally important, Brazil's complex multiracial history imbued eugenic theories of national improvement with additional emphasis on issues such as miscegenation and immigration (De Lacerda 1911: 377–82). According to the popular writer Monteiro Lobato, Brazil may have been 'condemned by race', but he hoped – together with many eugenicists and health reformers – that 'experimental medicine', as well as 'public campaigns related to sanitation, public health, vaccination, and medical care', would eventually provide the country with a much-needed national welfare system that encompassed all Brazilian races (Caulfield 2000: 149; Hochman, Lima and Maio 2010: 498). Not surprisingly, then, Paz Soldán's oft-quoted 1916 *Un programa nacional de política sanitaria* (A National Programme of Sanitary Politics) was considered in Brazil to be a 'fundamental eugenics text' (Stepan 1991: 53).

The acceptance of eugenics by Brazilian physicians, sociologists, anthropologists, and educators, therefore, was not only a reflection of the 'medicalization of race' by these professionals (Maciel 1991: 121–43), but also an acceptance of ethnic and cultural diversity within a discursive framework where the health of the nation took on a renewed significance (Lima 2007: 1168–77). According to Jerry Dávila, there are two reasons for 'the unique public role' enjoyed by eugenics in Brazil. On the one hand, eugenics 'provided the emerging scientific, medical, and social scientific authorities with a shorthand for explaining ideas of racial inferiority and defining strategies for managing or ameliorating that inferiority'; on the other hand, it 'armed this group with a scientific solution to what was basically a social problem' (Dávila 2003: 26). This belief in science, coupled with a conception of national progress based on improving the health of the population, informed early eugenic activities in Brazil. Moreover, attempts were made to popularize eugenic ideology for medical professionals and the general public (Marques 1994). For instance, in 1912 the journalist Horácio de Carvalho introduced the readers of the journal *O Estado de São Paulo* (*The State of São Paulo*) to eugenic developments in England. A year later, Alfredo Ferreira de Magalhães, director of the Institute for the Protection and Assistance of Children (*Instituto de Proteção e Assistência à Infância*) in Salvador organized the first conference on eugenics in Brazil. Magalhães conceptualized eugenics in terms of childcare, maternal assistance, and family education (Nisot 1929b: 207; Munareto 2013: 72–73).

Many of these ideas resurfaced at the First Brazilian Congress on the Protection of Children (*Primeiro Congresso Brasileiro de Proteção à Infância*), held in 1922 in Rio de Janeiro. They also provided Dr. Arthur Moncorvo Filho with ideological and institutional models for child welfare (Moncorvo 1926), such as the Museum of Childhood (*Museu da Infância*), created on the occasion of the First Brazilian Congress of Eugenics in 1929 (Wadsworth 1999: 103–24). Although Moncorvo and others promoted various interpretations of human improvement during the early 1920s, it was Renato Kehl (see Figure 5.2) who became Brazil's most important eugenicist. Like Delfino, Kehl left the First International Eugenics Congress of 1912 a convinced eugenicist and a life-long admirer of Francis Galton.

Kehl gave his first public lecture on eugenics in 1917, in which he argued for the 'regeneration of the Brazilian population'. Kehl had been impressed

Figure 5.2 Renato Kehl, (front row seated, second left)
Source: Fundação Oswaldo Cruz, Departamento de Arquivo e Documentação, Rio de Janeiro.

with the achievements of American and German eugenicists, and envisioned similar programmes of biological improvement in Brazil (De Souza 2006: 32–3; Hochman, Lima and Maio 2010: 499). He believed his hopes could come to fruition only by means of a eugenics organization. Thus, later that year Kehl and Arnaldo Viera de Carvalho, director of the Faculty of Medicine of the University of São Paulo, organized a meeting to discuss Brazil's civil marriage code in the context of eugenics. During the meeting Kehl and Carvalho proposed the organization of a eugenics society devoted to heredity, social, and racial health (De Souza 2006: 35). This materialized one month later as the Eugenics Society of São Paulo (*Sociedade Eugénica de São Paulo*), which is notable for being Latin America's first eugenics society. Arnaldo Vieira de Carvalho was chosen as its president, and Kehl as its secretary.

Once established, the Eugenics Society of São Paulo organized conferences, began publishing its own journal, *Annaes de Eugenia* (*Annals of Eugenics*), and sponsored a number of other eugenic publications. Victor Delfino and Carlos Enrique Paz Soldán became corresponding members (Stepan 1991: 48). Notwithstanding such auspicious beginnings, the Eugenics Society of São Paulo dissolved after the death of Carvalho and Kehl's relocation to Rio de Janeiro at the end of 1919. By then, however, eugenics had already become a familiar topic in Brazil, not least due to Kehl's numerous publications. In his short note on 'Eugenics in Brazil' published by the journal *Eugenical News* in 1921, Kehl cited two other eugenic societies, one in Manáos in the state of Amazonas in northern Brazil, led by Dr. João Coelho de Miranda Leão and another in Rio de Janeiro, affiliated to the Society of Neurology and Psychiatry, founded by the pioneer of Brazilian psychiatry, Juliano Moreira (Kehl 1921: 18).

The League of Mental Hygiene (*Liga de Hygiene Mental*), established in 1923 by the psychiatrist Gustavo Reidel, also carried out eugenic work in Brazil in the 1920s (Stepan 1991: 51–2; Hochman, Lima, and Maio 2010: 499–500). In its journal *Archivos Brasileiros de Hygiene Mental* (Brazilian Journal of Mental Health), as well as its public campaigns, the League concerned itself with the eugenic implications of social degeneracy: crime, delinquency, prostitution, and alcoholism. The League also recommended prenuptial medical examinations and institutional segregation as the most acceptable means to prevent further social and biological degeneration (Reis 1999: 29–55).

As in Argentina and Mexico, Brazil was also subjected to Italian influences. Authors like Giacinto Viola, Nicola Pende, Mario Barbàra, and Marcello Boldrini inspired research on the 'normal Brazilian type'. A Laboratory of Biotypology (*Gabinete de Biotipologia*) was established at the Faculty of Medicine in Rio de Janeiro, and during the 1930s physicians such as Isaac Brown, Waldemar Berardinelli, and Para Roha Vaz, employed biotypology as a scientific method to define the physical and phenotypic characteristics of the Brazilian nation (Gomes 2012: 705–19). Moreover, Corrado Gini visited Brazil in 1927, and delivered lectures in Rio de Janiero and São Paulo. Gini's tour perhaps encouraged Brazilian eugenicists to hold their first national Eugenics Congress two years later. Many Brazilian eugenicists were impressed with Fascist Italy's demographic policies. They also admired Italy's pronatalist programmes, which rewarded large families, honoured prolific mothers, created a national network of maternity and child clinics, and provided eugenically friendly summer camps for children.

Brazilian eugenics experienced a number of important changes in the 1920s. Renato Kehl's change of heart concerning the most efficacious direction of eugenics was one of the most significant. His enthusiasm for positive eugenics gradually gave way to a fascination with negative eugenic measures, such as sterilization. During much of the 1920s, Kehl headed the Propaganda and Hygienic Education Department of the Leprosy and Venereal Disease Service (*Inspetoria da Lepra e Doenças Venéreas*) in the National Department of Public Health (*Departamento Nacional de Saúde Pública*). There he continued his efforts to improve rural sanitation, preventive medicine, and public health (Hochman, Lima, and Maio 2010: 500), as illustrated by his 1923 book *Eugenia e medicina social* (*Eugenics and Social Medicine*).

Kehl changed his ideas about eugenics following a research trip to Germany in 1928, where he met the German eugenicists Hermann Muckermann and Eugen Fisher (Wegner and de Souza 2013: 266–7). Apparently, they encouraged him to reconsider the merits of German racial hygiene and genetic determinism. After returning to Brazil, Kehl began to advocate negative eugenic measures, such as the sterilization of 'degenerates and criminals'. The next year he publicized these views in his new periodical, *Boletim de Eugenia* (Bulletin of Eugenics), and in a book also published that year, *Lições de Eugenia* (Eugenic Lessons) (Kehl 1929a). Yet these issues were phrased within the Brazilian context, which was strongly influenced by Catholicism. Not surprisingly, perhaps, Muckermann's article 'Eugenics and Catholicism' ('Eugenia e catolicismo') was among the first translations to be published in the *Boletim de Eugenia* in 1929 (Wegner and de Souza 2013: 273).

Slightly different, both in tone and in substance, were the writings of other Brazilian eugenicists, who although 'favoured Mendelism over sanitation eugenics, distanced themselves from negative eugenics' (Hochman, Lima, and

Maio 2010: 501). For instance, the anthropologist Edgard Roquette-Pinto, director of the National Museum, argued in 1927 that 'alongside fatalistic eugenics that preaches that outside inheritance there is no salvation, another eugenics has been established, concerned with favouring the acquisition of the best somatic characters on the part of those who are living' (quoted in Hochman, Lima, and Maio 2010: 501). That same year Brazilian eugenicists attempted to mark the centenary of the death of the Brazilian Empress Maria Leopoldina by organizing a 'Concurso de Eugenia' (Eugenic Contest), during which three children representing the ideal eugenic child were selected (Kehl 1929b: 26–7). Even if the growing acceptance of Mendelism among Brazilian eugenicists was apparent by the late 1920s, adherence to a eugenics emphasizing preventive medicine, puériculture, and social hygiene remained widespread.

These various interpretations of eugenics coalesced at the Brazilian Congress of Eugenics (*Primeiro Congresso Brasileiro de Eugenia*), organized between 30 June and 7 July 1929 at the National Academy of Medicine in Rio de Janeiro. This was part of an ambitious series of conferences that included the Tenth National Congress of Medicine and the Fourth Pan-American Congress of Hygiene, Microbiology and Pathology. Roquette-Pinto served as president of the Eugenics Congress, and Renato Kehl as its general secretary (Kehl 1937a: 104). This proved to be the most important gathering of Brazilian eugenicists in the movement's history. Government officials, medical specialists, and foreign visitors discussed many of the typical eugenic topics: marriage, education, race, and the protection of nationality, racial typology, genealogy, immigration, venereal diseases, biometry, and puériculture.

Immigration was one of the most contentious subjects discussed at the Congress. Some participants, such as Renato Kehl, Miguel Couto, and the journalist Azevedo Amaral, proposed immigration restriction along eugenic and racial lines – a position rejected by Roquette-Pinto and others in attendance who favoured racial egalitarianism. Rather, Roquette-Pinto argued that racial integration was eugenic, rather than dysgenic, and that Brazil would find racial harmony and national unity in miscegenation. '[M]an in Brazil', he reminded his colleagues, 'needs to be educated and not substituted' (Roquette-Pinto 1929, vol. 1: 147).

Roquette-Pinto endorsed the government's active involvement with eugenics. The physician Levi Carneiro agreed, reaffirming the importance of the environment and education to eugenic and racial improvement. Echoing Gini's theory of regenerative eugenics, Carneiro envisioned an extensive programme of national rebirth based on revitalizing Brazilians' innate racial qualities

(Carneiro 1929, vol. 1: 112–21). Maria Antonieta de Castro, the president of the Association of Health Educators (*Associação de Educadoras Sanitárias*), and one of the few women at the Congress, also expressed concern with the racial quality of the future generations. Castro highlighted the importance of puériculture and neo-Lamarckist environmentalism in shaping the activities of the São Paulo State Inspectorate for Sanitary Education (*Inspetoria de Educação Sanitária do Estado de São Paulo*), particularly its Prenatal Hygiene Service (*Serviço de Hygiene Pré-natal*), and the Course for Young Mothers (*Curso das Mãezinhas*). The Inspectorate promoted programmes on childcare, infant health, prenatal hygiene, and maternal education, as well as baby beauty contests (see Figure 5.3) (De Souza et al. 2009: 769–70). Ultimately, the majority of participants at the First Brazilian Congress on Eugenics 'drew on a more optimistic vision of the future of a racially intermixed people: the population did not need to be replaced, but rather educated, and given proper infrastructure for hygiene' (Hochman, Lima, and Maio 2010: 504).

Renato Kehl was rather critical of this direction of Brazilian eugenics, as he was increasingly attracted to German racial hygiene. After the Congress, he gathered together nine other like-minded eugenicists and established the

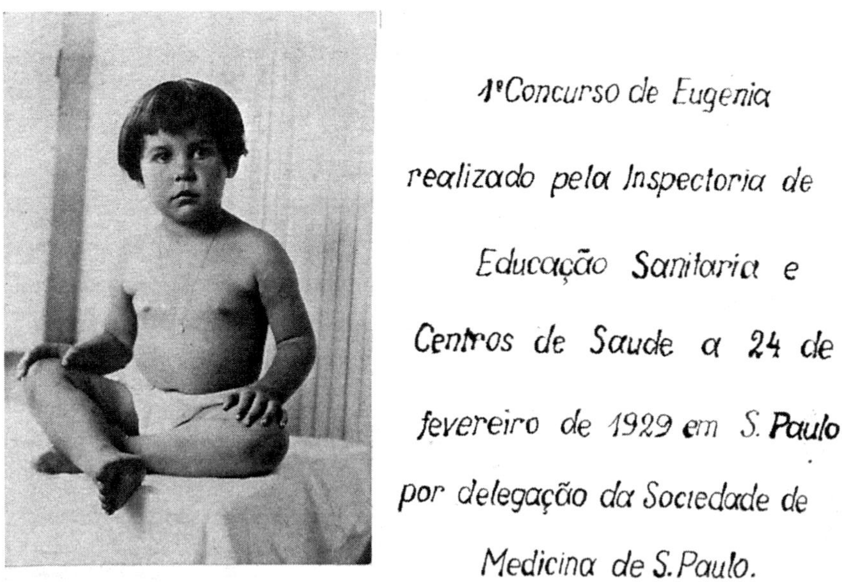

Figure 5.3 Eugenic Baby Prize Winner, São Paulo, 1929
Source: Maria Antonieta de Castro, 'A influência da educação sanitária na redução da mortalidade infantil'. Arquivo de Antropologia Física, Museu Nacional/ Universidade Federal do Rio de Janeiro.

grandiose-sounding Central Brazilian Commission on Eugenics (*Comissão Central Brasileira de Eugenia*) in Rio de Janeiro (Wegner and de Souza 2013: 265). The Commission's manifesto, published in September 1931, proposed both positive and negative eugenic measures, from financial assistance for orphans of worthy racial quality, to eugenic education in the schools, to prenuptial marriage certificates. Without these measures, Kehl feared that Brazil's degeneracy, as manifested in crime, vice, and mental defects, would only intensify (Kehl 1937a: 94–6 and 103–5). Most importantly, the Commission was determined to present its case to the government for restricting the immigration of non-Europeans into Brazil.

In Brazil, however, as in other Latin countries, both culture (Latinity) and religion (Catholicism) served as powerful obstacles to the dissemination of negative eugenic measures such as sterilization. As Ernani Lopez, president of the Brazilian League of Mental Hygiene (*Liga Brasileira de Higiene Mental*) remarked in 1933, 'Latin peoples remained skeptical about compulsory surgical sterilization' (quoted in Wegner and de Souza 2013: 272). As elsewhere in Latin America, the Catholic opposition to sterilization did not preclude the League – and some Brazilian psychiatrists – to promote negative eugenic measures aimed at the racial purification of society next to its proposals for hygienic education, preventive medicine, and positive eugenics (De Souza and Boarini 2008: 273–92; Wegner and de Souza 2013: 274–84).

The growing importance of eugenics within Brazilian scientific communities coincided with Getúlio Vargas' authoritarian *Estado Novo* (New State) (1937–1945). Emulating European states like Italy and Spain, Vargas promoted an exclusive nationalism that clearly appealed to eugenicists, particularly to those – like Edgard Roquette-Pinto and Gilberto Freyre – who had long campaigned for the creation of a racially harmonious Brazilian political community. In both its political and biological sense, '[f]usion through racial and cultural means, enabling blackness to disappear and the nation-state to form a new homogeneity, was itself taken to be "eugenic"' (Stepan 1991: 164). Tellingly, a member of the Commission on Eugenics, Belisário Penna, was appointed director of the Department of Health within the new Ministry of Education and Public Health (Stepan 1991: 163). Impressed by the apparent successes of Italy and Germany in empowering their countries, and under pressure from the far right, the pro-Nazi party *Acção Integralista Brasileira* (Brazilian Integralist Action), Vargas allowed Kehl's Brazilian Eugenics Commission to influence relevant parts of the new Constitution, then being written, which included the introduction of maternity and child benefits, restrictions on the work of mothers and infants, and the introduction of eugenic education. Arguments were also presented in the Constituent Assembly for obligatory prenuptial eugenic examinations and the racial selection of immigrants. As Nancy Stepan pointed out, '[t]he result of the various arguments was a eugenic and racial immigration law that established racial quotas as well as economic and other tests of fitness for entry into Brazil'. It clearly demonstrated the regime's 'commitment to whitening, eugenization, and homogenization as the official policy of the national state' (Stepan 1991: 166).

The main tenets of Latin eugenics – puériculture, preventive medicine, and social welfare – remained an integral part of Brazilian eugenics, even though Mendelism and population genetics were proving increasingly popular among Brazilian biologists, anthropologists, and sociologists. As one of the world's fastest-growing nations during the 1940s, Brazil remained committed to a welfare programme and eugenic paternalism that strove to avoid strict biological determinism and rigid scientific notions of racial difference (Griffing 1940: 13–16).

Homiculture and 'Anglo-Saxon eugenics' in Cuba

As in Brazil and Mexico, the Cuban population was highly multiracial. Consequently, racial issues played an important role in shaping Cuban eugenics, and by extension its influence on international Latin eugenics. Moreover, due to their geographical proximity to the USA, Cuban eugenicists were directly influenced by North American models of eugenics in ways not experienced by other Latin eugenicists, particularly with respect to immigration and sterilization.

As seen in Chapter 1, by the 1910s Cuban eugenicists like Eusebio Hernández and Domingo F. Ramos were inspired by French ideas of puériculture and championed a specific theory of eugenic improvement they termed '*homiculture*'. Both believed that homiculture was the expression of national progress and was fully compatible with the improvement of the Cuban 'race'. Pinard encouraged these two Cuban eugenicists to establish homiculture as a synthesis of eugenics and puériculture, which reflected Cuba's social and medical-specific conditions (Hernández and Ramos 1911). These eugenic postulates were also accompanied by a strong belief in the superiority of the European race and culture, and by concerns about 'undesirable' immigration.

As elsewhere, Cuban medical professionals were the most eager to embrace progressive ideas of national health based on eugenics, puériculture, and homiculture. For instance, at Cuba's Third National Medical Congress (*Tercer Congreso Médico Nacional*) held in 1914, Enrique Núñez highlighted the importance of Hernández' and Ramos' work on the development of child protection services and puériculture in Cuba. Physician José A. López de Valle similarly projected a eugenic ideal for Cuba centred on the protection of the family, child health, and hygiene (González and Peláez 1999: 132). In this context, medicine appealed to and embodied the Cuban elite's progressive views on the country's transformation into a modern nation. Diego Tamayo, a physician and editor of Havana's popular journal *Vida Nueva* (New Life), re-emphasized the physician–eugenic nexus when he wrote in 1921 that 'many of the syndromes that disturb our country could be addressed from within our profession, and therefore, we feel obligated to study them, so as to understand the pathologies and propose those remedies we determine to be most effective' (quoted in Bronfman 2004: 62). Cuba's widespread social and medical problems thus prompted progressive physicians and health reformers to engage in eugenic and homicultural projects, at the same time justifying state intervention to regulate and protect the health of the population.

Homiculture shared with puériculture and traditional eugenics a number of values, principles, and preoccupations with collective improvement. In his broad interpretation of homiculture, Ramos was equally influenced by Francis Galton and Adolphe Pinard. This dual eugenic heritage was apparent in the stream of papers he gave at various conferences, such as the Third International Congress for the Protection of Infancy held in Berlin, the American Public Health Association meeting held in Havana in 1911, and most notably at the Second International Congress of Eugenics of 1921 (Gonzáles and Peláez 2007: 29–33).

By the early 1920s Ramos had become one of the leading Cuban eugenicists (see Figure 5.4). Ramos was appointed to represent his country at the Second International Congress of Eugenics held in New York City between 22 and 28 September 1921. He was elected as one of the Congress' vice-presidents, along with Victor Delfino, Albert Govaerts, Georges Schreiber, Corrado Gini, and others. Ramos arrived several months in advance of the Congress, spending time at the Eugenics Record Office (ERO) at Cold Spring Harbor, New York. The experience was, for Ramos, a revelation. He became a close friend of Harry Laughlin, the vice director of the Office and one of the United States' most influential eugenicists. For their part, Laughlin and Charles Davenport, the ERO director, saw Ramos as essential to the expansion of their influence among Latin American eugenicists (Harry Laughlin to Walter Gilbert, 6 September 1934).

Figure 5.4 Domingo F. Ramos, on the right
Source: Courtesy of Claude Moore Health Sciences Library, University of Virginia.

Ramos's paper at the Congress, entitled 'Homiculture in its Relations to Eugenics in Cuba', synthesized his early ideas regarding eugenics. First and foremost, he validated Pinard's conception of puériculture – namely, 'the science which has its object the research and application of knowledge concerning the reproduction, conservation and improvement of the human species' – which placed the individual at the centre of eugenic research. Ramos highlighted two overlapping biological relationships: the individual to the species, and the individual to the environment. Homiculture served a unifying principle connecting all three. He argued that homiculture was a unified system of thought whose essence he articulated through a variety of biological discourses, from neo-Lamarckism to Mendelism, which informed 'sexual hygiene, school hygiene and prenatal and postnatal child hygiene'.

According to Ramos, it was finally 'the proper time to put Eugenics to work jointly with Public Health, as Eugenics has a great importance for a country [such] as Cuba'. The goal of homiculture was, therefore, to enable 'the scientific betterment of man, making the human species of the future the outcome of a scientific artificial selection and providing the environment in which it is going to live, artificially modified by the efforts of science' (Ramos 1923: 432–3). In this way, Ramos believed, homiculture transcended the eugenic dichotomy between the individual and the community: 'When we shall have a Eugenics Department organized, which I earnestly hope will be a reality in the near future, and we shall work for the betterment of old age, we shall have completed the programme of Homiculture connecting Public Health and Eugenics' (Ramos 1923: 434).

As elsewhere, improving the racial quality of the population through eugenics, puériculture, and homiculture relied on state funding and institutional support. It was also dependent on public displays of eugenic achievements. Exhibitions were one exemplary medium for informing the general public about eugenics and puériculture, such as the one organized on the occasion of the 6th Latin American Medical Congress (*VI Congreso Médico Latino-Americano*) held in November 1922 in Havana. The ERO provided 'demographic charts, maps, photographs, family history records, and scientific papers', while eugenicists such as Ramos and Arístides Mestre, a physician and assistant professor at the University of Havana, lectured on various eugenic topics ('The Sixth Latin-American Medical Congress', *Eugenical News* 1922: 113). The 'beautiful baby contests' or – as they were formally called – the 'National Motherhood, Homiculture and Eugenic Fertility Competitions' (*Premios de Maternidad, Homicultura y Fertilidad Eugénica*) introduced in Havana during the 1920s (Bronfman 2004: 119), is another excellent example. In 1924, speaking at one such event, Ramos reaffirmed the importance of hygiene and health for the racial improvement of future generations of Cubans (González and Peláez 1999: 162). This public lecture and the event at which it was delivered 'shared a Lamarckian approach, with an emphasis on enhancing environmental conditions to bring about racial "progress"' (Bronfman 2004: 119).

Throughout the 1920s, Cuban eugenicists remained in close contact with other Latin American eugenicists, as demonstrated by the consecutive Pan-American Sanitary Conferences organized in Havana in 1922 and 1924. However, after his visit to the USA, Ramos fell under the influence of Charles Davenport. As a result, '1921 marked the waning of "positive" eugenics, oriented towards environmental and reformist projects, and the rise of "negative" eugenics, with its focus on Mendelian genetics, racial purity, and prevention of reproduction for the genetically unfit' (Bronfman 2004: 117). During the 1920s and 1930s, Ramos devoted considerable effort to spreading the gospel of North American eugenics to his Latin American colleagues (Schell 2010: 478). He admiringly told Davenport after the 1921 visit that 'all [eugenic] works made in Cuba shall be under your inspiration', encouragingly adding that Cuba could be an 'outpost for good propaganda, under your inspiration in the Latin American countries'

(Ramos to Davenport, 27 June 1922). Interestingly, in May 1924 Ramos and José A. López del Valle participated in the First International Conference on Emigration and Immigration in Rome (González and Peláez 2007: 53), but neither attended the First Italian Congress of Social Eugenics held in September that year.

Characteristically, Ramos showed more interest in gaining power in international eugenic circles than he did in establishing a firm institutional basis for eugenics at home. Thus, he preoccupied himself for years with the establishment and direction of a Pan-American Office of Eugenics and Homiculture (Ramos to Davenport, 12 September 1922 and Ramos to Davenport, 28 December 1922). He first proposed this association at the Sixth Latin American Medical Congress in 1922, and again at the Fifth Pan-American Conference of Hygiene (*Quinta Conferencia Panamerican de Higiene*) held in 1923 in Santiago de Chile (González and Peláez 2007: 48–51). The Office's primary mission was to recommend 'measures that should be put in practice to prevent the propagation of disease and of physical, mental or moral defects of hereditary origin and to secure physical, mental and moral qualities of the same [hereditary] origin' ('Programme for the Constitution of the Pan-American Association of Eugenics and Homiculture, Santiago de Chile, March 1923').

Ramos's ambition to promote Cuba as a leader in international eugenics eventually materialized between 21 and 24 December 1927, when the first Pan American Conference on Eugenics and Homiculture (*Primera Conferencia Panamericana de Eugenesia y Homicultura de las Repúblicas Americanas*) was held in Havana. Besides Ramos, the organizing committee included Dr. Francisco Maria Fernández, State Secretary of Health and Welfare; Dr. Aristides Agramonte, a well-known pathologist and member of the Cuban Academy of Sciences; Dr. José A. López del Valle, president of the recently established Finlay Institute in Havana, and the prominent eugenicists Eusebio Hernández and José E. Sandoval. Representatives from Argentina, Bolivia, Colombia, Chile, Uruguay, and Venezuela attended the Conference. The well-known eugenicist Rafael Santamarina represented Mexico; and Carlos Enrique Paz Soldán spoke for Peruvian eugenicists. Charles Davenport was the North American delegate. Finally, Leonard Darwin, Adolphe Pinard, and Harry H. Laughlin were elected honorary members of the organizing committee ('Pan-American Conference on Eugenics and Homiculture', *Eugenical News* 1928: 17–9; González and Peláez 1999: 176–7).

Immigration, racial crossing, medical certificates, and, most importantly, the 'Code of Eugenics and Homiculture' (*Códico Panamericana de Eugenesia *y Homicultura*), written by Ramos with Davenport's support, were the main topics at the Conference. This Code was meant to serve as the foundation of a new eugenic organization, the Pan-American Office of Eugenics and Homiculture, to be established in Havana. In turn, with Ramos' encouragement, Charles Davenport hoped to persuade the Latin American representatives to see eugenics from the North American viewpoint. In his paper, Davenport 'enlightened' the delegates about the dangers of miscegenation as determined by his own investigations. Racially mixed individuals, he claimed, suffered mental disharmonies that threatened social stability. Moreover, these individuals tended to have 'the need of alcohol' which might have resulted from 'a feeling of insufficiency – a reflection of an internal conflict of instinct' ('Race Crossing', Folder 1). Davenport also advertised the benefits of North American immigration laws, which stemmed the flood of 'new and inassimilable elements' and eliminated some of the 'carriers of inferior physical, mental and moral qualities'. He encouraged the Latin American delegates to adopt similar laws ('The Eugenical Principle in Immigration', Folder 2).

The reaction of other delegates no doubt surprised Davenport and Ramos. In general, quite a few in the audience seemed disturbed by Davenport's interpretation of eugenics and his practical policy recommendations. Most did not appreciate Davenport's condescending and overtly racist remarks. For instance, the Mexican delegate, Dr. Rafael Santamarina, criticized North American psychologists for asserting that Mexican-origin school children were mentally inferior to their Anglo classmates, when the differences in the two groups' school performance were clearly due to environmental factors.

Nevertheless, Davenport and Ramos pressed ahead with their goal of promoting their 'Code of Eugenics and Homiculture' ('Códico Panamericana de Eugenesia y Homicultura', reproduced in González and Peláez 1999: 501–8; for an abridged version see Ramos 1934: 84–6). The Code recommended the establishment of national eugenic organizations that would classify *all* Latin Americans with regard to their biological potential. If judged eugenically fit, they would be granted the right to emigrate anywhere within the western hemisphere, subject to the immigration laws of the recipient country. Article 12 of Section III (on 'Migration'), for instance, stipulated that 'All nations of the Americas shall enact and apply immigration laws forbidding the entrance of those individuals somatically irresponsible or of bad germinal conditions, or coming from nations refusing either to accept this Code or comply with its provisions'. With respect to race (Section IV), Article 13 specified that each nation would be allowed to take the necessary measures to ensure 'the racial purity of the descendants', while Article 14 entrusted nations with the 'right to select which new races' could enter its territory and 'form part of its population' (González and Peláez 1999: 504).

Furthermore, the Code recommended the issuance at marriage of a eugenic health certificate (Article 15 Section V). This would serve as a guide to the desirability of the individual as an immigrant, should the occasion arise. Divorce was to be allowed in cases of insanity, criminality, syphilis, chronic alcoholism, and 'morphinism', [i.e. drug addiction] – no doubt for fear that offspring of such marriages might inherit the defects of their parents (Article 17). Failure to surrender records pertaining to one's biological history could also be punished by annulment of the marriage.

Section VI of the Code focused on pregnancy. Pregnant women would be required to follow medical advice with regard to their pregnancies, or face criminal penalties. 'Eugenically marginal' individuals would be educated about eugenics, and, if found receptive, allowed to have children as long as they followed the regulations of the eugenics authorities. Those with genetic defects considered too severe to be given the privilege to procreate would be segregated or sterilized, depending on their particular circumstances. Any individual who violated these regulations would lose their classification as 'eugenically fit'. Sections IX and X of the Code outlined the institutional relations between eugenics, homiculture, public health, and other scientific disciplines such as genetics, human embryology, anthropology, obstetrics, gynaecology, and paediatrics (González and Peláez 1999: 507).

Most delegates opposed the first version of the Code. Rául Cibils Aguire called it 'frightful', while Carlos Enrique Paz Soldán described it as a racial 'fantasy'. Many participants generally accepted the notion that eugenics was central to racial improvement. They also accepted that state intervention was needed to control the reproduction of the unfit, thus introducing marriage certificates. Eugenic sterilization was, however, rejected (González and Peláez 1999: 204–20). With the exception of Ramos and other radical eugenicists, the participants at the Conference 'wanted study, education, propaganda'; they wished to retain the sovereign right of each country to determine its own demographic and migration policies (Stepan 1991: 179).

The grand reconstruction of Latin American eugenics envisioned by Ramos and Davenport failed to materialize. 'Ultimately,' as Alejandra Bronfman argued, 'Ramos's vision of a genetically purified future, with its refusal to take into account the heterogeneous and racially mixed present, may have proved too impractical to a state interested in social reform and palpable results' (Bronfman

2004: 123). The only proposal that did generate universal enthusiasm was the neo-Lamarckian suggestion to bolster each country's physical fitness programme – a reaffirmation of Latin American eugenicists' commitment to environmentalist and positivist theories of human improvement ('Primera session de la Pimera Conferencia Panamericana de Eugenesia y Homicultura' – Jueves 22 de Diciembre de 1927, Folder 3).

Ramos gave the closing speech of the Conference. He said nothing about the 'wonderful and Christian work towards improving the species', which Francisco M. Fernández so optimistically highlighted in his opening address. Instead, Ramos claimed that eugenicists had the power to force the evolution of the human species, to create a 'new biological type' that would 'exceed in power' current human beings. In such a society, the eugenically superior would direct operations, and 'utilize' the labour of the inferiors. Each child would be mentally and physically evaluated, and then given the education and training appropriate to his or her abilities. Ramos insisted that 'human germs' be purged from the population 'by means of eugenics'. If individuals were found to be 'useless or noxious', they would face institutional segregation or sterilization. Borrowing from Galton, Ramos hoped that even future spiritual values would be based upon eugenics, the 'religion of life'. The priest of this religion would be the physician, 'in his office as cultivator of the human species' (Domingo F. Ramos' Closing Speech, Folder 4).

One tangible outcome of the Conference was the creation of the Pan-American Office on Eugenics and Homiculture (*Oficina Panamericana de Eugenesia y Homicultura*) under Ramos's leadership (González and Peláez 1999: 220–21). The first international activity involving the new Office, albeit in an advisory capacity, was participation in the Second Emigration Conference organized in Havana between 31 March and 17 April 1928. Harry H. Laughlin replaced Davenport at this Conference and, together with Ramos and Francisco M. Fernández, unsuccessfully argued for the introduction of eugenic immigration laws ('The Second Emigration Conference: Havana, 1928', *Eugenical News* 1928: 73–5). The Pan-American Office was also invited to join the International Federation of Eugenic Organizations at its 7th Meeting, held in September 1928 in Munich. José Enrique Sandoval, general secretary of the Pan-American Office, attended the meeting in Munich as well as the subsequent one held in Rome in September 1929. Together with Charles Davenport and Jon Mjøen, Sandoval also participated in the Second Italian Congress of Eugenics and Genetics (*Secondo Congresso Italiano di Genetica ed Eugenica*), which convened in Rome at the end of September 1929. His paper, entitled 'L'Eugénésie en Amerique'

(Eugenics in America) displayed an unmitigated support for Ramos's ideas of negative eugenics, such as sterilization and the control of immigration, which were gradually being embraced by the countries of Latin America (De Sandoval 1932: 201–5).

Ramos himself presented a concise version of the eugenic programme promoted by the Pan-American Office of Eugenics and Homiculture, with a special reference to immigration, at the Third International Congress of Eugenics held in New York between 21 and 23 August 1932. Unrestricted immigration was deemed to be dysgenic, and Ramos advocated placing it under strict eugenic observation. Race and environment were intimately connected, and both contributed to the eugenic quality of the immigrant (Ramos 1934: 80). According to Ramos, the future of national health in Latin America depended on the admittance of carefully selected racial elements. Ramos's eugenic interpretation of immigration was predicated on racial differences and on racial incompatibility. Eugenic policies, he claimed, 'should be based on the different biological constitutions of each individual and each race', with individuals classified as 'desirable and undesirable immigrants from the viewpoints of race and individuality and also in relation to sanitary conditions' (Ramos 1934: 83). Ramos then included excerpts from the 'Code on Eugenics and Homiculture' he presented in Havana in 1927 (Ramos 1934: 85–6), no doubt hoping to attract a wider international audience to his proposals.

During this period, there were also attempts to establish a Cuban eugenics society. Thus, in 1928 at the First National Congress of the Child (*Primer Congreso Nacional del Niño*) a proposal was made to create a National Society of Eugenics and Homiculture (*Sociedad Nacional de Eugenesia y Homicultura*), designed to assist with the successful application of a number of eugenic projects in Cuba, including premarital medical examination, sexual education, social assistance for 'incapacitated children', preventive education, and so on (González and Peláez 1999: 307). The Society did not materialize, not least due to the internal disagreements prevalent among Cuban eugenicists.

Political stability in Cuba deteriorated with the onset of the Great Depression. The ensuing political crisis also affected the Pan-American Eugenics Office. Antonio F. Barrera, head of the Cuban Secretariat of Infant Hygiene, came into conflict with Ramos, which apparently reinforced a decision by Machado to cut off funds for the Pan-American Office in 1931 (González and Peláez 1999: 221–2). In a letter to Laughlin, Ramos lamented the chaotic conditions in Cuba, and protested that it had become quite impossible to live in a country in which the government was using 'squads of Negroes and half breeds' to terrify respectable people. He was himself arrested and detained for a short time, then took refuge in the United States from March 1933 to 1935, seeking sponsors for his Pan American Eugenics Office. Eusebio Fernández also fled to Miami, where he lived out the rest of his life as a physician in private practice (Fernández to Laughlin, 19 October 1934).

While in the USA, Ramos sought to encourage Laughlin's help with a vision of a radical eugenic utopia, suggesting that the promotion of the eugenically purified white race in Latin America would 'constitute a sound base for the defence in general of said race in the whole world'. He warned that 'unquestionably the white race is surrounded by dangers today', especially the high birth rates of non-white peoples. The only solution was for the white race to 'stop, or control the advance of the rest, in an efficient and effective manner'. Ramos warned that, if the 'inferior Indian and Negro races' were allowed to dominate Latin America, this would embolden non-whites the world over. One fearful consequence would be that, '[an] America, dominated to the south of the Rio Grande by half-castes, would not be so strong to defend itself from a possible Yellow invasion, or to limit the ambitions of this race in the ways necessary, as an America of complete white domination'. Ramos even imagined 'a world-wide intellectual political movement' united by their conviction that biological inequalities between families and races were simply facts of nature. The culmination of Ramos' utopia would be, as he confided to Laughlin, 'the production of the new species, which, until now we have referred to as: – "Superman"' (Ramos's Letter to Laughlin, n.d.).

Undoubtedly, Ramos made good use of his time in the United States. He met with a number of US government officials, including Hugh Smith Cumming, the US Surgeon General and director of the Pan American Sanitation Board. He unsuccessfully pleaded for the Pan American Eugenics Office to be transferred to the Pan American Sanitary Bureau in Washington (Treasury Dept Memorandum: Pan American Conference on Eugenics and Homiculture, by H. S. Cumming, Surgeon General, 23 August 1934).[1] Ramos also met Clarence Campbell, a wealthy radical eugenicist, president of the [North] American Eugenics Society, and a convinced admirer of Hitler's eugenic programme in Germany. Around October 1934, Campbell wrote a letter to Sumner Wells, the US Assistant Secretary of State, in which he lauded Ramos – who, 'more than any other man', had been critical in stimulating the 'great interest' now shown by Latin Americans 'in their own racial improvement'. Ramos was ideally placed to serve as an intermediary between eugenicists in the United States and Latin America, because he 'thoroughly understands and sympathizes both with the United States position and that of Latin-American countries' (Clarence Campbell to Sumner Wells, Assist. Sec. of State, n.d.). In another letter, to Secretary of State Cordell Hull, Campbell encouraged the United States to forge an agreement among the American nations to maintain 'the purity and the dominance of the white elements, and the exclusion of undesirable immigrant racial elements' (C[larence] Campbell to US Secretary of State [Cordell Hull], 6 October 1934).

Ramos' exile from Cuba proved to be temporary. The country stabilized after Ramos' friend, Dr. Ramón Grau San Martín, emerged as Machado's successor, in September 1933. Ramos was quickly restored to his former positions, and was ready by November 1934 to attend the Second Pan American Conference of Eugenics and Homiculture (Report of the American Delegation to the Department of State, 14 August 1935). The official US delegation consisted of Ramos' friend, Hugh Smith Cumming; Bolivar Lloyd (the medical director of the US Public Health Service); and Kendall Emerson (the executive director of the National Tuberculosis Association and the executive secretary of the [North] American Public Health Association). Though Laughlin was not present, he and Campbell had close ties to Emerson (Laughlin to Dr. Kendall Emerson, Director, National Tuberculosis Association, 16 January 1935).

Ramos and the North American delegates once again argued for immigration restriction, matrimonial certification, and the sterilization of defectives. The main arguments for these recommendations were well known, namely, that 'the sane, the feeble minded, the indigent and even the criminal in prisons' were being cared for in 'luxurious' institutions at government expense. The financial burden on the taxpayer of supporting these 'non-productive members of society' threatened to become 'greater than he can bear', and potentially leading to 'revolt'. To avoid this, those members of the institutionalized population capable of work should be made to do so, or 'means must be found' to reduce the numbers of people dependent on government support (Hugh Smith Cumming, Bolivar Lloyd, Kendall Emerson, 15 August 1935). Ramos, in turn, suggested that sterilization was no different from forcing a population to receive vaccinations, an argument borrowed from the United States Supreme Court decision in *Buck v. Bell.*

Ramos returned to the United States several months later, again to resume his schemes on behalf of the white race. Ramos met with a variety of public officials, eugenicists, and racial organization leaders on the East Coast (Ramos to Laughlin, 18 January 1935; Ramos to Clarence Campbell, 11 February 1935). He addressed an excited letter to his new friends, laying out his and Laughlin's eugenic goals. He claimed he was struggling to 'assure the American freedom from the Yellow Man['s] ambition'. There were, he said, 'two yellow biological perils in America: the Yellow Fever and the Yellow Man'. Ramos claimed that he and Laughlin had, in fact, made important strides in spreading the North American eugenic programme in Latin America. They had awakened interest in eugenics in those Latin American countries populated mainly by whites. Ramos claimed that Brazil had 'a very good white stock; which was informed about eugenics and in a very good disposition to fight in favour of the purity of the race'. Northern Brazil was 'hopeless' as a site for white settlement, but could be used to solve 'the black problem in America'.

According to Ramos, the Caribbean countries (which also included Cuba) were the most promising regions for further white settlement. Ramos again advocated the same racist and eugenic strategies to accomplish these ends that he had been promoting for more than a decade. About the same time as he wrote the above letter, Ramos arrived in Miami. There, he hoped to convince University of Miami officials to establish a eugenics institute ('Memorandum: Subject: Preliminary conversation between President Bowman Ashe, Professors and Doctors Pearson and Zamora of the University of Miami and Dr. Francisco M. Fernandez, Past Director, and Dr. D. F. Ramos, Delegate, of the Pan American Office of Eugenics and Homiculture', 9 February 1935).[2]

By the early 1940s, the political situation in Cuba seemed to have developed in a particularly favourable direction for Ramos. In terms of practical eugenics, there was consensus among Cuban medical professions about the introduction of premarital medical certificates (González and Peláez 1999: 374–84). Ramos, moreover, became increasingly involved in politics, assuming the post of director of the Health Service (*Servicio Técnico de Salubridad*) and that the Secretary of Defence (*Secretario de Defensa*) in 1938. Due to his new position he was not able to attend the Third Pan-American Conference on of Eugenics and Homiculture (*Tercera Conferencia Panamericana de Eugenesia y Homicultura*) held in Bogota, Colombia, from 24 July to 7 August 1938 (Gonzáles and Peláez 1999: 261–3) and the First Peruvian Conference on Eugenics (*Primera Jordana Peruana de Eugenesia*), held in Lima between 3 and 5 May 1939. Yet the influence of his ideas was felt, particularly at the latter event, where ideas of negative eugenics and racial policies enjoyed widespread support (Benavente 1940: 26–32).

In 1937, Ramos was also instrumental in proposing that the Committee of Sanitation (*Comité de Sanidad*) was renamed the Committee of Sanitation, Eugenics and Homiculture (*Comité de Sanidad, Eugenesia y Homicultura*). The Child Hygiene Division of the Committee (*División de Higiene Infantil*) was renamed the Division of Eugenics (*División de Eugenesia*). In a letter to Harry H. Laughlin, the Division's director, Dr. Victoriano D. Agostini, commended Ramos's work and the role he played in having eugenics and homiculture 'recognized by the legislative and executive branches of the government' (quoted in Gonzáles and Peláez 2007: 260).

With this position, and subsequent appointment as director of the Finlay Institute (*Instituto Finlay*) in 1944, Ramos received the national and international recognition he so craved. This was also a period that witnessed Ramos' return to the eugenic ideals of his youth. For instance, in a lecture delivered to the American Public Health Association on 17 October 1941, Ramos signalled out the crucial importance of public health to a common Pan-American strategy of biological protection. 'As the result of new biological discoveries and the work entrusted with their application in the conservation of health and the improvement of man', Ramos noted, 'public health has become of extended importance and plays a greater role in the solution of the different problems of the individual, the family, the community, and the nation' (Ramos 1942: 629). Furthermore, in 1942, he organized a series of conferences on eugenics at the Finlay Institute. The participants – who included the lawyer José Agustín Martínez and physicians José A. Bustamante and Estela Agramonte de Rodríquez – unanimously criticized negative eugenics, promoting puériculture and social hygiene instead (Gonzáles and Peláez 2007: 78).

During this period, Ramos also became interested in Alexis Carrel's work on eugenics, as outlined in his 1935 bestseller, *L' Homme, cet inconnu* (*Man, The Unknown*). Ramos adopted Carrel's eugenic vision and henceforth took it upon himself to preach the benefits of a eugenic utopia in which human beings were re-engineered to serve in eugenic castes. Ramos' writings from this period illustrate his own eugenic 'Brave New World' set within the framework of homiculture. He hoped for a world ruled according to the principles of homiculture and 'biocracy' (that is, a biological aristocracy). For Ramos, this apparently meant that each individual would be assigned a eugenic value that would predetermine their role in society. Appropriately, he now promoted a new form of eugenic 'humanism' (Ramos 1940: 35–44), which corresponded with the general welfare policies followed by the Second Cuban Republic.

With the onset of the 1950s, Ramos was joined by other Cuban eugenicists, such as José Chelala Aguilera, in promoting an interpretation of eugenics centred on sexual and hygiene education, and the protection of the traditional Cuban family (Arvey 2012: 93–120). The positive proclivities of homiculture were now displayed most compellingly by a Cuban political elite preparing for wide social and national change. And – as in Romania during the 1950s and the 1960s – no group manipulated the eugenics discourse of the family and its corollary, national improvement, better than the communist political elite who, after 1959, concentrated its attention on social welfare, sanitation, puériculture, and preventive medicine as it attempted to build a socialist Cuban Republic.

***

The various interpretations of eugenics, puériculture, and homiculture proposed by eugenicists in Latin America during the interwar period demonstrate several important points emphasized throughout this book. Latin Americans had an interrelated eugenic agenda, heavily influenced by France, Italy, and Spain. In accordance with eugenicists in Latin Europe, eugenicists in Latin America emphasized a relatively open, assimilationist immigration policy, demographic growth, puériculture, social hygiene, and opposition to birth control and sterilization. Although negative eugenic methods of racial improvement, such as those inspired by the United States and Nazi Germany, were advocated – as for example at the First Peruvian Conference on Eugenics (*Primera Jornada Peruana de Eugenesia*), convened in Lima between 3 and May 1939 – such views were in minority. Positive eugenics, biotypology, puériculture and a Catholic interpretation of eugenics remained dominant in Latin America (Burga 1940: 33–8 and Solano 1940: 96–101). In Venezuela, for instance, a Society for Puériculture and Pediatrics (*Sociedad Venezolana de Puericultura y Pediatría*) was established as late as 1939. What the eminent social hygienist Carlos Enrique Paz-Soldán called 'the racial tradition of the Peruvian Medical School', in full display at the Second Peruvian Conference on Eugenics (*Segunda Jornada Peruana de Eugenesia*), held in Lima between 25 and 29 May 1943, did not however adumbrate the popularity of neo-Lamarckist theories of social and biological improvement so characteristic of Latin eugenics. Moreover, and as elsewhere in Latin America during the 1940s, political elites in Peru adopted a pronatalist population policy favourable to population growth, a policy endorsed both by the Catholic Church and the eugenicists (Cueto, Jorge and Carol 2009; Orbegoso 2012: 230–43).

This is not to deny, however, the powerful influence exercised by North American and German models of eugenics among eugenicists in Latin America. During the late 1930s, the Cuban eugenicist, Domingo F. Ramos, was – as discussed in this chapter – strongly impressed with sterilization and the control of immigration, both methods promoted by North American eugenicists like Charles Davenport and Harry H. Laughlin. This was a complex problem, however, and not only for Ramos, but for all Latin eugenicists at the time. During the 1940s, Ramos' position on the matter was carefully nuanced, and he too (like most Latin eugenicists in Argentina, Mexico, Brazil, Cuba, Peru, and Venezuela) remained ultimately committed to the language of positive eugenics, homiculture, and puériculture, which well served their progressive views of national improvement, blending secular rhetoric about the importance of a strong state with Catholic virtues of family and reproduction.

6

The Latin Eugenics Federation

By the mid-1930s, eugenicists in the Latin countries had formulated their own general model of human improvement. Yet Latin eugenics was far from monolithic. On the contrary, as discussed in the previous chapter, it was premised on a number of theories, including the critique of Anglo-Saxon or Nordic racial superiority, neo-Lamarckism, puériculture, homiculture, biotypology, Catholicism, and the opposition to interventionist reproductive practices, such as sterilization. As such, Latin eugenics often dovetailed with various programmes of preventive medicine, public health, and social hygiene.

Latin eugenicists regarded the Anglo-Saxon and Nordic eugenic narratives of human improvement as increasingly problematic, for they centred mostly on race, the control of reproduction, the elimination of the 'unfit', and to a certain extent, the lowering of government expenditures on public health by eliminating the need for state care of individuals classified as hereditarily and socially 'defective' (Briand 1938a: 55). Latin eugenicists believed their methods to improve the nation and race were more 'humane', arguing that eugenic goals were best realized by voluntarist approaches, improving environmental health, encouraging population growth, educating the public about sexual hygiene, and cultivating moral behaviours linked to heath and reproduction. Their eugenic vision of a healthy nation stressed the protection of mothers and children within the family, as well as provision of eugenically sound social welfare programmes. However, this is not to say that Latin eugenicists were somehow immune from the more radical eugenics measures practiced in the United States and Nazi Germany. Many Latin eugenicists believed that certain methods of population management and control, such as medical examination before marriage, were required to ensure the health of future generations. In a few cases, governments in Latin countries even experimented with still more radical practices: compulsory eugenic sterilization was authorized in Vera Cruz in 1932, while Catalonia and Romania legalized abortion for eugenic reasons in 1936.

In addition to formulating a distinct theory of human improvement, Latin eugenicists were also able to establish their own international eugenic organization in 1935. This significant moment in the history of Latin eugenics represented the convergence of decades of eugenic developments in the Latin countries, while also resulting from the deterioration of relations between eugenicists in the Latin countries and eugenicists who championed negative eugenics combined with Nordic racial science, especially in Nazi Germany. The intellectual characteristics and broad ideological functions of Latin eugenics, particularly after 1933, must therefore be understood within the larger context of international rivalries. Given the powerful support to the negative eugenic measures provided by the Nazi dictatorship in Germany, Latin eugenicists reacted similarly, with marked aversion. They realized how their own national eugenic ideologies were fundamentally concordant; they also feared the likely consequences for their own nations of a possible German domination, in science as in all other aspects of national life. Until the profound disjuncture brought about by outbreak of the Second World War in 1939, the elaboration of common Latin eugenic thought remained infused with a sense of Latinity, based on a common cultural and scientific heritage.

Parting ways

During the 1930s, the growing dissention between Anglo-American, Nordic, and Latin variants of eugenics was apparent in the strained relations between members of international eugenic organizations. As noted already, French, Italian, and Belgian eugenicists such as Lucien March, Corrado Gini, and Albert Govaerts had played important roles in the international eugenics movement since its inception in 1912. However, they did not succeed in changing the directions of the international eugenic movement as did the British, the American, and the German eugenicists. In 1925, the Permanent International Eugenics Committee became the International Federation of Eugenic Organizations. Its leadership reflected the dominance of Anglo-Saxon and Nordic eugenics: the British eugenicist Leonard Darwin was its first president, followed by Charles B. Davenport, an American, and then Ernst Rüdin, from Germany. During the 1920s, Lucien March, Corrado Gini, and Albert Govaerts also served in the organizational leadership of the IFEO, although never as presidents.

By the early 1930s, the influence of the French eugenicists within the IFEO was in decline, not least due to their unfailing commitment to Latin eugenic principles, including neo-Lamarckism and natalism. The merging of the

French Eugenics Society with the Eugenics Section of the French Office of the International Institute of Anthropology in 1926 further reinforced the differences between the goals of French eugenics and those of many other eugenicists outside the Latin sphere (Kühl 2013: 63–7).

Italian eugenicists, and Gini, in particular, were increasingly apprehensive about their participation in the IFEO, as well as the difficulties in having their interpretation of eugenics that reflected their interests in demography and biotypology accepted by the other eugenicists from Anglo-Saxon and Nordic countries. In his inaugural speech to the Second Italian Congress of Eugenics and Genetics in 1929, Corrado Gini made a clear distinction between the version of eugenics promoted in 'the Latin countries' and that in the 'Anglo-Saxon world'. According to Gini, Anglo-Saxon eugenics was dominated by three interrelated theories: (a) 'the primacy of heredity over the environment'; (b) 'the superiority of the Nordic race'; and, finally, (c) 'the progressive degeneration of modern nations due to the increased fertility of the lower classes' (Gini 1932: 18–9). Despite what some Anglo-Saxon eugenicists alleged, the sterilization of those considered to be 'defective' and racially 'unworthy' could not prevent further social and biological degeneration. Instead, Gini proposed his version of eugenics, one he termed 'regenerative eugenics' ('eugenica rinnovatrice'), premised on an explicit communion between heredity and the environment, the regenerative effects of racial crossing, and demographic changes in the population. He also emphatically connected a nation's 'hereditary patrimony' with its fecundity, and argued that eugenicists were not justified in assigning degrees of relative superiority or inferiority to complete populations or entire races. Such extremist eugenic views contradicted those of the Latin countries, which stressed 'Latin equilibrium'.

This assertive exegesis on the superiority of Latin eugenics prompted the 'Anglo-Saxon' and 'Nordic' eugenicists at the Congress, such as Cora B. S. Hodson, Felix Tietze, Eugen Fischer, Charles B. Davenport, and Jon A. Mjøen, to aggressively counter by expressing their support of eugenic sterilization. In a heated exchange along these lines, Cora B. S. Hodson, the London-based general secretary of the IFEO, insisted on the need to sterilize 'degenerates'. Ernesto Pestalozza, probably the most passionately opposed to sterilization of all the Italian eugenicists, reacted angrily: 'To Signora Hodson I would say that I would save my enthusiasm for operations that remove illness from the sick or dying, and not for mutilations, which I as a surgeon would sternly oppose' ('Prima Seduta', *Atti del Secondo Congresso Italiano di Genetica ed Eugenica* 1932: 37). Pestalozza also stressed that compulsory sterilization was contrary to the ideals of Italy and of the Catholic Church. Only such positive eugenic programmes as infant assistance, prenatal care, and the moral and physical education of children could 'perfect the individual and prepare him for a race superior in numbers and quality'.

The disagreements that erupted during the 1929 Italian Congress over negative eugenics echoed wider tensions within the IFEO. At the next meeting in Munich in 1928, Gini complained to the president, Charles B. Davenport, about the London Bureau, which he said was 'only a source of delay, confusion and bad intentions'. In particular, Gini seemed to be at loggerheads with Cora B. S. Hodson. Gini advocated restructuring the IFEO in such a way that Hodson's position would be eliminated (Gini to Davenport: 29 March, 9 June 1929, and June 1931). However, Gini's recommendations were not implemented.

The situation deteriorated further after the 1929 IFEO meeting in Rome. In her official report on the meeting, published in *Eugenical News* (the official journal of the IFEO), Hodson inserted a long speech defending sterilization that was never actually given, and added a declaration claiming the IFEO delegates agreed on the necessity of sterilization, which was in fact never made ('The Meeting of the International Federation of Eugenic Organizations', *Eugenical News* 1929: 153–7). Upon reading this, Gini sent letters to Hodson correcting the record, which she duly ignored. In retaliation, Gini declared that the Italian Eugenics Society would no longer have any contact with her.

The dispute continued at the 9th Meeting of the IFEO, held in Farnham, England, in September 1930. The IFEO delegates accepted Gini's earlier proposal that the organization's secretariat be restructured, but postponed implementation until 1932. In turn, Gini refused to authorize the Italian Eugenics Society to pay dues to the IFEO so long as Hodson remained in her post. As a result, in November 1931 Hodson declared that the Italians had forfeited membership in the organization on account of their unpaid fees (Gini to Davenport, 16 June 1932; Cora Hodson to Davenport, 9 August 1930). In a letter to the current IFEO president, Charles B. Davenport, Hodson complained in exasperation that the 'Latins' defaulted 'with impunity' on their dues to the IFEO, yet still expected to be treated exactly like those nations who paid their membership (Hodson to Davenport, 17 June 1932). She also claimed that Gini refused to cooperate with any organization (such as the International Union for the Scientific Study of Population), which did not reflect his views (Hodson to Davenport, 28 July 1932).

Despite Davenport's efforts to placate Gini, the Italian Eugenics Society withdrew from the IFEO in August 1932 (Gini to Davenport, 20 August 1932). The official departure of the Italians was eventually confirmed on 26 September 1935, in a letter sent by the Italian Ministry of Education to the President of the Council of Ministers. 'Italian scholars', it was noted, 'must abstain from collaborating with the International Federation of Eugenic Organizations, from which our representatives have distanced themselves in consideration of its programme, which evidently contrasts with the Italian direction regarding the qualitative population policy' (quoted in Cassata 2011: 177).

These disagreements paralleled those expressed at the World Population Congress in Geneva, held between 29 August and 3 September 1927. Once again, the organization brought together eugenicists from around the world, including prominent Latin demographers and physicians such as Lucien March, Corrado Gini (see Figure 6.1), Alfredo Niceforo, Livio Livi, Marcello Boldrini, Léon Bernard, Jean Bourdon, and Eugène Dupréel (Sanger 1927).

Following this Conference, a meeting was held in Paris at the Musée Social in July 1928 to establish an International Union for the Scientific Study of Population (IUSSIP). Belgium was represented by Valère Fallon and Eugène Dupréel; France by Léon Bernard, Adolphe Landry, and Lucien March; Italy by Corrado Gini, Franco Savorgnan, and Marcello Boldrini; Spain by Severino Aznar and Joaquín Espinosa. Other founding countries included Denmark, Great Britain, Greece, Germany, Sweden, Argentina, Brazil, the Soviet Union and the USA ('International Population Union', *Eugenical News* 1928: 131–2). In this case Raymond Pearl, an American biologist, was elected president.

Gini was, however, determined to make Italy a leading member of the organization, given the Italians' prominence in population research. Although

Figure 6.1 Lucien March (first on the left) and Corrado Gini (centre) at the 1927 World Population Congress in Geneva
Source: Wellcome Library London. Image reproduced by permission of Alexander C. Sanger.

his efforts to promote Italian population policies were thwarted in the IFEO, he expected to have even greater influence in the IUSSIP. Not surprisingly, Gini demanded that the 1931 meeting of the Population Union be held in Rome. American and British members opposed Gini's proposal, regarding him simply as a spokesman for the Italian fascist regime. Refusing to accept this decision, Gini organized a rival Population Union meeting in Rome. The outcome was two congresses in 1931 devoted to the study of population: one held in London between 15 and 18 June, the other in Rome between 7 and 10 September (Pitt-Rivers 1932).

The Italian Congress managed to attract most of the prominent Latin eugenicists, anthropologists, and demographers at the time, including Lucien March, Henri Briand, Eugène Pittard, Armand Siredey, Sabin Manuilă, Severino Aznar, Octavio V. López, Enrique Ruiz Guiñazú, Javier Ruiz Almansa, and Albert Govaerts. Prominent Italian eugenicists and demographers were also in attendance, including Agostino Gemelli, Giacomo Acerbo, Nicola Pende, Alfredo Niceforo, Sergio Sergi, and Giuseppe Genna. The eugenics topics covered at the Congress included biotypology, pronatalism, the relationship between constitution and fertility, the demographic characteristics of large families, the expansion of the white race, miscegenation, and so on (Gini 1934a). The Italian contributions, more explicitly, 'celebrated the "positive" and pronatalist consensus within Italian eugenics and fascism and defined prolific people as the most select and desirable group within a biologically well-endowed and gifted Italian race' (Quine 2012: 117).

The Congress in Rome marked the zenith of Gini's international career as a demographer, and domestically as one of Italy's foremost social scientists. Mussolini did not attend the Congress, as initially announced. Yet the Duce self-assuredly read and corrected Gini's opening speech, requiring that a paragraph praising Malthus be removed, and countermanding Gini's decision to invite Marie Stopes (Ipsen 1996: 205; Cassata 2011: 142). Soon after the Congress, in early 1932, Gini resigned under pressure as director of the Central Institute of Statistics, prompting speculations that Mussolini had forced him out.

Gini's campaign for the cause of Latin eugenics continued unabated, however. He was determined to assert the 'Latin perspective' at the Third International Congress of Eugenics. On 21 August 1932, Gini gave an official 'Response to the President's Address' delivered by Charles B. Davenport. Gini's lecture stressed the importance of international congresses in enabling those interested in 'the problems of heredity and the improvement of the human races' to meet and discuss their views (Gini 1934b: 25). One of the most central eugenic topics, he continued, was the fundamental relationship between the quantity and quality of the population. They were 'indissolubly connected',

> [N]ot only because in practice it is difficult to think of a measure affecting the number of inhabitants which does not also affect their qualitative distribution, or of a measure hindering or encouraging the reproduction of certain categories of people which does not also modify, directly or indirectly, the number of population, but also and above all because population is a biological whole. (Gini 1934b: 25–6)

Gini argued that entire nations followed demographic cycles of birth, aging, and death, much as did individuals. However, these cycles could be modified through a constant process of adjustment to both heredity and the environment.

Gini adapted Francis Galton's definition of eugenics to emphasize the genetically restorative power of a healthy environment in a neo-Lamarckian context, and the importance of statistics, demography, and medicine in relation to genetics. According to Gini, eugenics focused on controlling those factors 'apt to improve the racial qualities of humankind not only from the view-point of their causes, as does Genetics, but also from that of their practical consequences, of their history, of their diffusion, of their economic, political, moral, cultural reflexes' (Gini 1934b: 26). He warned that it would be a mistake to consider eugenics only from 'a narrow point of view as a chapter of Genetics applied to man, or worse still, of experimental Genetics applied to man'. This would only lead to a regrettable neglect of 'all the other problems, so vast, complex and delicate', which could influence human improvement (Gini 1934b: 27). More broadly, this was an attempt to restore eugenics to its nurtural, environmentalist context.

Gini's celebration of social and cultural factors informed his interpretation of 'regenerative eugenics'. He claimed that eugenics had only reached a stage of development that warranted investigation, whose aim was to study 'through series of successive generations, how new stocks arise, what circumstances determine their formation in the midst of the obscure mass of the population – a formation which can hardly be explained by the heredity of superior factors heretofore non-extant' (Gini 1934b: 27). By understanding the complex relationship of natural selection, mutual adaptation, and 'the change of environment caused by emigration', eugenicists could come to understand how modern societies and cultures evolved. Indeed, this had long been the aim of his cyclical theory of the population, which sought to discern the factors affecting fecundity and 'demographic metabolism' – a theory which was an integral component of his eugenic thought since the early 1920s.

Although at first it might appear that Gini was merely attempting to readjust the priorities of a scientific theory, in reality his redefinition of eugenics carried an unmistakable political message. The hereditarian and racial outlook of the Anglo-Saxon and Nordic eugenicists was the result of faulty assumptions about human improvement that grew out of their ideological world view rather than scientific research, Gini claimed. The task of understanding the social and biological programme of human improvement and making it accessible to the general public and political elites was delegated to the eugenicists. Only they had the determination and the scientific knowledge needed to make eugenics the 'religion of the future'. Perhaps more importantly for our discussion here, the neo-Lamarckian basis of Gini's eugenic beliefs demonstrate that he was now much more preoccupied with asserting a common identity for Latin eugenics in the scientific world than with making futile efforts to dispute the dominance of Anglo-Saxon and Nordic eugenics.

Twenty years had passed since Adolphe Pinard first defined puériculture as different from eugenics at the First International Eugenic Congress in London. At the Third International Eugenic Congress in New York it was Corrado Gini's turn to offer a programmatic statement of Latin eugenics. If in 1912 it was puériculture that gave Latin eugenics its distinctive character, in 1932 it was Gini's 'regenerative eugenics' and the Latin countries' opposition to sterilization.

The ideological differences between Latin and Nordic eugenics widened at the next population congress, held between 26 August and 1 September 1935 in Berlin. Now the world was treated to a full display of Nazi racial and eugenic arrogance, which further confirmed the politicization of German population research and racial hygiene. The most prominent German eugenicists and demographers presented papers, including Alfred Ploetz, Ernst Rüdin, Eugen Fischer, Otmar Freiherr von Verschuer, Fritz Lenz, and Friedrich Burgdörfer. Marie Stopes, Harry H. Laughlin, Herman Lundborg, G. P. Frets, and Cora B. S. Hodson (representing the IFEO), were also present, together with a number of participants from Poland, Czechoslovakia, Hungary, Yugoslavia, and Bulgaria. Numerous eugenicists from the Latin countries also participated. The vice-presidents of the Congress included Corrado Gini, Adolphe Landry, Sabin Manuilă, Luis de Hoyos Sáinz, and Severino Aznar. Stefano Somogyi, Franco Savorgnan, Livio Livi, and Carlo Alberto Biggini represented Italy; René Martial, Jean Dalsace, and Fernand Boverat came from France; Norbert Ensch represented Belgium; and Argentina sent Victor Delfino (Harmsen and Lohse 1936: viii–xix and 932–59).

In his welcoming address, Wilhelm Frick, the German Reich's Minister of Interior and the Congress' honorary president, attempted to dispel some of the 'misconceptions' about his country's racial and eugenic programmes. Nazi leadership, he complained, has been 'reproached' for 'following a special racial cult and as injuring the commands of Christian brotherly love by our eugenic measures' (Frick 1936: 16). This population congress, he hoped, was an opportunity for all participants 'to assist in bringing about a better understanding of National Socialist Germany', and to 'be a valuable building stone in the worthy work of peace and construction of our Führer, Adolf Hitler' (Frick 1936: 17).

Eugen Fischer, as the acting president of the Congress, reiterated the ideological importance of eugenics and its essential role in building a racially strong Germany. The adoption of qualitative eugenic measures, he noted, was 'incontestably the credit of the National-Socialist leadership in Germany'. The purposeful eugenic re-engineering of the German nation by the Nazis filled Fischer with both national and professional pride. 'I am glad to be able to recognize unreservedly', he rejoiced, 'that today almost all the other nations have realized the seriousness of the qualitative aspect of the movement of population and that the science of human heredity and racial hygiene or eugenics are now receiving the attention due to them' (Fischer 1936: 45).

Dr. Jean Dalsace, head of the laboratory at St. Antoine Hospital in France, was one Latin eugenicist who did not hesitate to oppose negative racial hygiene. In his paper, suggestively entitled 'A propos de la stérilisation' (About Sterilization), Dalsace thoroughly criticized the use of sterilization and castration to eliminate the transmission of hereditary diseases. In a language familiar to the Latin eugenicists at the Congress, Dalsace argued instead for the improvement of the environment, corrective education, and the establishment of Centers of Birth-Control and 'Conscious Maternity' as means to prevent the proliferation of criminals and social problems associated with criminality (Dalsace 1936: 706–12).

Dalsace's paper wonderfully exemplified the fundamental opposition of Latin eugenicists to interventionist reproductive policies. His views may have surprised most participants at the Congress, but they undoubtedly echoed wider eugenic developments in the Latin countries across the world. Only a month after the Berlin Congress, the eugenic societies of Latin America met in Mexico City with the purpose of creating an international organization to promote their eugenic vision of social and biological improvement. This new organization was the Latin International Federation of Eugenic Societies.

Latin eugenics *à rebours*

As discussed in the previous chapter, in Latin America biological determinism was tempered by a commitment to neo-Lamarckist and environmentalist eugenics, which allowed for the nation's social and biological renewal through puériculture, biotypology, homiculture, and a host of preventive and public health measures. Unlike Europe, where Latin eugenicists were in the minority, in Latin America this interpretation of eugenics was dominant. Given this, it is perhaps not surprising that the movement to create an international Latin eugenics organization first gained decisive support in Latin America.

In the early 1930s, preparatory meetings were held in Argentina and Mexico to create a Federation of Latin Eugenic Societies (*Federación de Sociedades Latina de Eugenesia*). In November 1934, at the Second Pan-American Congress on Eugenics and Homiculture in Buenos Aires, the Argentinian eugenicist Josué A. Beruti had already announced the creation of such a federation to be based in Rome and to include the Argentinian Association of Biotypology, Eugenics and Social Medicine, the Belgian Society for Eugenics and Preventive Medicine, and the Italian Society of Genetics and Eugenics (Beruti 1934: 169).

The 7th Pan-American Congress of the Child (*VII Congreso Panamericano del Niño*) was organized in Mexico City between 12 and 19 October 1935. A preliminary meeting (*Reunión Preliminar*) of the Latin International Federation of Eugenic Societies (*Federación International Latina de Sociedades de Eugenesia*) was held on this occasion, organized by the president of the Mexican Eugenics Society, Adrián Correa, with the assistance from the eugenic societies of Argentina and Peru, and the approval of eugenic societies in Brazil, Belgium, France, and Italy. One of the chief goals of the new Latin Eugenics Federation was to 'educate those Latin countries that did not have a eugenic society, of the importance of the biological and social studies, particularly with respect to child welfare' (*Memoria del VII Congreso Panamericano del Niño*, vol. 1: 80–82 and vol. 2: 631).

Corrado Gini was elected president of the new organization. As he was not able to attend the Congress, Alfredo A. Saavedra, the secretary of the Mexican Eugenics Society, read his inaugural lecture. It began by noting the enthusiastic support received for the new organization from the Latin countries of Europe, including France, Romania, Spain (Catalonia), Belgium, and Romandie (the French-speaking cantons of western Switzerland). Gini then reiterated the classic traits associated with Latin eugenics, beginning with 'human dignity and personal integrity', and continuing with the Latin countries' 'long tradition of civilization' and their 'more balanced and fair-minded attitude' (Gini 1936: 77–8).

Gini recognized that there were economic and cultural differences among the Latin nations. Some had 'a past superior to the present', while others were 'experiencing a phase of renewal with hopes for a grand future'. Yet they all confronted similar eugenic problems, especially the management of demographic growth and the improvement of their nations' biological wellbeing. In terms of practical 'Latin eugenic commonalities', the first to be mentioned was the rejection of neo-Malthusianism. Most Latin countries were known for their high birth rates, exceeding those of many Nordic and Anglo-Saxon countries. Since demographic growth was essential to the biological renewal of nations, Gini explained, the most populated Latin countries were encouraged to favour emigration to their counterparts with lower birth rates.

Gini then connected migratory patterns to the 'the relationship between quantity and quality' of a people, especially as it affected social equilibrium: this was one of the Latin Federation's 'fundamental eugenic problems' that could be 'objectively studied, in all its complexity' (Gini 1936: 78). The quantitative development of the population also involved questions of race, ethnic hybridism, and cultural assimilation. As was the norm in their culture, the Latin nations could deal tolerantly and dispassionately with such issues, '*sine ire et studio*' – quite unlike (it was implied) the inflexibility provoked in the Anglo-Saxon and Nordic cultures (Gini 1936: 79). Latin eugenicists were not 'blinded by nationalism to the point of believing, against history, that we can speak of a superiority of race, across time and place'. Rather, 'all races were absolutely equal from the point of view of their intellectual attitudes' (Gini 1936: 79).

The issue of how to reach eugenic perfection also involved deep-rooted cultural norms. Indeed, as Gini put it, 'Latin scientists will never forget that the object of their research is not flies, or rabbits, or cows, but men; individuals, that is, who have a personality and rights, and who might be coordinated and subordinated, but not completely sacrificed to the collective interests of society'. Because Latin culture always privileged individual over collective interests, the ancient Romans abolished human sacrifice and slavery. It was 'only natural', therefore, for 'the descendants of Rome' to 'feel reluctant to consider measures which deprive human beings of the most essential attributes of their personality, and which sacrifice one of the most salient manifestations of life' (Gini 1936: 80).

Rejection of neo-Malthusian birth control measures was one of the central 'Latin eugenic commonalities'. Given the nature of Gini's 'regenerative eugenics' – a synthesis of neo-Lamarckism and demography – radical measures such as sterilization, the method preferred in the Anglos-Saxon and Nordic countries to prevent the reproduction of 'defectives' and 'degenerates', had no place in Latin countries. This categorical denunciation of sterilization concluded Gini's inaugural lecture.

With the mission of the International Federation of Latin Eugenic Societies now established, the focus turned to strengthening institutional ties between its constituent members. The first 'constitutive' meeting of the Federation followed two months later, on 18 December 1935, once again in Mexico City. The Italian Society of Genetics and Eugenics; the Argentinian Association of Biotypology, Eugenics and Social Medicine; the Peruvian League of Hygiene and Social Prophylaxis; and the Mexican Society of Eugenics sent delegates, together with representatives from Columbia, Cuba, Costa-Rica, Chile, Guatemala, San Salvador, Nicaragua, Haiti, Uruguay, Honduras, Panama, Paraguay, and Brazil. Romania (Sabin Manuilă, the director of the Demographic, Anthropological and Eugenics Department of the Romanian Social Institute); Romandie (Eugène Pittard of the Anthropological Laboratory in Geneva); and Catalonia (Hermenegild Puig i Sais from the Catalan Eugenics Society) also joined the Federation.

The resolution adopted in Mexico City outlined the following course of action:

> 1. To assist with the creation of eugenic societies in Latin American countries, which will then join the International Federation of Latin Eugenic Societies
> 2. To organize the first Congress of the International Federation of Latin Eugenic Societies in Paris in 1937 on the occasion of the Universal Exhibition
> 3. The Organizing Committee of the first Congress of the International Federation of Latin Eugenic Societies consists of the President of the Federation and Presidents of the member societies; it is recommended that the Committee begin its work immediately
> 4. Copies of this resolution should be sent to all societies represented in the International Federation of Latin Eugenic Societies, as well as to those joining afterwards. ('Informations', *Revue anthropologique* 1936: 190–1)

As the Latin eugenic organization turned its attention to Europe, it cemented the leading role attributed to French medicine and Italian demography, while at the same highlighting the international importance gained by a much smaller Latin country: Romania.

Ex oriente lux?

As discussed in Chapter 3, Romanian eugenics was first institutionalized in Transylvania during the 1920s, alongside developments there in social assistance, preventive medicine, and public health. Romanian eugenicists such as Iuliu Moldovan were concerned as much with improving the sanitary conditions of the population as with preserving the Romanian national character in Transylvania. To this effect, Moldovan promoted an interpretation of national eugenics ('igiena națiunii') tailored to express Romania's ethnic realities and its biological future. Moldovan used the term 'biopolitics' to describe this hybrid national and eugenic programme. The dominant principles underlying Romanian biopolitics during this period were the country's territorial integrity and the doctrine of the homogeneous ethnic nation premised upon racial affiliation. This combination of nationalism with eugenics, and the particularly localized work carried out by Moldovan and his students in Transylvania, may explain why the Biopolitical and Eugenics Section of the 'Astra' Association he had established in Cluj remained virtually unknown outside Romania ('Eugenics in Romania', *Eugenical News* 1936: 84–85).

During the 1930s, two new directions of research into heredity and population health in Romania complemented Moldovan's broad interpretation of national eugenics. The renowned neurologist and endocrinologist Gheorghe Marinescu (see Figure 6.2) represented the first one. During the 1930s, Marinescu articulated his vision of eugenics particularly in relation to his field of research, neuropsychology and endocrinology. 'The nation', he argued, was a 'synthesis: race + culture'. Without denying the importance of 'environmental hygiene and education' for social and biological improvement, Marinescu also endorsed interventionist eugenic policies (Marinescu 1936: 50). Medicine was essential to eugenics. 'All [medical] cases', he noted, 'must be examined from a eugenic viewpoint' (Marinescu 1936: 54). Reiterating many of the observations on the nature of human degeneration proposed by French, German, and American physicians, Marinescu proposed an interpretation of eugenics that not only accounted for the influence of biological factors over hereditary diseases, consanguinity, and the predisposition of 'inferior races' to certain maladies, but also explained the lack of sexual education, poverty, and malnutrition caused by social factors (Marinescu 1936: 62–3). According to Marinescu, the contribution of eugenicists to the improvement of the nation's health needed to be conveyed to a larger audience, not restricted to the medical profession.

Figure 6.2 Gheorghe Marinescu
Source: Library of the Institute of Medicine, Institute of Public Health, Bucharest.

On 7 January 1935, under Marinescu's leadership, the Romanian Royal Society of Eugenics and Heredity (*Societea Regală Română pentru Eugenie și Studiul Eredității*) was established in Bucharest. The demographer Sabin Manuilă was one of the vice-presidents; the social hygienist Gheorghe Banu was elected secretary general. Among its members were prominent Romanian politicians and scientists, including the paediatrician Mihail Manicatide, the psychiatrist Grigore Odobescu, the endocrinologist Constantin Parhon, and the forensic pathologist Nicolae Minovici ('Autorizarea de funcționare, actul constitutiv și statutele Societății Regale Române pentru Eugenie și Studiul Eredității', *Revista de Igienă Socială* 1936: 271–3). The main goal of the new eugenics society was 'to formulate and publicize the principles and the methods by which to select and improve the most robust and better endowed individuals of the Romanian nation, while at the same time ensuring the reproduction of those individuals'. It also endeavoured to pass the legislation needed to achieve this goal ('Autorizarea de funcționare, actul constitutiv și statutele Societății Regale Române pentru Eugenie și Studiul Eredității', *Revista de Igienă Socială* 1936: 273).

Introducing the activities of the Society to the Romanian Academy on 20 May 1935, Marinescu explained that the Society would undertake research into what he termed, following the French school of hereditarian medicine, the 'normal and pathological heredity' of the population, especially the study of the 'superior and normal hereditary personality' (Marinescu 1936: 84). The Society would suggest social hygienic measures to reduce infant mortality, the spread of venereal diseases, and alcoholism. It would promote puériculture, the protection of working mothers, and the improvement of sanitary conditions in rural areas. While Marinescu shared with other Romanian eugenicists this preoccupation with public health and social hygiene, he also advocated more aggressive eugenic methods, such as the introduction of prenuptial health certificates and even the 'sterilization of degenerates, idiots, imbeciles, mentally ill, criminals and so on' (Marinescu 1936: 84).

Sabin Manuilă, the director of the Central Institute of Statistics (*Institutul Central de Statistică*) in Bucharest, was another promoter of eugenic thought in Romania.[1] On 14 March 1935, the Demographic, Anthropological and Eugenics Department (*Secţia de Demografie, Antropologie şi Eugenie*) was established at the prestigious Romanian Social Institute (*Institutul Social Român*), with Manuilă as president and Dumitru C. Georgescu, a statistician and anthropologist, as secretary. The creation of this special Department devoted to eugenics confirmed the official importance afforded to biological research into the health of the Romanian population, which also defined the ideological parameters of the Romanian government's involvement in social and racial politics during the 1940s.

The eugenic research promoted by the new Department was located at the interstices of Moldovan's biopolitical eugenics and Marinescu's constitutional eugenics, with a particular focus on sociology, statistics, and demography. The focus was on the nation's 'social body'; the ultimate aim was to provide a totalizing interpretation of all social, cultural, economic, biological, and political elements that formed the Romanian nation. The new Department aimed to research the 'morphology, the structure, the evolution and the biological value' of the Romanian population. No modern state, it was inferred, could survive without a rational population policy and demographic management (Georgescu 1936: 56–7). Eugenics offered the appropriate conceptual framework for those Romanian statisticians and demographers envisioning a total control of the national body through intimate knowledge of its social functions.

By the late 1930s there were three functioning eugenic societies in Romania, and in 1937, they formed the Union of the Eugenic Societies in Romania

(*Uniunea Societăților de Eugenie*), under the presidency of Gheorghe Marinescu. These societies and departments at various state institutions brought together Romanian eugenicists from different disciplines as well as provided a platform for international collaboration.

In addition to the domestic proliferation of eugenic societies in Latin countries during the early 1930s, Nicola Pende's personal endeavours to promote biotypology, and his capacity to inspire 'disciples', contributed to the growing interest in establishing connections with other Latin eugenicists throughout the world, thus helping to bring about the International Federation of Latin Eugenics Societies. Already in 1931, Marinescu presented at the Romanian Academy a lengthy account of the Italian constitutional school, extolling both its scientific relevance in relation to other branches of biology and medicine, such as endocrinology, but also its positive eugenic ambitions. As Marinescu explained, biotypology prioritizes the health of the individual within a corrective rather than eliminationist tradition of eugenics (Marinescu 1934: 1–34).

In February 1935, Pende himself visited Romania at the invitation of the Italian Cultural Institute, giving lectures at the Royal Foundation 'Carol I' in Bucharest and at the University of Cluj. His first public lecture in Bucharest, on 25 February, dealt with 'Biotypology and the Improvement of Human Constitution' ('Biotipologia şi înbunătățirea constituției umane'), providing a unique opportunity for the general Romanian public to become familiar with his theories, and their practical application to the field of pre- and post-natal orthogenesis, maternity, and the 'biological and political problem of race'. With its focus on the pregnant woman, the child, and the young adult, orthogenesis 'was preferred by the Latin races'; in contrast to negative eugenics, seen as 'anti-biological and anti-human'. Moreover, Pende argued, it was 'utopian to want to preserve a pure race, because all nations consisted of a number of races'. As Latins, the Romanians were encouraged to adopt biotypology, and thus harness 'the values of the nation' to the 'necessities of the state'. Individual and collective improvement was conterminous (Pende 1935b: 67–8).

During his stay in Bucharest, Pende also participated in a meeting of the Romanian Society of Neurology (*Societatea Română de Neurologie*) presided over by Marinescu; visited the Colțea and Colentina Hospitals; and toured the Institute of Legal Medicine, led by Nicolae Minovici. As expected, the proverbial Italian–Romanian cultural and linguistic affinity was often invoked during these meetings. Pende described this special relationship as the 'permanent reality of the Latin genius' ('Visita profesorului Pende', *România medicală* 1935: 68). Although such pronouncements certainly contained an element of hyperbole, the cultural and scientific connections between Romania and Italy at that time were remarkably strong (Iorga 1936).

Pende was equally praised for his achievements in the capital of Transylvania, Cluj. His lecture at the University focused on the 'orthogenetic physical education' (*Universitatea 'Regele Ferdinand I' Cluj: Anuarul* 1934–1935: 456–7). The Cluj-based journal *Endocrinologie, Gynecologie, Obstetrică* (Endocrinology, Gynaecology, Obstetrics) devoted the April 1936 issue to Pende's visit. In his introductory article, the prominent social hygienist and former Minister of Health, Iuliu Haţieganu, praised Pende's role in establishing the centrality of constitutional medicine and endocrinology to individual health, thus creating a 'new discipline, human biotypology, whose overwhelming contribution to the main questions in racial hygiene and eugenics was widely recognized today' (Haţieganu 1936: 193). Aristide Grădinescu, president of the Society of Endocrinology, Gynecology, Obstetrics of Cluj offered, in turn, an overview of Pende's contribution to biotypology and constitutional medicine, and its lasting influence among Romanian physicians (Grădinescu 1936: 196–202). The journal also published Pende's article on 'Syphilis and the Endocrino-Sympathetic System' ('Sifilisul şi sistemul endocrino-simpatic') (Pende 1936: 203–13).

What this meant in practical terms was a substantive transformation of the eugenic repertoires in Romania, to include biotypology, constitutional medicine, and endocrinology (Râmneanţu 1937: 257–61 and 1939). Equally important, it was a highly revealing example of Latin collaboration in medicine in general and in eugenics in particular (Andronescu 1936). The importance of 'Latinity' as a unifying scientific philosophy was further displayed at the 4th Congress of the Latin Medical Press (*4me Congrès de la Presse Médicale Latine)* held in Venice between 29 September and 3 October 1936 (see Figure 6.3).

The Congress brought together physicians from most Latin countries, including France, Belgium, Argentina, Romania, Cuba, Portugal, Spain, and Italy. Davide Giordano, an Italian surgeon and president of the *Fédération de la Presse Médicale Latine* (Federation of Latin Medical Press), served as president of the Congress. The first session reviewed 'the history of medical press in Latin countries', with contributions from Enrique Noguera (Madrid), Maxime Laignel-Lavastine (Paris), Augusto da Silva Carvalho (Lisbon), and Valeriu Bologa (Cluj). The second session discussed 'the social influence of the medical press', while the third concentrated on 'medical education and practice in Latin countries'. Nicola Pende provided the final lecture of the Congress, a fine-tuned argument in favour of the 'Latin-Mediterranean Spirit in Education and the Practice of Medicine'. It elucidated what by then was a familiar and powerful

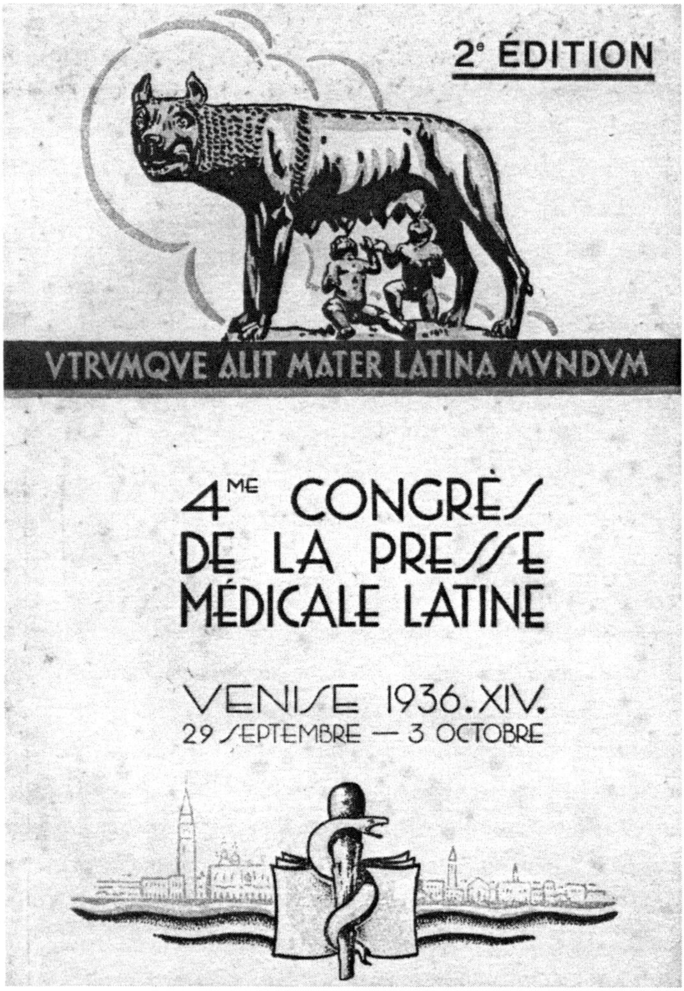

Figure 6.3 The Programme of the 4th Congress of the Latin Medical Press
Source: Marius Turda's personal collection.

trope: that medical thinking in the Latin countries valued the relationship between individual and collective health rather than subsuming the former into the latter, and recognized the dynamic interaction between heredity and the environment (*Programme: 4me Congrès de la Presse Médicale Latine* 1936; '4me Congrès de la Presse Médicale Latine', *Revue Médicale Française* 1936: 480).

By early 1937, Nicola Pende's and Corrado Gini's decade-long efforts to explicitly politicize their vision of eugenics by defining it in terms of Latin cultural norms had borne fruit. Latin eugenics had finally become officially codified and was advocated through its own international organization. It was as a result of international scientific collaboration such as that carried out by Corrado Gini and Nicola Pende during the 1920s and early 1930s, in Latin America and Europe, as well as the creation of the Latin International Federation of Eugenic Societies in Mexico in 1935, that Latin eugenics received its formal international recognition. For almost five years, until 1940, Latin eugenics became the common platform upon which eugenicists from Latin countries articulated and validated their concerns about social and biological improvement, and guided their assumptions in population and medical research, as well as anthropology and sociology. Four consecutive international congresses – three of them held in Paris and one in Bucharest in 1937 – allowed for the short-lived but meaningful existence of a formalized international Latin eugenics movement.

Latin eugenics in action

Some of the themes of this mature phase of Latin eugenics were apparent in the First International Meeting of Biotypology (*Réunion Internationale de Biotypologie*), organized in Paris on 24 July 1937 by the French Society of Biotypology. The topic of the meeting was 'the study of human personality from the point of view of biotypology'. Nicola Pende and Giacinto Viola attended, as did Gheorghe Marinescu ('Comptes rendus de la Réunion Internationale de Biotypologie', *Biotypologie* 1934: 73–5). As expected, Pende outlined the importance of biotypology, particularly in relation to eugenics and biology (Pende 1937: 76–84), while Viola discussed the constitutional evaluation of the individual (Viola 1937: 85–92).

The importance of biotypology to French demographers and eugenicists was confirmed just five days later, on 29 July, when the International Congress on Population (*Congrès International de la Population*) began in Paris. It was organized under the auspices of the French presidency, together with the Ministries of National Education and Public Health. The Congress was presided over by Adolphe Landry, the prominent champion of Latinity and pronatalism in France. Vice-presidents of the Congress included the Catholic historian Édouard Jordan; an Italian statistician, Livio Livi; and a number of prominent German demographers and eugenicists, including Friedrich Burgdörfer, Eugen Fischer, and Ernst Rüdin.

The Congress reflected Latin demographers' preoccupations with quantitative population policies. Sections were devoted to 'the general population theory'; 'historical demography'; 'statistics'; and 'special research' on birth-, marriage-, and mortality rates, natural fertility, and so on. Emphasis throughout was placed on education, the environment, and, of course, positive measures to encourage population growth. As would be expected, German eugenicists (Otmar von Verschuer, Karl Thum, and Ernst Rüdin) discussed qualitative population policy in terms of hereditary diseases, genetics, racial pathology, and negative racial hygiene. They were in a minority, however. Authoritative counter-arguments came from the German-American anthropologist Franz Boas; two Jewish anthropologists from Prague, Ignaz Zollschan and Maximilian Beck (Zollschan 1938, vol. 8: 93–105; Beck 1938, vol. 8: 106–16); and even Frederick Osborn, the leader of 'moderate' eugenicists in the United States (Osborn 1938, vol. 8: 117–22). Boas presented a paper on 'Heredity and Environment' (Boas 1938, vol. 8: 83–92), in which he criticized 'the incredible amount of amateurish work, produced for more than a century, but particularly by modern race enthusiasts' (Boas 1938, vol. 8: 83). By positing the importance of cultural diffusionism, social interaction, and adaptation to the environment, Boaz provided a powerful endorsement of the critique of biological determinism and racial difference. His commitment to the belief that culture was determined by the environment rather than genetics was eloquently expressed in his closing remarks: 'The existence of a unity of bodily build in even the most homogeneous population known to us can be disapproved, and the existence of a cultural personality embracing a whole "race" is at best a poetic and dangerous fiction' (Boas 1938, vol. 8: 92).

Boas' commentary was revealing. French, Belgian, and Romanian participants at the Congress shared his anxiety with the growing tide of racism in Europe during the 1930s. Their critique of racial determinism reinforced similar arguments made by Latin eugenicists: that culture and environment, not innate racial qualities, led to human improvement. In this sense, race was a questionable analytical category. Human society was highly differentiated and constantly evolving, and could not be reduced to simple racial stereotypes.

The French and Romanian papers dealing with biometry and biotypology also expressed this nuanced and multi-layered conception of social and biological improvement. The French anthropologist Eugène Schreider described a 'biotypological mission' under the supervision of the French ethnologist, Paul Rivet, to study an indigenous Mexican tribe, the Otomi people (Schreider 1938, vol. 8: 8–13). Petru Râmneanțu, a Romanian anthropologist and eugenicist from Cluj, used Giacinto Viola's and Nicola Pende's methodologies to study the relationship between biotype and fertility, providing a complete inventory of medical, social, and biological causes affecting the life of the peasant population in two Romanian villages in the Banat (Râmneanțu 1938, vol. 8: 14–20).

Similarly, the French social hygienist Grégoire Ichok discussed the application of biotypology to the assessment of physical aptitudes in the industrial professions, the regulation of which was seen as a prelude to a wider rational organization of social life and reproductive biology (Ichok 1938, vol. 8: 33–7).

Finally, on 1 August 1937, the French psychologist and biotypologist Dagmar Weinberg gave the Congress's closing plenary lecture on biometry and biotypology (Weinberg 1938, vol. 8: 237–48). Weinberg reiterated the scientific importance of classifying individuals and human groups according to biometry, defined as 'the quantitative study of characteristics that differentiate individuals or human groups', and biotypology, which visualized 'those differences as a total unit of analysis and comparison' (Weinberg 1938, vol. 8: 238). To reaffirm the biotypological and environmentalist interpretation of human improvement at such an important international event was to assume, again, that the racial imperative invoked by the German racial hygienists was not acceptable to Latin eugenicists as a justification for negative eugenics and social selection.

As the Congress on Population was drawing to an end, another one was just beginning: The First Congress of Latin Eugenics (*Ier Congrès Latin d'Eugénique*). The Congress of Latin Eugenics took place at the Faculty of Medicine of the University of Paris, from 1 to 3 August 1937. It was organized under the patronage of the French Ministry of Public Health. Louis Martin, president of the International Institute of Anthropology, and Gustave Roussy, dean of the Faculty of Medicine, were honorary presidents. The organizing committee included Georges Schreiber, Henri Vignes, and Georges Heuyer. The Latin International Federation was now composed of 22 representatives from 11 countries (see Table 6.1).

The outgoing president of the Federation, Corrado Gini, gave the opening address. Gini praised the objectivity, moderation, and humanity of the eugenic sciences of the Latin countries, the outcome of 'their most ancient civilization' (Gini 1937a: 5–6). Next to follow was the new Federation president, Eugène Apert, who spoke on 'the social importance of eugenic studies'. Apert also stressed the eugenic significance of morality, and claimed eugenicists were an enlightened elite whose role was to actively pursue solutions to their nations' health problems. Instead of relying exclusively on heredity and genetics, as was the tendency among Anglo-Saxon and Nordic eugenicists, Latin eugenicists promoted eugenic improvement through preventive medicine, social hygiene, public health, and education ('L'importance sociale des études eugéniques') (Apert 1937: 7–12). Finally, Gheorghe Marinescu welcomed the participants. He highlighted the common medical tradition of the Latin countries, in

Table 6.1 The Latin International Federation of Eugenic Societies (1935–39)

Representatives	Eugenic society	Country
Mariano R. Castex Arturo R. Rossi	Argentinian Association of Biotypology, Eugenics and Social Medicine	Argentina
Norbert Ensch Albert Govaerts		Belgium
Renato Kehl	Central Brazilian Commission on Eugenics	Brazil
Hermenegild Puig i Sais Josep A. Vandellós	Catalonian Eugenics Society	Catalonia
Eugène Apert Raymond Turpin A. Brousseau	Eugenic Section of the International Institute of Anthropology	France
Corrado Gini Agostino Gemelli M. Saibante	Italian Society of Genetics and Eugenics	Italy
Adrián Correa Alfredo Saavedra	Mexican Society of Eugenics	Mexico
Carlos A. Bambarén Ursula Ch. de Schmitt	Peruvian National League of Hygiene and Social Prophylaxis	Peru
Almerindo Lessa		Portugal
Gheorghe Marinescu Sabin Manuilă	Federation of Romanian Eugenic Societies	Romania
Eugène Pittard Hersch Liebmann		Romandie

Source: Ier Congrès Latin d'Eugénique: Rapport 1937: 381–3

particular 'Pasteur's doctrine of hygiene' (Marinescu 1937a: 13–4). In forging such a creative synthesis of history, biology, and medicine, these three leaders of the Latin eugenic movement considered this Congress a unique opportunity to discuss the 'vital problems' preoccupying the Latin nations.

The first cluster of such problems was grouped under the heading 'Miscegenation and immigration' ('Métissage et immigration'). Here René Martial, a French physician and anthropologist, chastised recent racial mixing and uncontrolled immigration into France, which he claimed was weakening the racial vitally of the French nation (Martial 1937: 16–35). The Italian statistician Dino Camavitto presented a report of the research expedition conducted by the Italian Committee for the Study of Population Problems in Mexico in 1933 (Camavitto 1937: 40–59). Étienne Letard, a professor of veterinary medicine and agronomy, viewed the issue of racial miscegenation from his own area of expertise in animal husbandry (Letard 1937: 61–8). During the discussions that followed, the 'racial problem' highlighted opposing views, as Martial attempted to define the 'racial question' within a conceptual framework that others, such as Marinescu, found deeply flawed.

The second panel of the Congress compared the 'qualitative and quantitative growth of population' ('La valeur comparée des accroissements qualitatifs et quantitatifs d'une population'). Papers were given by the French statistician Raul Husson; the Brazilian eugenicist Renato Kehl; the Italian statistician Paulo Fortunati; the French eugenicist Georges Schreiber; and the zoologist Philippe L'Hèritier. These speakers recognized that the relationship between qualitative and quantitative measures to ensure population growth constituted one of the defining features not only of eugenics in general but of demographic movements in certain countries (such France and Italy) in particular. As part of their strategy to improve the nation's health, eugenicists like Kehl defined their work as 'political statistics, based on social and eugenic measures to protect and improve the national population' (Kehl 1937b: 78).

Demographic growth and the related topics of marriage rates, fecundity, and 'social metabolism' was also discussed in the third section, comprised exclusively by Italian statisticians and demographers: Vincenzo Castrilli, Giorgina Levi della Vida, and Cesare Padovan. In turn, the fourth section was devoted to dissertations on means to cure various maladies of the germ plasm ('Les maladies du plasma germinatif et leur guérison') and attracted contributions from French, Italian, and Romanian participants: Edmond-Alexandre Lesné, Christian Champy, Raymond Turpin, H. Rogier, Alexandru Caratzali, Oddo Casagrandi, Gustave Roussy, René Huguenin, and A. Brousseau.

The Latin eugenicists' predisposition for neo-Lamarckism did not exclude research on genetics. One paper, in particular, needs to be mentioned here: a study on the aetiology of 'mongolism' (later termed 'Down Syndrome') based on 104 cases. It was the result of the collaboration between French and Romanian geneticists Raymond Turpin and Alexandru Caratzali, assisted by H. Rogier, in the paediatric unit of the Armand-Trousseau Hospital in Paris (Turpin, Caratzali, and Rogier 1937a: 154–63). The authors offered a multi-factorial causation for 'mongolism': (a) 'the substitution of a single gene', whose expression was 'polymorphic, influenced by environmental conditions'; (b) 'a Mendelian combination of several genes', particularly with respect to twins; (c) and finally 'an anomalous chromosome' (Turpin, Caratzali, and Rogier 1937a: 158). It was important to consider, they concluded, the influence of both heredity and the environment in causing the malady, especially the age of the mother at the moment of conception (Turpin, Caratzali, and Rogier 1937a: 163).

The fifth section was devoted to the relationship between 'constitutional type and eugenics' ('Type Constitutionnel et Eugénique'), with Corrado Gini as the first speaker (Gini 1937b: 200–11). At the outset, Gini questioned the conceptual clarity attached to the word 'type'. In statistics, he noted, it was used in a number of contexts, with various meanings, to indicate normal (or Gaussian) distribution, but also the 'dominant type' of a character of personality (Gini 1937b: 201). Such critical remarks, however, were not meant to dissuade the support enjoyed by biotypology among Latin eugenicists, but to spell out the 'usefulness of the statistical studies of constitutions' (Gini 1937b: 204).

Next, Gini addressed the relationship between biotypology (he preferred the term 'Science des Constitutions') and eugenics, beginning with the thorny issue of the nature of individual, constitutional differences. Some were due to heredity, he claimed, others due to the environment. This was common knowledge but worth repeating, Gini believed, as it influenced the 'inter- and intra-racial character of individual differences'. According to biotypologists, the interaction between heredity and the environment was central to the creation of constitutional types, but did not favour one 'hereditary patrimony' over another. While some might describe this attitude as scientific eclecticism, Gini called it 'agnostic', deploring the fact that most biotypologists 'studied constitutional forms without being preoccupied with their inter- and intra-racial origin' (Gini 1937b. 206–7). This statement, and the discussion that followed Gini's paper claiming the need for more scientific accuracy for biotypology (Gini 1937b: 211–2), reaffirmed the conceptual importance of race for eugenics so far as he was concerned. This was clearly in disagreement with other participants' profession of the universality of human nature.

After Gini, the French physiologist Raoul Husson and Alfred Thooris, a scientific consultant for the French Athletics Federation, gave papers. The final two papers in this section were delivered by Marcello Boldrini, an Italian statistician and fellow of the Pontifical Academy of Sciences, and by the French psychiatrists, Georges Heuyer and Andrée Courthial. They dealt with 'Constitution and Eugenics' ('Constitution et Eugénique') (Boldrini 1937: 228–31; Heuyer and Courthial 1937: 232–8). In his paper, Boldrini distinguished between 'passive' eugenics, which was concerned with collecting information about the 'social influence on genetic characters and phenomena, such as the [human] constitution and constitutional characters', and 'active' eugenics, which sought 'the improvement of future generations thorough the use of [their] genetic qualities' (Boldrini 1937: 230). 'Active eugenics' relied heavily on the practical application of theories of heredity to society, which Boldrini considered to be untimely.

More importantly, Boldrini viewed any 'programme of active eugenics as impossible and dangerous', since there was still too insufficient an understanding of genetics to attempt to 'improve humanity by manipulating the qualities of the species'. Instead, he recommended that eugenicists and biotypologists draw on the 'sciences that influenced the improvement or impairment of individual qualities' for centuries: namely, hygiene, medicine, and education (Boldrini 1937: 231). In essence, Boldrini believed that Latin eugenicists could attain social and biological improvement without the need for genetic engineering and racial selection. The rejection of 'active measures' of eugenics, such as sterilization, was also emphasized by the Polish-French psychiatrist Franziska (Françoise) Minkowska, who gave a paper on 'Eugenics and Genealogy' ('Eugénique et généalogie') (Minkowska 1937: 341–50).

The extensive professional collaboration between the Latin scientists was one distinguishing features of this Congress. The work on 'mongolism' by Turpin and Caratzali has already been mentioned. They presented another four papers together: one with M. Gorny, detailing the research they conducted on 1100 families to determine the influence of the parents' age and health on the offspring as well as on fertility rates (Turpin, Caratzali, and Gorny 1937b: 240–61); one with Nicolae Georgescu-Roegen, deputy director of the Romanian Institute of Statistics, on the influence of maternal age on stillbirths (Turpin, Caratzali, and Georgescu-Roegen 1937c: 271–7); and, finally, two papers on stillbirth in twins (Turpin and Caratzali 1937d: 283–5 and 1937e: 286–90). In the same section, Gini's close collaborator Nora Federici presented her research on 'numerous Italian families' (Federici 1937: 278–82).

The last section of the Congress dealt with the relationship between eugenics and demography, pathology and pedagogy. It included three Romanian eugenicists: Mihail Manicatide, Gheorghe Banu, and Gheorghe Marinescu. Manicatide discussed infant mortality in Romania (Manicatide 1937: 292–5); and Marinescu examined the relationship between gigantism and acromegaly from the point of view of histology and genetics (Marinescu 1937b: 351–68).

Banu offered what was arguably the Romanian delegation's most important contribution to the Congress, both in terms of theory and practical solutions. Banu's paper, entitled 'The Dysgenic Factors in Romania. Principles of a Practical Eugenic programme' ('Les facteurs dysgénique en Roumanie. Principes d'un programme pratique d'eugénique'), revealed a complex understanding of eugenics that owed much to the French and Belgian eugenic ideals of the 1920s (Banu 1937: 296–319). Banu believed that eugenics was particularly relevant to Romania, due to its widespread medical problems, which included venereal and contagious diseases and alcoholism, as well as its social and economic problems. Nevertheless, Romanians were endowed with healthy racial qualities: they had 'high fertility rates, the capacity to adapt and assimilate, innate strength, resistance to adverse circumstances', and so on. As such, Banu argued that 'a rational eugenic programme' was needed to protect Romanians from disease and poverty, as well as to assure the biological future of their nation (Banu 1937: 297).

The establishment of synergies between preventive medicine, social policies of health, and the biological screening of the population was central to Banu's eugenic programme. Reflecting this, the main principles around which Banu constructed his eugenic programme were the 'introduction of prenuptial medical certificates'; 'a scientific study of heredity'; 'the organization of school hygiene and the introduction of the medical card for each individual'; 'the fight and propaganda against venereal diseases'; the 'reorganization of the maternal and infant assistance'; and finally 'the partial introduction of preventive sterilization' (Banu 1937: 309–10). He then presented each principle in detail (Banu 1937: 310–9). Like other Latin eugenicists, Banu understood social and biological improvement as a synthesis of social hygiene, puériculture, and biotypology on the one hand, and practical measures to improve the quality of the population, on the other hand.

The Italian pedagogue Giacomo Tauro gave the final paper at the conference. Tauro believed that eugenicists' 'main task' was to educate the nation. He summed up the message of the Congress in distinctly neo-Lamarckian terms: 'educational activity constitutes, in our opinion, the best means to attain eugenic goals' and would 'stop the decay of individuals, of nations and of races, and [thus] favour their continual betterment and progress' (Tauro 1937b: 379–80).

In addition to these presentations, a eugenic exhibition was set up in one of the rooms of the Congress, consisting of the usual charts of pedigrees and inheritable traits, genealogical maps, illustrations of various hereditary issues, and so on. The Congress also sponsored a visit to the School of Puériculture (*École de Puériculture*) in Paris, led by Adolphe Pinard's successor, Benjamin Weill-Hallé ('Première Réunion de la Fédération des Sociétés Latines d'Eugénique', *Revue Anthropologique* 1938: 83–5).

Reviewing the Congress for the journal *Revue Anthropologique*, Henri Briand considered that the Latin Federation's 'efforts towards positive eugenics and to oppose the Anglo-Saxon negative eugenics were successful' (Briand 1938b: 309). The president of the Congress, Eugène Apert, was commended for his 'victory' in organizing an event that was 'remarkable for the quality of its participants', including Corrado Gini and Gheorghe Marinescu (Briand 1938b: 309). Furthermore, Briand noted, French 'national disposition' was against 'eugenic practices devised by the Anglo-Saxons'. It was possible, however, to study 'heredity and demographic phenomena' from vantage points other than biological and racial determinism. In France and elsewhere in the Latin world, 'the methods of positive eugenics' were no longer just 'good-natured remarks' and 'philosophical digressions' but were deemed of 'paramount practical importance' (Briand 1938b: 310).

By fusing historical and cultural commonalities with medicine, biology, genetics, demography, and biotypology, this first congress on Latin eugenics powerfully illustrates that eugenics was a discipline and a movement from which the Latin countries could theoretically and practically benefit. As is apparent, Latin eugenicists were also engaged in 'cutting-edge' genetic research. Equally important, the Congress legitimized the international status of Latin eugenics, and of its message of morality and ethics. In this respect, and in tune with the previous congresses on biotypology and population, the Latin Eugenics Congress endeavoured to provide a scientific narrative of human improvement that was distinguished from, and opposed to, ideas of Nordic superiority. Although dispersed geographically, Latin eugenicists worked on a common corpus of topics, including biotypology, demography, racial morphology, social psychology, as well as eugenics and genetics. That they were also unified by their efforts to demonstrate that their Latinity was the foundation of their common scientific agenda and internationalism became clear at the *XVIIe Congrès International d'Anthropologie et d'Archéologie Préhistorique* (17th Congress of Anthropology and Prehistoric Archaeology) organized in Bucharest between 1 and 8 September 1937.

The official patronage of the Romanian king, Carol II, coupled with extensive media coverage, brought renewed attention to the importance of Latin science. Romanian anthropologists, demographers, physicians and eugenicists, and their Latin colleagues used the Congress as a platform to announce to the international community their achievements in such diverse disciplines as palaeontology, archaeology, folklore, serology, and eugenics, as well as to strengthen the institutional network between Latin nations.

On this occasion Constantin I. Angelescu, the Romanian Minister of National Education, was chosen as president. The Organizing Committee included the most important Romanian anthropologists, demographers, geographers, and eugenicists at the time, such as Ion G. Botez, Sabin Manuilă, Simion Mehedinţi, Iuliu Moldovan, Victor Papilian, Francisc Rainer, and Nicolae Minovici, together with Louis Martin from France, Eugène Pittard from Switzerland, and Louis Vervaeck from Belgium (*XVIIe Congrès* 1939: 13–4). Other officials included the anthropologist and a founding member of the French Society of Biotypology, Georges-Paul Boncour, the philosopher Henri Berr, and the Spanish ambassador to Romania, Manuel López Rey (see Figure 6.4). It is worth noting that there were no German participants at the Congress.

The third section of the Congress, devoted to heredity, eugenics, and selection ('Hérédite-Eugénique-Sélection'), was presided over by the Dutch psychiatrist and eugenicist Gerrit Pieter Frets, and included the participation of the British medical geneticist, Lionel S. Penrose, who spoke on 'Maternal Age and Congenital Disease', with a special reference to Down Syndrome (Penrose 1939: 600–5).

With the exception of Iordache Făcăoaru, all Romanian eugenicists who contributed to this section belonged to the Romanian Royal Society of Eugenics and Heredity and to the Demographic, Anthropological and Eugenics Department of the Romanian Social Institute.[2] Some of them, like Gheorghe

Figure 6.4 Participants at the 17th Congress of Anthropology and Prehistoric Archaeology in Bucharest, 4 September 1937. Manuel López Rey (third from the left), Georges Paul-Boncour (third from the right) and Henri Berr (second from the right) are seated on the first row.
Source: *În amintirea profesorului Fr. J. Rainer (1874–1944)*, Bucharest, 1945, p. 16.

Marinescu, Nicolae Georgescu-Roegen, and Alexandru Caratzali, presented papers similar to those they had delivered at the Latin Congress in Paris. Gheorghe Banu, however, submitted three papers, expanding on the arguments in favour of a practical eugenic programme for Romania he had outlined in Paris (Banu 1939a: 615–20; 1939b: 620–6 and 1939c: 635–40).

Reflecting the cross-generational relationship within the eugenic community in Romania, this section featured the research of a new generation of Romanian eugenicists, educated in Bucharest under the leadership of Gheorghe Marinescu, Sabin Manuilă, and the notable Romanian anatomist and anthropologist Francisc Rainer. Maria Veştemeanu discussed the influence of fertility and early marriages on demographic growth (Veştemeanu 1939: 587–90); Dumitru C. Georgescu also dealt with differential fertility in Romania (Georgescu 1939: 640–67). Veturia Manuilă and Maria Domilescu examined the eugenic and demographic consequences of illegitimate marriages and children (Manuilă and Domilescu 1939: 683–94). Eugenia Petrovici analysed the relationship between heredity, syphilis, and mental disorders (Petrovici 1939: 605–14). She especially highlighted the growing importance of endocrinology, particularly in relation to reproduction physiology and biotypology, expressing her conviction that it could contribute decisively to the control of pathological heredity, particularly with respect to mental diseases.

There was an important debate on eugenic sterilization at the Congress. Petrovici argued for the need to introduce eugenic sterilization in Romania as the most efficient method to prevent the further degeneration of the race (Petrovici 1939: 614). Her senior colleagues, Gheorghe Banu and Iordache Făcăoaru, concurred. Such views were expressed in connection to arguments in favour of state intervention in Romania in order to accelerate the application of eugenic ideas of social and biological improvement. As Făcăoaru put it, 'It is the state that should ensure that biological responsibilities and the eugenic ideal are achieved' (Făcăoaru 1939a: 634).

Ion Vasilescu-Bucium, president of the juridical section of the Royal Romanian Society for Heredity and Eugenics in Craiova (the capital of Oltenia), decried the lack of eugenic legislation, such as sterilization, in Romania (Vasilescu-Bucium 1939: 678–81). All civilized countries, Vasilescu-Bucium believed, 'pursued the same eugenic goal – to preserve the value and the strength of their race'. Accordingly, many of them had introduced 'prophylactic measures to protect the moral and physical health of their social collective' (Vasilescu-Bucium 1939: 679). In Romania, Vasilescu-Bucium noted, the eugenic transformation of society was in infancy. It was only with the newly introduced Penal Code

(17 March 1936) – 'whose eugenic leanings were inspired, if somewhat timidly, by the principles [outlined] by Galton's science' – that for the first time in Romania contributing to 'the degeneration of the race' was considered a criminal offence (Article 446). The Penal Code also criminalized the marriage of a healthy person to someone suffering from a venereal or contagious disease, and condemned early marriages. Article 484 permitted eugenic abortion as a preventive measure when one of the parents suffered from insanity and 'there was certainty that the offspring will suffer from severe mental defects' (Vasilescu-Bucium 1939: 679–80). In conclusion, Vasilescu-Bucium hoped that other 'eugenic desiderata' such as sterilization would soon be achieved in Romania, not least due the Royal Romanian Society of Eugenics and Heredity's growing popularity and its lobbying efforts (Vasilescu-Bucium 1939: 681).

Vasilescu-Bucium's paper was succeeded by a lively discussion between those in favour of voluntary sterilization and those insisting on the need for compulsory sterilization. Marinescu reaffirmed what he had said in his 1935 article on eugenics, agreeing with voluntary sterilization in the case of hereditarily transmitted diseases, but rejecting compulsory sterilization for 'social and religious reasons' (Vasilescu-Bucium 1939: 681). His former student, the psychiatrist Gheorghe Stroescu, opposed this view. Stroescu argued that 'compulsory sterilization is the only way of preventing the reproduction of those with hereditary illnesses and of improving the race' (Vasilescu-Bucium 1939: 682). Basing his arguments on his medical experience in France and Germany, and contrasting the eugenic practices in these two countries, Stroescu preferred the latter:

> While voluntary sterilization can be applied to intelligent patients, it cannot be carried out on the mentally ill and imbeciles. In our case, compulsory sterilization is the only means to prevent the continual increase of the feeble-minded, especially in isolated villages in the mountains. Voluntary sterilization proved inefficient in countries where it has been applied. (Vasilescu-Bucium 1939: 682)

Făcăoaru also maintained that 'voluntary sterilization was ineffective'. He therefore suggested that, in order to reconcile the two perspectives, the following motion should be submitted to the participants: 'The third Section presided over by Mr. Frets, having declared that voluntary sterilization produced no effect in the countries where it has been applied, proposes that eugenic sterilization be made obligatory and coercive'. Marinescu found this formulation too drastic. Instead, he suggested a resolution which stated that 'The third Section presided over by Mr Frets proposes that eugenic sterilization is applied with prudence and only with the consent of the patient or his family' (Vasilescu-Bucium 1939: 682). Ultimately, Marinescu's position on voluntary sterilization prevailed.

Although all participants agreed that eugenic sterilization in some form was essential to any national programme of biological improvement, the moderate view triumphed. Despite Stroescu's and Făcăoaru's escalating rhetoric, positive measures to improve the nation's health continued to dominate the thinking of the leading Romanian eugenicists. Notably, Marinescu did not specify which category of individuals would be subjected to voluntary sterilization, much less the legal and medical reasons required to permit it.

Făcăoaru's lengthy historiographic discussion of the work produced by Romanian authors during the period between 1920 and 1936 was the last paper included in this section on eugenics. It was and remains the first substantial overview of Romanian publications related to eugenics, heredity, and biopolitics (Făcăoaru 1939b: 718–99). In addition to discussing the most important Romanian contributions to eugenics, Făcăoaru also put forward his own ideas of the biological improvement of the Romanian nation. These included education and educational reform – for instance, he wanted to introduce the study of human genetics, eugenics, anthropology, ethnography, and demography into the school and university curricula. Eventually, the state and society would experience a biopolitical transformation according to the principles of positive and negative eugenics (Făcăoaru 1939b: 788–90). Echoing eugenic developments elsewhere, particularly in Nazi Germany, Făcăoaru proclaimed the primacy of the 'ethnic hereditary patrimony of the nation' over the interests of the individual (Făcăoaru 1939b: 791). The Romanian nation, according to Făcăoaru, had to become the object of eugenic and biopolitical knowledge. This was a view that other Romanian eugenicists shared. But in accordance with the main tenets of Latin eugenics, the Romanian attitude remained predominantly oriented towards positive measures to improve the health of the population, alongside child and maternal welfare, and family protection.

By 1937, the need for the state's eugenic transformation, reflecting a characteristic mixture of national emergency with confidence in the general applicability of eugenics to society amassed considerable authority in most Latin countries. The lecture delivered by the anthropologist Eusébio Tamagnini at the inaugural meeting of the Portuguese Society for Eugenic Studies (*Sociedade Portuguesa de Estudos Eugénicos*) in Coimbra on 9 December 1937, is another important example. Tamagnini's eugenic programme fused positive environmental measures aimed at improving the health of the Portuguese families with an emphasis on education. Nonetheless, his vision of eugenic health and welfare did not preclude the endorsement of more interventionist measures such as the interdiction to marry for individuals afflicted by hereditary diseases or the ban on consanguineous marriages. Such efforts to control reproduction and thus improve the racial quality of the nation were conscious attempts to translate into practice the eugenic management of the population without resorting to negative methods such as sterilization (Tamagnini 1938: 554–9).

Like his Romanian counterparts, Tamagnini wanted to reverse what was perceived to be the racial degeneration of his nation, but his approach to eugenics emphasized improvement and protection of the health of the population rather than elimination of its 'defective' members. In endorsing this view, Portuguese eugenics, like other national eugenic movements in the Latin countries, had another distinctive advantage over Anglo-Saxon eugenics (with its emphasis on birth control and segregation) and German racial hygiene (with its predilection for racial superiority and hereditary selection): it accorded with a humanitarian and Christian insistence that true human improvement was possible only once eugenics became the product of culture and civilization rather than of state-orchestrated genetic engineering.

***

The Second Congress of the International Federation of Latin Eugenic Societies was planned to take place in Bucharest between 25 and 30 September 1939, organized by the Union of the Eugenic Societies in Romania. The endocrinologist Constantin I. Parhon was elected the third (and last) president of the Federation (*see* Table 6.2), while the geneticist Gheorghe K. Constantinescu was appointed secretary general of the Congress ('Al II-lea Congres al Federaţiunii Societăţilor Latine de Eugenie', *Revista de Ştiinţe Medicale* 1939: 247–8).

The themes proposed for the Congress included the prevention of hereditary diseases; natalism and demographic growth; heredity and infectious diseases;

Table 6.2 Presidents of the Latin International Federation of Eugenic Societies (1935–39)

President	Country	Period
Corrado Gini	Italy	1935–37
Eugène Apert	France	1937–38
Constantin I. Parhon	Romania	1939

heredity and intelligence; and heredity and endocrinology ('Al II-lea Congres Internațional al Federațiunii Societăților Latine de Eugenie' *Revista de Științe Medicale* 1939: 247–8). The paper Gini had prepared for the Congress, for instance, dealt with the Albanian colonies in three villages in the province of Calabria (Gini 1941: 75). But the German invasion of Poland on 1 September 1939, and the subsequent outbreak of the Second World War, prevented the Latin eugenicists from meeting again. Eventually, the Congress was 'postponed indefinitely' (SA/EUG/E.12: 1932–1940).

Although there were no international eugenic congresses after 1940, the collaboration between Latin eugenicists in Europe continued during the War. For instance, Arturo Rossi, Renato Kehl, Corrado Gini and Gheorghe Banu were among the Latin eugenicists listed as honorary members of the Mexican Eugenics Society for the Betterment of the Race in 1944 (López-Guazo 2005: 265). Further examples include Gregorio Marañón's participation in the First Congress of Endocrinology (*Premier Congrès d'Endocrinologie*) organized in Bucharest in June 1939; Corrado Gini's lecture to the Faculty of Medicine at the University of Brussels, also in June 1939, at the invitation of the Belgian Medical Society of Physical Education and Sports (*Société Médicale Belge d'Education Physique et des Sports*); and Corrado Gini's and Fabio Frassetto's participation in the National Congress of Population Science (*Congresso Nacional de Ciências da População*) held in Oporto in 1940. Both Frassetto's paper on biotypology and Gini's paper on denatality 'continued to develop what was a formative principle of Latin eugenics – namely, that "hyperfecundity" was a positive force for the race and that "prolificy" was a product of a superior racial constitution' (Quine 2010: 319).

Eugenicists in Latin America also strengthened their collaboration during the 1940s. Most notably, in 1942, the Argentinian Social Museum (*Museo Social Argentino*) established a Commission for the Scientific Study of Population (*Comisión para los Estudios Científicos de la Población*), headed by Nicolás Besio Moreno, a civil engineer and statistician. The Commission included most prominent Latin American eugenicists at the time, and former members of the Latin Eugenics Federation: José Vandellós (Venezuela); Jorge de Romaña and Susana Solano (Peru); Alfredo M. Saavedra (Mexico); Roberto Berro, Juan Pou Orfila, and Augusto Turenne (Uruguay), César Jacome (Ecuador); José Albuquerque and Renato Kehl (Brazil); and finally, Francisco Ealker Linares (Santiago de Chile). Like the Latin Eugenics Federation before it, the Commission was concerned with improving the 'quantity' and 'quality' of the population; immigration and colonization; urban and rural depopulation; the protection of women and children; public health and social prophylaxis; ethnological and anthropological research; and 'vital statistics' relating to birth rates, marriage, morbidity, and mortality (Miranda 2005: 215).

Reflecting the political changes in Europe and the specter of another world war, Latin eugenics gradually became more aggressive during the late 1930s and early 1940s, immersed as it was in an increasingly militarized and divisive world. As we shall see in the next chapter, many Latin eugenicists were drifting towards nationalist and ethnic radicalism, and increasingly voiced support of national racial biology and racial protectionism.

7

Latin Eugenics and Scientific Racism

Like their counterparts elsewhere, eugenicists in the Latin countries proposed social and biological theories of human improvement that characterized, classified, and ultimately utilized racial taxonomies. Stefan Kühl has suggested that one 'useful way to distinguish between strands in the eugenics movement is to emphasize their differing conceptions of race improvement. All eugenicists', he argued, 'held the idea that it was possible to distinguish between inferior and superior elements of society, but not all traced inferiority directly to an ethnic basis' (Kühl 1994: 70). As was so common at the time, most Latin eugenicists promoted a world view that included human races as a defining element: for them, the white races were at the apex of their racial hierarchies. They regarded miscegenation between the white and non-white races as a form of 'racial pollution' that would in some way harm the ostensibly superior white race. Yet, largely because of their own national histories, intellectuals from Latin countries also tended to downplay biological differences *between* the white races, often ascribing white ethnic variation to cultural, linguistic, religious, and spiritual differences, rather than rigid biological ancestries. Beginning in the late 1930s and into the Second World War, biological determinism gained ascendency in Latin eugenics, due in no small part to the influence of Nazi Germany. The palpable tension between the racial 'tolerance' that characterized most of the history of Latin eugenics, and the racial chauvinism of its last phase, was a significant factor in the disintegration of the international Latin eugenics movement by the end of the Second World War.

Races, immigration, and miscegenation

While often deploring miscegenation between whites and non-whites, Latin eugenicists praised inter-European racial mixing whenever it served their purposes. For example, Mussolini attempted to use the argument that racial mixing between various Mediterranean nations (Italians, Spanish, and French) was eugenically beneficial as one element in his campaign to increase Italian influence in the 'Latin world'. Indicative of this, in 1926 the Fascist Italian deputy Carlo Barduzzi announced his country's desire to strengthen ties with Spain and South America, to counter the 'German, Slavic, and Moslem block'. In terms of racial unity, there were a number of reasons why such an alliance made sense. Spain, Barduzzi exclaimed, had undergone a racial 'awakening' similar to that in Italy; there was a 'great ethnic affinity' between the Italian and Spanish races. This now included South America, given the millions of Italian settlers there. He expected that France would naturally wish to join such an Italian–Spanish–Latin American alliance. Barduzzi predicted that 'The union of Latin people will carry to triumph the Roman idea, which will minimize the danger of a new European conflagration' ('Fascist Asks Latin Opposition', *New York Times* 1926: 25).

In 1938, Nicola Pende attributed the unique superiority of Italians to the benefits of miscegenation. He valued the 'matrimonial eugenics' resulting from the ethnic fusion that had occurred in the Italian peninsula since Roman times, claiming that Italians 'must value the principle *Italiani con Italiani*, in order to preserve and further improve the pure civilized characteristics of the progeny of Rome and the different ethnic components that in one sense or another have made a contribution of indisputable value to our supremacy' (quoted in Cassata 2011: 211–12).

To some extent, these examples reinforce Victoria de Grazia's argument that 'behind Italian fascist policy, there was a very different conception of population engineering' than in Nazi Germany, and that 'among Italian eugenicists, the FrenchAustrian Benedict Morel, with his emphasis on genetic mixing as a source of strength, was more popular than the "German" Mendel, who emphasized selection for pure strains' (De Grazia 1992: 53). However, very similar attitudes prevailed when Latin eugenicists considered relations between the white and the non-white races. While racial borders were blurred between Latin nations, they became visibly enhanced when applied to Europeans mixing with other races, especially those from Africa. Debates about immigration and miscegenation, coupled fears of social and biological degeneracy, informed the Latin eugenic vocabulary during the 1920s. In March 1919, only several months after the end of the First World War, the Italian Society for Genetics and Eugenics adopted a resolution proposed by Vicenzo Giuffrida-Ruggeri demanding a ban on marriages between Europeans and Africans. It 'would be fitting', Giuffrida-Ruggeri believed, 'that the different eugenic societies endeavour to obtain from their respective governments legislative enactments where they do not yet exist, tending to prohibit marriages between Europeans and representatives of African races with the exception of Mediterranean races (Berbers, Egyptians) and Arabs of the white race' ('The Italian Society of Genetics and Eugenics', *American Journal of Sociology* 1920: 638). Enthusiastic about the proposal, Corrado Gini advocated its adoption by the Eugenics Education Society in an August 1919 letter to Leonard Darwin. However, Darwin considered the proposal too premature and politically controversial. Undaunted, the Italian Society for Genetics and Eugenics continued to campaign for similar measures during the first five years of its existence (Cassata 2011: 71).

At the same time, French eugenicists also debated their views on race, basing their conclusions on 'eugenic principles'. Considering that the French Army had deployed almost half a million non-white colonial soldiers (*troupes indigènes*) in Europe during the War (Fogarty 2008), French eugenicists fretted over issues such as marriages between the 'white' and 'black' races. For example, during the 1920s Eugène Apert was preoccupied with the effects of racial mixing on the 'health of the French nation'. Apert operated within an established European racial tradition that construed 'whiteness as a category which could be fragmented and internally hierarchized' (Camiscioli 2009: 87). While Apert acknowledged that racial fusion between various populations was an historical reality, he nevertheless insisted that post-war circumstances in France and elsewhere in Europe required heightened racial protection against further degeneration, coupled with the eugenic elimination of 'undesirable elements'. While mixed marriages between the Latins (*inter alia*, Spaniards, Italians, and Belgians) were beneficial, *métissage* between the French and the Africans was 'racially detrimental' (Apert 1923: 1565–9; 1924: 149–67).

Apert was also influenced by a racialized eugenic discourse, most notably represented in France by Charles Richet, the esteemed Nobel laureate for physiology and vice-president of the French Eugenics Society. In his book *La sélection humaine*, written in 1912 but only published in 1919, Richet proposed a radical eugenic programme of human breeding geared towards the creation of a superior Western European race. He condemned the 'confusion of ethnic types' ('la confusion des types ethniques') characterizing Europe in general and France in particular (Richet 1919: 28). Not surprisingly, then, Richet wished to prevent the 'mixing between superior and inferior humans' (Richet 1919: 58). Indicative of the tendency of Latin eugenicists to ascribe racial difference to factors other than biological determinism, Richet claimed that the superiority of the Europeans over the Africans and Asians was not a matter of *nature* but of *nurture* and *culture*. A 'superior race', Richet argued, was 'an intelligent race' (Richet 1919: 64–5). In light of this, Richet developed a racial hierarchy placing:

> *[T]he black race* – incapable of thinking and innovating and unable to constitute themselves as a nation – at the bottom of the hierarchy of the human races; then, and very far above, *the yellow race* – not very inventive or creative, but brave, hard working, and capable of assimilating quickly; finally, above these two races, *the white race* which has accomplished the most in the world, has created an intelligent society, has invented thousands of industries, has used natural resources and animals at will, and has conquered through sciences the entire planet. (Richet 1919: 78. Emphasis in the original)

Such racial taxonomies were certainly not new, having characterized European discourses on race since the Enlightenment. Richet employed these taxonomies to define France's eugenic priorities. 'The first principle of human selection', he declared, 'was to officially prohibit the marriage between whites and women of another race, yellow or black' (Richet 1919: 84). However, racial exclusion and the eugenic control of marriage between different races was only the preliminary phase of a much more comprehensive process of human selection. The future of the race was to be determined by biological, social, and cultural boundaries separating those who belonged to the community from foreigners and outsiders who were viewed as aliens or potential enemies. Richet also advocated a complementary system of eugenic cleansing of the French race itself, according to which those members of society deemed 'degenerate', 'abnormals', 'unhealthy', and 'diseased' were to be separated from the 'healthy' majority (Richet 1919: 163–8). As Richet envisioned it, human selection was predicated on guarding the racial patrimony of the nation from further contamination and decay.

Richet's book reflected wider anxieties over European hegemony and associated post-World War I debates about the racial constitution of the French, Portuguese, Italian, and Romanian nations. For instance, in his 1919 *Raça e Nacionalidade* (Race and Nationality), António Mendes Correia (see Figure 7.1) asserted the existence of a racial nucleus for the Portuguese nation, which survived unaltered throughout time, notwithstanding repeated migrations in and out of the Iberian peninsula. The 'Portuguese race' had built a world-wide colonial empire, and was the source of the biological and cultural prosperity of the nation (Mendes Correia 1919). Mendes Correia believed that these achievements of the 'Portuguese race' were the fruitful products of a fusion of hereditary and environmental traits. All European cultures, he maintained, were equally justified in their colonial projects because of their superior hereditary

Figure 7.1 António Mendes Correia
Source: Patrícia Ferraz de Matos, *Mendes Correia e a Escola de Antropologia do Porto*, Lisbon 2012, p. xxxiii.

qualities and their cultural competencies and practices, which indigenous colonial communities could not match (Mendes-Correia 1933: 142–8).

During the 1930s and 1940s, such interpretations of the Portuguese imperial mission echoed broader eugenic concerns about miscegenation, coupled with António Salazar's doctrine of a political and cultural renewal of the *Estado Novo* (De Matos 2010: 89–111). Thus, for Gonçalves Pereira, a professor at the Technical University (*Universidade Técnica*) in Lisbon, 'The Portuguese transformed nomadic races into sedentary ones, warlike and anarchic populations into peaceful, hardworking peoples. They saved thousands of human beings [and] showed themselves always superior to differences of race, caste or religion' (quoted in De Matos 2013: 37).

This new vision of race, eugenics, and empire was widely discussed at the First Congress on Colonial Anthropology (*Congresso Nacional de Antropologia Colonial*) in 1934 and at the National Congress of Population Science (*Congresso Nacional de Ciências da População*), organized in the summer of 1940 within the impressive 'Exposition of the Portuguese World' (*Exposição do Mundo*

Português). It also permeated Mendes Correia's 1943 *Races of Empire* (*Raças do Império*), which highlighted the racial diversity of the Portuguese empire and the transatlantic bonds within it (De Matos 2012: 253–62).

The renewed anxieties about racial purity and racial survival often led to the proposal of more radical eugenic measures. An example is provided by Enrico Morselli's 1923 book *L'Uccisione Pietosa (L'Eutanasía) in rapporto alla Medicina, alla Morale ed all'Eugenica* (*Compassionate Killing – Euthanasia – in Relation to Medicine, Morality and Eugenics*). Like Richet, Morselli not only condemned racial mixing between white and black races, but also recommended the strongest possible measures – including euthanasia – to prevent the 'reproduction of those degenerates' (Morselli 1923: 24). The white races were 'superior' to the blacks, as they exhibited 'intelligence and the spirit of inventiveness' together with 'social solidarity, a sense of individual duty, the consciousness of high morals and social work, [and] the formation of an intellectual aristocracy solely dedicated to the development of science, art, and religion' (Morselli 1923: 78).

Immigration issues were a natural corollary to these concerns regarding racial homogeneity. This was especially obvious at the International Conference on Emigration and Immigration (*Conferenza internazionale dell'emigrazione e dell'immigrazione*), held in Rome between 15 and 31 May 1924 under Mussolini's patronage. The Conference participants asserted the importance of eugenics in providing guidelines for social and biological improvement, and demanded the introduction of racial screening and selective immigration policies. French participants advocated medical measures to protect the nation's racial health from unwanted immigration; the Italians proposed that citizenship be defined by racial lineage (Aliano 2008: 66–7; Camiscioli 2009: 89).

Two years later, in October 1926, the French Society of Public Medicine (*Société de Médicine Publique*) devoted its Thirteenth Annual Congress, held at the Pasteur Institute, to the topic of immigration. In an alarmist tone, physicians Georges Dequidt and Georges Forestier warned their colleagues not to 'forget that the first waves of Orientals and Slavs that are breaking on France presage the invading flood which threatens to submerge that which is left of our civilization and the health of our race' (quoted in Schneider 1990: 236).

Mussolini held views regarding miscegenation that were consistent with those of Morselli and, especially, Gini. In 1928, Mussolini contrasted 'the reproductive vigor and demographic expansion of the Yellow and Black races' to that of the 'White races'. In cases in which these races come into contact, 'one of which was young, vital, fecund, and the other old, overly mature, near senility' their 'destiny' seemed clear: the younger race had the right to replace the older (Mussolini 1928: 13). This argument proved especially useful whenever Mussolini felt the need for theoretical support for his demographic policies. For instance, at the onset of the 1940s, Italian racial discourse did not hesitate to claim that the young, vigorous Italian race had supplanted the decrepit French race in southern France – with the obvious implication that, from a scientific-demographic perspective, this region ought to be turned over to Italy (Landra 1938: 8–10; Businco 1939: 15–17; Interlandi 1939: 7).

After the Italian conquest of Ethiopia, Mussolini criticized the cohabitation of Italian soldiers with native women. He moved swiftly to ban interracial marriages, in 1937. Two years later, another law condemned long-term sexual relations between Italians and Africans as detrimental to the Italian race. Mussolini believed that his imperial ambitions could only be realized if Italians acquired a new 'colonial consciousness' attendant on acknowledgement of their superior racial status, 'to eradicate the "plague" of miscegenation and to loathe its most "monstrous" product: the *meticciato*' (Sòrgoni 2003: 419). This racialization of national belonging gained rhetorical and political power in the late 1930s, although as we have seen it had already been a concern of Italian eugenicists for decades.

Similar developments can be observed in other Latin countries. In Spain, *Generalísimo* Francisco Franco led a revolt between 1936 and 1939 to overthrow the Spanish Republic. He united behind him diverse political forces such as the fascists (the Falange); conservative Catholics; ultra-nationalists; and militarists. However, Franco only succeeded in launching a viable military challenge to the Republic by accepting substantial military aid from Mussolini, who had renewed hope that Spain would fall under his influence. After three years of horrific civil war, the remnants of the Republican forces surrendered or fled into exile. Franco set up an authoritarian Catholic-Nationalist dictatorship that ruled Spain for the next thirty-seven years. Franco's regime stood for the iron-manacled unity of the nation based on the revival of an aggressive, expansionist Spanish race that had given the world the Golden Age of the conquistadors. Francoists were determined that Spain would once again reclaim its place as a great nation. Such a profound rejuvenation required the guiding hand of the 'Caudillo' ('Leader') assisted by the Roman Catholic Church (Peláez 2005: 87–114).

From the outset, the new regime was committed to wiping out 'degenerate modernism', embodied in birth control, abortion, communism, and feminism. As in fascist Italy and Nazi Germany, significant efforts were directed at educating the youth according to a new morality and eugenic culture that glorified the virility and vitality of the Spanish race (Boalick 2013: 32–40). Equally important, motherhood was described as both a biological and religious experience (Morcillo 2010: 47–9). After the alleged family estrangement promoted by the Second Republic, Spanish women were now reintegrated into their natural state as the 'angeles del hogar' (angels of the home), and entrusted with the domestic management of the family.

Such eugenic rhetoric coloured Franco's demographic and social policies during the 1940s. Already in July 1938, nine months before completely gaining control over the country, Franco began introducing legislation designed to raise the birth rate, aiming for forty million Spaniards within a few decades (never one to be outdone, Mussolini set his own goal for Italians at sixty million). Many of these new eugenic measures were borrowed from neighbouring countries. The Spanish state centralized and made obligatory the previously voluntary family allowances of employers. The government increased the severity of punishments for abortion. The murder of an innocent foetus, according to the regime, was a more serious offence than the murder of a sinful person; based this justification, abortion was decreed to be a crime against the state (Fernández-Ruiz 1939; Aznar 1943: 97–110).

Like Mussolini in Italy and Salazar in Portugal, Franco promoted Spain's former imperial *grandeza*, coupled with policies designed to encourage population growth and the protection of the nation's biological patrimony. The attempt to realize Franco's goals borrowed the language and ideology of racial hygiene (Nash 1992: 746). This is well demonstrated by the ideas of Antonio Vallejo-Nágera, one of the main racial ideologues of the new political regime. His 'racial hygiene' was 'peculiarly Catholic and authoritarian' (Richards 2012: 197), permeated more by 'spirituality and morality than hereditarianism'. Broadly speaking, Vallejo-Nágera's conservative racial hygiene represented a major departure from the anarchist, leftist and liberal eugenic discourses voiced in Spain during the 1920s and early 1930s, which had supported contraception and in some cases even abortion. In contrast to the general understanding of eugenics, which Vallejo-Nágera claimed was too limited in scope and supposedly tainted by liberal and communist ideology, the new Spanish racial hygiene embraced the physical and spiritual totality of the nation. His 1934 book, *Higiene de la raza* (Hygiene of the Race), reflected the growing popularity of scientific racism among Latin eugenicists. Similar to the Italians and the Germans, the Spaniards should 'demand the recuperation of [their] spiritual and racial valor' (Vallejo-Nágera 1934: 1–2).

Nevertheless, conservative maternalism, natalism and puériculture remained the central tenets of eugenics in Franco's Spain. In his 1938 book, *Política racial del Nuevo Estado* (*Racial Politics of the New State*), Vallejo-Nágera directly connected Spain's eugenic vitality to natalism, the protection of the family, and a heightened sense of religious duty. 'The ideal citizen in the New State', he maintained, 'will be married and a prolific parent' (Vallejo-Nágera 1938: 55). As Franco's new national project unfolded, race was coupled with newly recast eugenic prescriptions for maintaining social control over the body politic. The influence of Nazi Germany's aggressive programme of eugenic engineering during the late 1930s supported the growing importance of race and scientific racism in Spain and elsewhere in Latin Europe.

German racial hygiene in Latin Europe

During the late 1930s and early 1940s, German racial hygiene had a small but determined following in the Latin countries. In France, after Georges Vacher de Lapouge's death in 1936, the social hygienist René Martial and the Swiss-born anthropologist Georges Montandon promoted German racial hygiene and a racialised version of eugenics (on the latter, see Conklin 2013: 91–9).

Martial (see Figure 7.2) was a respected public health reformer, having served as director of the Bureau of Hygiene in the Parisian suburb of Alfortville in the 1920s; he was also a member of the French Academy of Medicine (Schneider 1990: 232). Martial was particularly interested in the anthropo-biology of race and immigration, subjects he taught at the Institute of Hygiene of the Medical Faculty in Paris (Martial 1931 and 1938). As mentioned in the previous chapter, Martial was also one of the participants in the First Congress of Latin Eugenics in Paris, where he discussed the relationship between *métissage* and immigration. Unlike Apert and Richet, however, Martial was interested not only in the effects of racial mixing between white and black races, but also in the migration into France from other 'white' European countries such as Poland and Russia. Martial did not condemn mixed marriages between the French and the Poles or Russians; these only proved the elastic nature of 'cross-breeding' and its racial adaptability (Martial 1935a: 83–92).

By the mid-1930s, Martial was recognized by French eugenicists as an authority on immigration, having published not only in specialized journals like *Revue Anthropologique* (Martial 1933: 352–69 and 447–67), but also in popular magazines such as *Mercure de France* (Martial 1935b: 267–94). Here, he summarized his interpretation of immigration as a two-pronged political programme: one external, the other internal. The foundation of the external

Figure 7.2 Réne Martial
Source: Bibliothèque de l'Académie nationale de Médecine, Paris.

policy was provided by 'the trilogy: history, biology, [and] psychology'. In terms of practical measures, this policy presupposed an interrelated and multi-layered internal eugenic selection, directed at 'the race, the nation, the family, [and] the individual' (Martial 1935b: 294).

To Vacher de Lapouge's arguments about racial immutability and the incompatibility between superior and inferior races, Martial added his own ideas derived from social hygiene, eugenics, serology and genetics (Larbiou 2005: 98–120). He thus argued that, while the French race had throughout history absorbed other races, its vitality and fecundity were increasingly under threat. In 1928 Martial proposed the introduction in France of a policy of 'interracial grafting' ('greffe interraciale'), a theme he fully developed in his 1931 book *Traité de l'immigration et de la greffe inter-raciale* (*Treatise on Immigration and Interracial Grafting*) and again in his 1943 book, *Notre race et ses aïeux* (*Our

Race and Her Ancestors). 'Interracial grafting' was 'a seductive metaphor that incorporated Martial's previous concern with assimilation (the ability of the graft to take hold), but it also placed a great deal more emphasis on the *selection* of the graft than just a simple concern for its health' (Schneider 1990: 241). The French were stronger because they were a mixed race, Martial claimed. However, without careful eugenic planning, uncontrolled immigration would undermine the race's biological strength and eventually jeopardise its future. The aim was not to achieve racial purity, as in Nazi Germany, but to promote appropriate and beneficial racial crossings – *le bon métissage*, in Martial's words (Martial 1934).

As noted in the previous chapters, after 1933 a minority of French eugenicists looked favourably upon the Nazi racial programme. The anthropologist Georges Montandon was certainly one of the most prolific in this respect (Conklin 2013: 179–86). He was also a notable inspiration for Guido Landra and other Italian racial scientists (Cassata 2011: 129). In contrast, a reaction against the presence of Nazi sympathizers among the French anthropologists, Germany's growing militarism, and the rise of Latin eugenics as an international movement seeking to oppose German racial hygiene contributed to the creation of the group *Races et Racisme* in Paris. It included historians Georges Lefebvre and Edmond Joachim Vermeil, the philosopher Célestin Bouglé, the legal scholar Louis Le Fur and the ethnographer Paul Rivet. The group also published a journal with the same name, which was 'scholarly in its anthropology, anti-Nazi in its ideology, and intensely patriotic' (Schneider 1994: 112). Leading French biotypologists such as Henri Laugier and Eugène Schreider contributed to this new journal, reaffirming the resilience of Latin eugenics and biotypology (Conklin 2013: 164–5).

Some Portuguese eugenicists were also attracted to German racial hygiene, particularly the physician José Ayres de Azevedo (Castanheira 2010). As with many other eugenicists in the late 1930s and 1940s, Azevedo was especially interested in racial anthropology and blood-group research. He believed that serology could offer more accurate means to ascertain the effects of racial mixing between European and non-European races. In two papers, presented at the 1940 National Congress of Population Science, Azevedo reasserted the civilizing, colonial mission of the Portuguese race (Azevedo 1940a: 61–75). He viewed racial miscegenation through a two-pronged analysis: one internal (classification and differentiation) and the other external (delineating relations to other races). He argued that research into the Portuguese blood groups proved that centuries of colonial expansion notwithstanding, they in fact, had retained their dominant European racial characteristics (Azevedo 1940b: 551–64). Like other Latin eugenicists attracted to German eugenics and racial science, Azevedo visited Germany during the early 1940s, and worked at various institutes, including the Institute for Race Biology and Race Hygiene in Frankfurt and the Kaiser Wilhelm Institute for Human Heredity Sciences and Eugenics in Berlin (Cleminson 2014: 156–61).

In Italy, Mussolini's rather sudden decision to embrace Aryan racism in 1938 had more direct consequences for the Latin eugenic movement. Some sixteen years after he first came to power, Mussolini found himself increasingly in Hitler's shadow as the leader of a European fascist revolution. The reason, he concluded, was that Italians had remained stubbornly immune to fascist culture; he had good reason to doubt that they would embrace a course of militaristic expansionism similar to that of Germany. Hence, Mussolini sought to 'brutalize' Italians by awakening their 'racial consciousness', a prerequisite for their new role as conquerors.

By the summer of 1938, Mussolini seemed convinced that the source of German discipline, sense of purpose, and military might lay in the glorification of their Nordic race, and hatred of the Jewish 'fifth column' that threatened it. Simply put, the Duce resolved that what had worked for Germany ought to work just as well for Italy. He would command the Italian intellectual establishment to reconstruct Italian racial identity in a manner that encompassed a pride in their 'Aryan' racial heritage, as well as a rejection of Italy's own (very small) Jewish population (Israel and Nastasi 1998; De Donno 2006: 394–412; Livingstone 2014: 22–74).

To advance his new racial ideology, Mussolini made use of a younger generation of Italian racial scientists who admired Nazi Germany, such as Giulio Cogni, Guido Landra, Leone Franzì, and Lidio Cipriani. Mussolini told them to convince the public that the 'Aryan heritage' of Italy was the source of Italian greatness. Even the Führer was willing to play along. 'From the cultural point of view', Hitler once commented, 'we are more closely linked with the Italians than with any other people. The art of Northern Italy is something we have in common with them: nothing but pure Germans. The objectionable Italian type is found only in the South, and not everywhere even there'; as evidence, Hitler expressed his admiration of 'lovely girls from the Campagna' (*Hitler's Table Talk* 2000: 268–9).

To announce this new racial paradigm, Mussolini ordered the publication of a statement supposedly endorsed by reputable Italian scientists affirming the existence of human races; it claimed that race was defined by biology and heredity; that the Italian people were of Aryan origin; that a pure 'Italian race' existed, based on 'the ancient purity of blood' rather than any cultural and linguistic heritage; that the Jews did not belong to the Italian race; and that interracial mixing between Italians and non-European races was biologically detrimental (for a full translation of the Manifesto see Gillette 2001: 318–20). This seminal document, the *Manifesto della Razza* (Racial Manifesto, also known as the Manifesto of Racial Scientists), appeared in the 14 July 1938 edition of the semi-official newspaper *Giornale d'Italia* (Israel 2007: 103–8). Notably, among the original signatories of the Manifesto, only Franco Savorgnan was a member of the Italian Society of Genetics and Eugenics.

The new doctrine was institutionalized through the creation of a propaganda bureau, the Office for Racial Study and Propaganda (*Ufficio studi e propaganda sulla razza*) of the Ministry of Popular Culture (*Ministero della Cultura Popolare*). The new Office was entrusted with establishing a library for racial studies; assembling a racial photographic archive; preparing articles and conferences on Italian racism; founding a racist journal, *La Difesa della Razza*; establishing contact with other nations on racial matters; and producing a film on race (Gillette 2002a: 361). Shortly after this, on 5 September 1938, the Central Office of Demography (*Ufficio Centrale Demografico*) was transformed into the General Directorate of Demography and Race (*Direzione Generale per la Demografia e la Razza*), followed in November by the introduction of 'measures to protect the Italian race' ('Provvedimenti per la difesa della razza italiana'). Like Nazi Germany, Fascist Italy, too, was gradually becoming a 'racial state' (Da Grazia 2000: 219–54; Cassata 2008: 56–103).

The Duce appointed Guido Landra, a young researcher from the Institute of Anthropology of the University of Rome, as director of the Racial Study Office. Several years before, at the beginning of his career, Landra endorsed the interpretation of the Mediterranean race put forward by Giuseppe Sergi, Italy's most famous anthropologist of the era. However, by the late 1930s, Landra became a devotee of Nordic racism. In a document he prepared for Mussolini in April 1938, 'Le questioni della razza in Italia nel 1938' ('The Question of Race in Italy in 1938'), Landra asserted that 'the new Italian race' was 'born from the long fusion of the three principal races that have populated the peninsula for thousands of years: the Mediterranean race, the Alpine race, and the Adriatic race, with the successive addition of Nordic blood'. Nevertheless, he viewed the Italian race as 'essentially Mediterranean', since 'in Italy the Mediterranean race has an ethnic predominance over the others, whatever physical type they have' (quoted in Gillette 2001: 310).

Eugen Fischer's visit to Italy in the spring of 1938, followed by Rudolf Frercks' (the vice-director of the Office of Racial Politics of the Nazi Party, *Rassenpolitisches Amt der NSDAP*) visit to Rome in October 1938, strengthened Landra's Nordic predispositions (Cassata 2008: 200). As director of the Italian Racial Office, Landra enthusiastically cultivated his connections with German racial scientists, culminating in the 'Italo-German Committee on Racial Questions' ('Comitato Italo-Tedesco per le questioni razziali') that met in Germany between 13 and 21 December that year ('Ministero dell Cultura Popolare, Gabinetto, Ufficio Razza, Guido Landra 1938'). The purpose of the meeting was 'to commence a preliminary examination of the opportune means to avoid in the respective racial propaganda those arguments that could harm the amicable spirit between the two nations' (quoted in Gillette 2002a: 365). Much like José Ayres de Azevedo, Landra also visited various racial hygiene institutes in Germany, and met the leading German eugenicists, anthropologists, and politicians – including Hitler, who awarded the iron cross second class to Landra for his efforts in promoting Italian and German cooperation on racial issues (Cassata 2008: 200–1).

In addition to his connections with German eugenicists, Landra was probably selected as director of the Racial Study Office because of widespread opposition to Mussolini's new racial directives by Italy's most important eugenicists: Nicola Pende, Corrado Gini, and Agostino Gemelli. These prominent scientists were sufficiently ensconced in their own power bases to cause Mussolini difficulties otherwise avoided by placing a relative unknown anthropologist such as Landra in charge of Italian racial propaganda. Landra was also easily replaceable. On 15 February 1939, he was removed from the direction of the Racial Office and replaced by Sabato Visco, director of the Institute of General Physiology (*Istituto di fisiologia generale*) at the University of Rome (Cassata 2008: 62).

The new Italian racial movement was also given its own journal, *La Difesa della Razza* (The Defence of the Race), founded in August 1938 and edited by Telesio Interlandi, a radical fascist editor (Michaelis 1998: 217–40; Cassata 2008: 5–55). Landra quickly became the real force behind the journal. It gave Landra the opportunity to expound on the many intersections between German racial hygiene and Italian racism. In one article, Landra expressed his hope that Italian eugenics would be directed to 'maximally stimulate the most gifted elements of the Italian people from a racial point of view'; to create favourable conditions for the development of the 'great mass of average elements'; and 'finally to diminish by eugenic methods, such as sterilization and castration, the grey mass of weak and asocial elements until they disappear' (Landra 1942a: 11).

Other contributors to *La Difesa della Razza*, such as Giuseppe Lucidi, Lidio Cipriani, Marcello Ricci, and Telasio Interlandi, advocated a similar interpretation of eugenics based on hereditarianism, racial anthropology, and the rejection of neo-Lamarckism (Cassata 2011: 233–45). For example, an October 1938 article by Marcello Ricci, an assistant professor of anthropology at the University of Rome, warned of the social and economic dangers that societies faced when allowing defective individuals to reproduce over the generations. Ricci also looked forward to the 'true improvement of the race' that would result from the 'effective diminution of defective genetic traits' by the 'application of opportune measures tending towards the elimination of the reproductive activities in individuals dangerous to the race' (Ricci 1938: 31).

In fact, *La Difesa della Razza* endeavoured 'to demonize and dismantle the neo-Lamarckian basis of Italian eugenics' (Cassata 2008: 197–8). In one instance, the journalist Elio Gasteiner published an article in the 20 August 1938 edition of *Difesa della Razza* rejecting the importance of the environment for the improvement of the race. He argued that the 'immense work for the physical education of the youth has no effect on quality or on a desired hereditary racial improvement. For the individual, there will certainly be constitutional advantages, but these improvements are paratypic; that is, they are not hereditary and therefore cannot change the race' (quoted in Cassata 2008: 227).

However, some contributors to the journal were not so ready to dismiss neo-Lamarckism, which after all had long been integral to their understanding of eugenics. In one case Renato Semizzi, professor of social medicine at the Universities of Padua and Trieste, argued that 'improvement of the environment' was indeed eugenically beneficial; he recommended that the state provide 'for the prophylaxis and the correction of the disabled in the fight against all social illnesses' (quoted in Cassata 2008: 232). *La Difesa della Razza* also published contributors from German, French, Romanian, Hungarian, and Greek eugenicists, such as Eugen Fischer, Georges Montandon, Nicolae Minovici, Vintilă Horia, Zoltán Bosnyák, István Milotay, and Ioannis Koumaris.

This propaganda campaign, designed to rapidly inculcate the Italian people with this new racial consciousness, also had important domestic and international repercussions, prompting a radicalization of the eugenic discourse in other Latin countries ('Acțiunea Italiei pentru apărarea rasei', *Buletin Eugenic și Biopolitic* 1938: 254–55; Vornica 1941: 179–84). Such views did not go unchallenged. For instance, Pope Pius XI denounced this apparent attempt to 'borrow racism from the Germans and deny Italy's Roman heritage'. Even King Vittorio Emmanuele II wondered about the wisdom of Mussolini's *volta face* (D'Aroma 1957: 275).

The advocates of Aryan racism in Italy faced three serious obstacles which they would never be able to fully overcome: the opposition of the vast majority of Italy's traditional medical and scientific community, which continued to support Latin eugenics to the extent possible during the last years of Italy's fascist regime; the Catholic Church's determined anti-sterilization doctrine, along with a certain distaste for excessive anti-Semitism; and the indifference, if not outright hostility, of large segments of the Italian population. Prominent Italian eugenicists such as Corrado Gini, Agostino Gemelli, and Nicola Pende would not allow themselves to be pried away from Italian intellectuals' centuries-long pride in their 'Mediterranean' Roman ancestors.

Attempting to counter this hostile reaction, Mussolini ordered some of Italy's most respected biologists and anthropologists to endorse the Manifesto. Nicola Pende was among them. He was called in unawares on 25 July 1938 by the Minister of Popular Culture, Dino Alfieri, and along with a cluster of his colleagues told in no uncertain terms that the Duce ordered him to sign the Manifesto. Pende was outraged. In a derogatory remark obviously aimed at Landra, Pende responded angrily: 'We cannot endorse the great stupidities written by youngsters whom we ourselves made the mistake of graduating one or two years ago!' (quoted in Toscano 1996: 892–3). Nevertheless, the newspapers announced Pende's adherence the next day. Pende would always maintain that he never actually signed the document.

For several months thereafter, Pende pleaded with Mussolini to allow him to publish a retraction of his supposed adherence to the Manifesto, and a reaffirmation of his earlier work endorsing the Latin identity of Italians. The Duce granted him permission to continue expressing his opinions on race in a limited number of cases. In October 1938 Pende published an article, 'La terra, la donna e la razza' ('Land, Woman, and the Race') in the official fascist journal *Gerarchia* (*Hierarchy*) in which he attacked Interlandi and the Italian Aryanists. Pende claimed that it was genetically impossible for Italians to be Aryans. The 'soil', he said, was an essential factor in the 'biological-spiritual characteristics of the race', which took millennia to develop. Italians were the 'children' of Italy, a land drenched by an 'abundant Mediterranean sun'. The new 'Nordic Aryan' model for Italy proposed by racial eugenicists and anthropologists was absurd: 'To want therefore to fix a racial or physical or especially psychological ideal type for the Italian people to strive for, which does not have its natural habitat in Italy, but in Scandinavia or Scotland for example, is as illogical as hoping that blacks raised for centuries in Italy will turn white!' (Pende 1938: 663–9).

Pende's endorsement of biotypology, orthogenesis, and Latin eugenics continued unabated, notwithstanding the radically different conception of the Italian race that was now official dogma. In another 1938 text, Pende again reiterated the Roman and Latin origins of the Italians:

> The biological Italic type, which has many original racial elements, in the course of its history, is physically and psychologically nothing less than the *progeny of Rome*, because it is mother Rome that for millennia knew how to assimilate and amalgamate peoples of European races that were morphologically and psychologically different, in order to form a *romano-italico type*, that persists from the time of Roman Italy, and that has an ethnic profile in a biological sense, which cannot be confused with other national types, even in the great sphere of Latin families. (Quoted in Knudsen 2010: 10)

Even though Pende insisted on retaining the traditional view of the Italian race and Latin eugenics, he did not fall out of Mussolini's favour, as did Gini in the early 1930s. Quite the contrary: in 1938, Mussolini finally approved the construction of the Institute of Orthogenesis and Racial Improvement (*Istituto di Ortogenesi e Bonifica della Stirpe*) in Rome, a project pursued by Pende since 1934. Pende reciprocated by proposing the organization of an imposing exhibition dedicated to Latin eugenics and fascist orthogenesis ('Mostra dell'Ortogenesi Fascista della Stirpe') for the Universal Exhibition planned to be held in Rome in 1942 ('Documenti inerenti l'Esposizione universale di Roma E42'. 24 June 1938). Moreover, in April 1939, Pende became a member of the Commission of National Education and Popular Culture (*Commissione dell'educazione nazionale e della cultura popolare*).

Pende's scientific prestige did not preclude Interlandi from attacking him in the pages of his newspaper, *Il Tevere*. In response, Pende complained to Mussolini that Interlandi had viciously 'covered my name with contempt based on overwhelmingly exaggerated falsifications and lies about my racial ideas' (Pende to Mussolini, 18 October 1938). The Duce enforced a truce: Interlandi refrained from attacking Pende, while Pende substantially toned down his complaints about the new racial order in Italy.

Regardless of the Duce's admonitions, Pende continued his relentless attack against the radicals now officially in control of Italian eugenics. In his 1939 book *La scienza dell'ortogenesi* (*The Science of Orthogenesis*), Pende criticized 'the infamous eugenics of certain eugenicists who believe that race can be improved or purified' by 'surgically sterilizing individuals of both sexes who have hereditarily transmittable illnesses' (quoted in Cassata 2011: 212). Once again, Pende asserted the 'moral, scientific and social value of the Italian science of orthogenesis'. What was needed, Pende argued, was not a 'utopia of creating better descendents through crossings with distant races or of selecting the fittest generators and excluding the unfit for the improvement of the race', but the 'scientific control' of the individual throughout his or her life (quoted in Cassata 2011: 213). Through the management of individual health, the science of orthogenesis provided a eugenic foundation encompassing heredity and environment within an ideal of the race and the nation. Even after Italy had joined Germany in fighting the war, in 1940, Pende openly criticized Nordic racism, denouncing the fantasies of 'racially pure utopias' prevalent in the 'Nordic' countries, such as the United States and Germany. In fact, the 'Nordics' were a disharmonious race, he alleged, set in a conflict between opposing forces (Pende 1940: 570).

Corrado Gini also objected to the Aryanization of Italian racial discourse, and consequently also had to endure his opponents' scorn. Giovanni Preziosi, a well-known fascist and one of Italy's foremost anti-Semites, chastised Gini for his participation in the 'infamous and antiracist eugenic congress in Paris', and described the Latin International Federation of Eugenics as an instrument in 'the hands of Jews and Masons' (quoted in Cassata 2011: 186). Preziosi's attacks did not intimate Gini, however. He moved ahead with the organization of the Third Congress of the Italian Society of Genetics and Eugenics (*Terza riunione della Società Italiana di Genetica ed Eugenica*) in Bologna between 5 and 7 September 1938 (*Genus* 1939: 1–371). At the Congress Gini took the occasion to announce the forthcoming Second Congress of Latin Eugenics to be held in Bucharest in September 1939 (Cassata 2011: 186). In fact, most participants to the congress in Bologna, including Agostino Gemelli, Marcelo Boldrini, and Nora Federici, continued to endorse traditional Latin eugenics, regardless of the tergiversations of fascist doctrine.

The proponents of Latin eugenics in Italy, Romania, and elsewhere tried unsuccessfully to prevent the growing racism engulfing their movement, due to Mussolini's endorsement of Aryan racism after 1938 and, more dramatically, the outbreak of the Second World War in 1939. However, within a year, Romania and Italy had voluntarily joined Germany's side; France and Belgium were invaded and occupied by June 1940. These events dealt a nearly fatal blow to the Latin eugenic movement, and broke the international links that held it together. Eugenicists in the Latin countries were now forced to deal with the most immediate and basic issues of national survival (Lackerstein 2012). Importantly, in some cases, these fears of national dissolution prompted the state to finally implement measures of social and biological improvement that had long been advocated by the Latin eugenicists. For the first time in the history of Latin eugenics, there was political consensus that the nation's demographic potential and racial strength would be fatally undermined unless the state decisively encouraged the growth of strong families along with the protection of mothers and infants.

French eugenics under the Vichy

The development of French eugenics under the Vichy regime is perhaps the clearest example of the transformation experienced by Latin eugenics during the Second World War. In essence, there was as much continuity with the pre-1940 period as there were significant departures. In medicine and anthropology, the disciplines closest to eugenics, most practitioners continued to be active and publish during the occupation (Grimoult 2012: 55–66). For example, the racial anthropologist Georges Montandon began editing the anti-Semitic and racial journal *L'Ethnie Française* and became the editor of the *Revue anthropologique* in 1941 (Jarnot 2000: 17–34); René Martial became the co-director of the Institute of Anthroposociology (*Institut d'anthroposociologie*), created in 1942 by the General Commission on Jewish Affairs (*Commissariat général aux questions juives*). Joseph Saint-Germes, a professor of law, shared the directorship with Martial. The Institute's president was none other than Vacher de Lapouge's son, Claude. Its executive committee was composed of the hygienist Louis Cruveilhier from the Pasteur Institute; Émile Charles Achard, secretary general of the Academy of Medicine; and Jules Renault, previously an 'inspecteur général' in the Ministry of Public Health. In the same year, Martial was also given the newly established chair in racial ethnology ('ethnologie raciale') at the Faculty of Medicine in Paris (Larbiou 2005: 111–12).

Being under direct German occupation, Paris had to endure a more intense Nazification of its cultural and scientific activities. This was also the case with respect to eugenic propaganda. Karl Epting's *Institut Allemand* (German Institute), a German propaganda centre in Paris, sponsored a series of talks in 1941 given by leading German eugenicists: Leonardo Conti, Otmar von Verschuer, and Eugen Fischer (Briand 1941a: 195–6). The latter, for instance, spoke on 'the problem of the race and racial legislation in Germany'. From the outset, Fischer remarked that these two topics caused 'the greatest incomprehension for the foreigner' (Fischer 1942: 83). He hoped that his lecture would offer some clarification in this respect. First, he dispelled any doubt about what race was. In short, '*Race is heredity and only heredity*. What is not hereditary is not racial' (Fischer 1942: 84. Emphasis in the original). Second, there was no hope for neo-Lamarckism in a world dominated by the Nazi racial ideology: hereditary racial characteristics were 'unalterable', Fischer announced (Fischer 1942: 95).

Since 'the history of civilization' was in fact 'the history of race', Fischer continued, and since 'cultural achievements' were actually 'racial achievements', the state had the 'enormous responsibility' to ensure the survival of the race. Racial mixture between superior and inferior races was completely rejected (Fischer 1942: 104–5). Fischer chastised the French for being negligent in their duties to the Aryan race by allowing a flood of black African blood into France, and thus into Europe. This led to a regression of the intellectual and cultural capacity of France, with dramatic consequences if racial mixing continued. Above all, however, the French needed to protect their racial essence from the Jews (Fischer 1942: 106–8). Fischer held up National-Socialist Germany as a model; the 'hereditary health and racial purity of the German people' were protected by the state through a number of racial laws and administrative measures (Fischer 1942: 108–9). The message was clear: France needed both quantitative *and* qualitative eugenic policies.

Fischer's lecture echoed many of the eugenic themes debated in the Vichy era. The regime championed a conservative, nationalist revival that aimed to correct the French nation's racial, cultural, and social degeneracy. In this it embraced many of the eugenic projects put forward in France since the 1920s. Eugenicists under Vichy were obsessed with pronatalism, halting the falling birth rate, and protecting the family. They also endorsed – at least rhetorically, if not in practice – the introduction of prenuptial birth certificates and negative eugenic practices, such as sterilization. Furthermore, unlike the pre-war history of French eugenics, a widely prevalent anti-Semitism now appeared in most French eugenic writings, attempting to justify the harsh anti-Semitic measures the state instituted in the early 1940s. Vichy's racial laws were much like those of Italy and Romania at the time, which professionally and socially segregated the Jews from the Italian and Romanian societies.

The association of ideas of biological renewal with the promise of social protection strengthened the regime's positive eugenic message as expressed through the triadic structure of 'family, mother and home'. This new formula was given the highest symbolic importance: Vichy replaced the republican French motto, 'Liberty, Equality, and Fraternity' with a more conservative and Catholic slogan: 'Travail, Famille, Patrie' ('Work, Family, Nation'). In many ways, Vichy saw the protection of the French family as the cornerstone of society to be as its ultimate purpose. To this effect, a consultative population committee ('Haut comité de la population') was established which in turn drafted a new 'family code' ('Code de la Famille'), introduced on 29 September 1939, accentuating the role of state intervention in family life and the state's control of motherhood (Sauvage 1941). Quite a few new laws granted entitlements to parents with large numbers of children, testifying to the Vichy regime's commitment to accelerating the birth rate and 'protecting the family'. In this sense, the regime's demographic and family policies answered the long-standing eugenic demand in France for the normalization of large families, coupled with the definition of the population in terms of its biological capital (De Luca 2008).

A number of administrative bodies were created to implement and coordinate the new policies. The *Secrétariat d'État à la Famille et à la Jeunesse* (Ministry for Family and Youth) was established in July 1940, followed in September by the *Secrétariat général à la jeunesse* (Secretary General of Youth) within the Ministry of Education. This was followed by a *Comité consultative de la famille française* (Consultative Committee of the French Family), created in June 1941 and composed of prominent members of pro-natalist and Catholic associations. One of its members, Emmanuel Gounot [president of the Lyon branch of the League of Numerous Families (*Ligue des familles nombreuses*)], succeeded in having his proposal for a national federation of family associations enacted into law on 29 December 1942 (Capuano 2009: 127–43). Another body, the *Commissariat general à la Famille* (General Commissariat on the Family), was established on 7 September 1941, and entrusted with promoting the official rhetoric of French natalism (Capuano 2009: 58–64). The Commissariat was tasked with promoting the family and infusing a 'familial mystique' into the French people, to 'make the motto *Travail, Famille, Patrie* a reality' (Childers 2003: 158).

The Vichy government implemented various additional measures to advance its family-centred ideology and programme of social renewal. For instance, Mothers' Day ('Journée des Mères') was elevated into a national rite (see Figure 7.3), with celebrations throughout France, and the usual 'medals for mothers' (Jennings 2002: 101–31). Although married women were discouraged from working, the continuation of the Third Republic's family allowances presumably compensated (Jackson 2001: 331–2). Fathers of large families were exempt from the labour draft imposed on the Vichy government by the Germans. Divorce laws were made much more stringent. Sentences for distributing birth control 'propaganda' were strengthened, and abortion became a crime against the state, the nation, and the race, punishable by death (Quine 1996: 74).

Figure 7.3 Journée des mères, 31 Mai 1942
Source: Archives Départementales du Loiret, Orléans.

All of these measures were consistent with the goals and methods advocated by Latin eugenicists since the 1920s. However, the new political circumstances caused some radical departures from French eugenic traditions, including the wider acceptance of negative eugenic proposals, such as sterilization. Nazi-inspired eugenic language became common during the Vichy period. For instance, René Leriche, president of the High Council of Physicians, wrote that 'it was a great crime' for those with a hereditarily transmissible illness to reproduce. 'Unproductive people', he continued, 'the sterile, the parasites – have no right to life; they must disappear'. Leriche was also an anti-Semite, claiming that the Jews were purposefully spreading various dysgenic practices and materials to genetically poison the French, so that they would bear only 'idiots and ill people'. Leriche explicitly connected his eugenics with his politics, advising France to take the hand of friendship proffered by Nazi Germany (Leriche 1941: 146–86).

Most of these elements appear in a detailed proposal for national eugenic renewal, namely André Langeron's *Recherche d'un Français nouveau* (*In Search of a New France*), published in 1942. Langeron had made a name for himself as a reactionary military officer and germanophile in the 1930s; during the Vichy era he was one of the leading members of the quasi-fascist *Parti populaire français* (French Popular Party). His ideas bore a clear similarity to those of other eugenicists in the Latin countries of the time, such as Julius Evola in Italy and Antonio Vallejo Nágera in Spain. Alexis Carrel, Vichy's leading eugenicist, also clearly influenced Langeron's understanding of eugenics.

Langeron believed that France could only be saved if the Vichy government subjected its people to a 'racial revolution' leading to the creation of a sort of New French Man, 'more virile, more beautiful, and stronger' than the degenerate race defeated by Germany. Signs of physical degeneration were ubiquitous; Langeron bemoaned the 'sad-looking bodies' and physical deformities he saw in French youth; their intellectual development, he claimed, showed a parallel degeneration (Langeron 1942: 15). Many of them were the children of unknown fathers or broken homes. In other instances, egoistical parents, caring only about themselves, had thrown their children out onto the streets. A large number of these parents were alcoholics whose children were afflicted with mental disabilities or severe hereditary illnesses (Langeron 1942: 26–7).

Echoing eugenic arguments put forward by Vacher de Lapouge, Martial, and Montandon, Langeron claimed that racially foreign elements had polluted France. The Third Republic had encouraged 'gangs of coloured labourers' and colonial soldiers to settle in France after the First World War. Studies conducted in the United States had 'proven beyond a doubt' that the mental aptitudes of such people were inferior to those of the European race. These foreigners fathered biracial children who were inevitably 'mediocre, unstable, and deceitful', and were rejected by both their maternal and paternal families (Langeron 1942: 60). The Jews in France were an equally dangerous 'foreign element' characterized by a hereditary culture and way of life at odds with French values. Langeron called for a strengthening of Vichy's anti-Semitic laws, meant to limit the 'influence' of Jews on French society (Langeron 1942: 61 and note 1).

Langeron also warned against France's depopulation. The lukewarm countermeasures of the feeble Third Republic, such as government subsidies for large families and the patriarchal Family Code, had done nothing to reverse the decline of the nation's birth rate, he argued (Langeron 1942: 71, 104). But he believed that Vichy's uncompromising eugenic reforms would succeed where previous measures failed. One of the 'supreme duties' of the Vichy government was to remould the French people according to 'five year racial plans' (Langeron 1942: 61 and 71). New racial and eugenic marriage laws would implement these plans. For example, no couple would be allowed to marry without the express consent of a 'racial tribunal'. Women seeking marriage certificates would first need to demonstrate sufficient knowledge of eugenics, puériculture, and paediatrics (Langeron 1942: 101). As in Nazi Germany, Langeron proposed that racial tribunals be instituted to ensure that only 'Aryan' French couples would be allowed to marry. Alcoholics, prostitutes, syphilitics, 'parasites', and the 'morally unclean' would be denied the privilege of marriage and procreation. Langeron also included the proscription of cohabitation in order to prevent degenerates from breeding outside of marriage (Langeron 1942: 87–9). Women's primary duty, not surprisingly, was to care for their husbands, and prolifically bear and raise children. Ruthless punishment for the promotion of birth-control techniques or abortion would ensure that women fulfilled their duty to the French state (Langeron 1942: 100–1).

Anxious to re-ruralize France, Langeron promoted relocating urban industry to rural areas, which would be connected to cities by modern transportation infrastructures. To accommodate the housing needs of large families, the government would build decent residential complexes in these rural areas. Such projects would be a far better use of government funds than the Third Republic's proclivity to pour money into 'sumptuous' insane asylums and hospitals (Langeron 1942: 79–82).

Influenced by Alexis Carrel, Langeron foresaw the creation of a new eugenic caste in France. School children would undergo periodic medical examinations; their athletic and physical prowess would be cultivated and assessed. The results would be recorded on racial-eugenic cards (Langeron 1942: 103). By such means, the state would determine the individual's biological role in the new France (Langeron 1942: 129–30). At the top of the social hierarchy would be a new ruling elite, a somewhat modernized, idealized version of the mediaeval French knighthood. Character and physical perfection, rather than intellectual ability, would be the primary qualifications for those who would rule France (Langeron 1942: 238 and note 1, 241).

Langeron expected that the new *Fondation Française pour l'Étude des Problèmes Humains* (French Foundation for the Study of Human Problems) would provide the scientific management of the French eugenic state. The Foundation was created on 17 November 1941, with Alexis Carrel, a Nobel Prize-winning physiologist and author of the classic eugenic text *L'homme, cet inconnu* (*Man, the Unknown*) (1935), as Regent ('Régent') (for the official document, see Drouard 1992c: 337). As Andrés Reggiani described it, 'The convergence of Vichy's ideological projects and the humanitarian emergency created by wartime restrictions gave Carrel the opportunity to carry out his long-cherished project' – an institute 'to regenerate the race' (Reggiani 2007: 111).

The Foundation was supplied with a massive budget, and was charged with scientifically studying 'the most appropriate measures to safeguard, improve, and develop the French population' (quoted in Reggiani 2007: 112). The Foundation had departments specializing in population biology, child and adolescent biology, biotypology, and bio-sociology. Its efforts reflected the main tenants of Latin eugenics, focusing on both qualitative and quantitative improvements of the population. Thus, the causes of low natality rates were investigated, and puéricultural methods were proposed to increase the birth of 'hereditarily gifted children'. The 'apparent lack of virility' of adolescent males was also a matter of concern, as was their health and physical fitness (Tumblety 2012: 216).

The Foundation also undertook anthropological and biotypological studies to compare the ethnic characteristics of sturdy French peasants with various immigrant groups, in order to rank them according to their compatibility with the 'true French'; it also created a genetic archive, containing data on desirable and undesirable hereditary traits. Although building on existing eugenic traditions in France, it seemed that the Foundation 'developed its own brand of eugenics largely independent from' the preceding French Eugenics Society and the *École d'anthropologie* (School of Anthropology) (Reggiani 2007: 137).

These various eugenic projects reflected the diversity of interests represented in the Foundation. Catholic pro-natalist eugenicists collaborated with opponents of immigration and supporters of sterilization, such as Félix-André Missenard, the vice-regent of the Foundation (Drouard 1992c: 224–5). Racial biologists also found a place in the Foundation. The anthropologist Robert Gessain, for example, promoted the notion of racial incompatibility between the French on one hand, and blacks and Jews on the other hand (Drouard 1992c: 230–1).

The Foundation's efforts towards national rejuvenation and consolidation of eugenic policies were ultimately translated into Vichy's only eugenic law, introduced on 16 December 1942. Entitled the Law Relating to the Protection of Maternity and Infancy, it required, among other measures, an obligatory health examination before marriage ('Loi 941 du 16 décembre 1942 relative á la protection de la maternité et de la première enfance', *La Presse Médicale* 1943: 5–6). Section II, Article 4 stipulated that 'the authorization of marriage' would be not be issued 'until each of the future spouses produce[d] a medical certificate, no more than a month old, stating only that the bearer has been examined with the purpose of marriage' ('Loi 941 du 16 décembre 1942 relative á la protection de la maternité et de la première enfance', *La Presse Médicale* 1943: 5).

Most of the French eugenicists who debated the topic of premarital medical examination during the 1920s (discussed in Chapter 3) were no longer alive, but their dream of improving the health of the population through the rational planning of marriage and reproduction had finally been realized. For example, the pre-marital certification law first proposed by Adolphe Pinard almost twenty years before was finally enacted. Above all else, this provision was meant to promote eugenic marriages (Tumblety 2012: 215). Eugenic puériculture, another of Pinard's legacies, was also enshrined in the new eugenic law. Pre- and postnatal assistance in public maternity wards was guaranteed, as well as a wide range of material support for pregnant and breastfeeding women. Finally, the law required that each newborn be provided with a health book (*carnet de santé*).

Although the French Eugenics Society ceased to exist after the capitulation in 1940, there were many holdovers from the Third Republic. Both Paul Vignes, who had once been a prominent member of the French Eugenics Society, and René Martial, were now associated with Carrel's Foundation. Raymond Turpin, one of the participants in the Latin Eugenics Congress in 1937, continued to advocate the idea of family subsidies and emphasize the importance of premarital medical examination. Henri Briand championed physical education as an essential measure to fortify the race. He was particularly supportive of the government's plans to introduce physical education in the French schools (Briand 1941b: 8–15).

Vichy's political project of crafting a new French nation dovetailed with its eugenic programme of biological improvement; both were centred on the ideal of a healthy national community. As a result, the nation was reconceived as a modern laboratory of social and biological engineering. Eugenics was envisioned as the transformative scientific instrument that would strengthen the biological bonds between the nation and the state, resulting in a politicized biology that would prepare the French people for the new biopolitical state.

The normalization of race in Romania

During the early 1940s, the themes of racial revival and national regeneration by means of racial hygiene also dominated Romanian eugenics. As we have mentioned, during the 1920s and 1930s, Ioan Manliu and Iordache Făcăoaru were among those Romanian eugenicists influenced by German racial hygiene. By the early 1940s, however, Gheorghe Banu (see Figure 7.4) developed the most original synthesis between Latin eugenics and German racial hygiene.

In the decade preceding the Second World War, Banu fused public health, social hygiene, and puériculture into one eugenic ideology. In 1931, he founded a journal, *Revista de igienă socială* (Review of Social Hygiene), which propagated his ideas and which he continued to edit until 1944. Banu also projected eugenics into the political realm: he became a deputy in the

Figure 7.4 Gheorghe Banu
Source: Library of the Institute of Medicine, Institute of Public Health, Bucharest.

Romanian Parliament in 1934; he served as Minister of Health between 1937 and 1938, and was then appointed as general-director of the Social Service a year later. In 1935, following the examples of France and the USA, an Academy of Medicine (*Academia de medicină*) was established in Bucharest; it sought to foster biomedical research in Romania. Banu joined the Academy on 16 June 1936, having first delivered a public lecture on the 'principles of racial hygiene' (Banu 1936: 835–69).

Inspired by the Vichy's eugenic measures, the Romanian Academy of Medicine established a number of committees after 1940 dealing with eugenics, the protection of mothers and infants, public health, and social assistance. Banu was a member of all these committees, promoting governmental measures to improve the racial qualities of the Romanian population. In 1940, he became a professor at the Institute of Hygiene and Public Health in Bucharest (*Institutul de Igienă și Sănătatea Publică*), and the first chair of social medicine at the University of Bucharest. Finally, in 1943 Banu became the director of the Institute of Hygiene and Public Health in Bucharest, and began the publication of his monumental *Tratat de medicină socială* (*Handbook of Social Medicine*). Although the Handbook was planned to include nine volumes, only four were published dealing with eugenics, demography, school medicine, social assistance, and infections and venereal diseases (Banu 1944, vols. 1–4).

Banu published his most important eugenic work, *L'hygiène de la race* (*The Hygiene of the Race*), in 1939. Guido Landra described the book as the 'most modern of treatise on racial hygiene in existence' (Landra 1942b: 48). Here, Banu offered both an informed theoretical discussion of heredity and proposed concrete solutions for the biological improvement of the Romanian race. Most notably, the sixth section of Banu's book focused on the 'principles and methods for the normalization of the race'. The foundation of Banu's eugenic philosophy centred on the notion that social hygiene and racial improvement were closely linked. To 'normalize the race' would require the maintenance and increase in the 'normal elements' of the race, and the elimination 'from the heart of the social organism elements which are physically and mentally deficient'. Congruent with this goal, Banu showed particular interest in 'practical and theoretical investigations of heredity; biological and hereditary statistics; the study of family genealogies; the biological and hereditary status of the population; and the demographic evolution of communities' (Banu 1939d: 256).

Like prenuptial certification and compulsory segregation, preventive sterilization was one of the 'socio-biological measures' needed to ensure this 'normalization of the race'. Banu refuted two arguments against his interpretation of eugenic improvement. Sceptics might condemn sterilization on rational grounds, he wrote, as an 'encroachment on the human rights of the individual'; others might assert the primacy of Christian morality, which 'opposed the control of heredity' (Banu 1939d: 290–3). While some of the objections raised by 'moralists and the representatives of the Church' were legitimate, Banu nevertheless contended that the benefits of preventive sterilization outweighed all else. Consequently, it was necessary for penal codes to be devised according to 'the principle of social protection', instead of reflecting the 'dogmas of liberal orthodoxy' (Banu 1939d: 294). Eugenic sterilization, by its very nature, bore significant implications for the state. It offered a means to cut expenditure and re-invest in other public sector areas, diverting funds that otherwise would have been 'wasted' on the treatment of dysgenic social groups. Moreover, Banu continued, preventive sterilization was, 'first and foremost, of biological importance: it [was] about the purity and the vital value of the race' (Banu 1939d: 297). His overriding aim was to work towards a programme of biological rejuvenation in which relationships between the individual and his or her racial community were mutually advantageous. He therefore stressed that sterilization of 'pathological individuals' such as 'imbeciles, idiots, epileptics, criminals, and those affected by diverse psychoses, syphilis, and tuberculosis as well as hemophiliacs and diabetics' was essential for the conservation and betterment of the race (Banu 1939d: 297–8).

Banu's discussion of racial hygiene was representative of the theoretical and ideological views propounded by eugenicists in Romania during the war. It also illustrated the professed need to reconsider previous proposals by conceiving of eugenics as a totalizing biopolitical strategy of national survival that was just as racial and political as it was medical. As Petru Râmneanțu heralded in 1940, there was only one institution capable of restoring Romania's racial vitality and rejuvenating the Romanian nation through a 'totalitarian demographic policy': the biopolitical state (Râmneanțu 1940: 29–52).

The growing interaction between Romanian eugenicists and German and Italian racial scientists also favoured the racialization of Romanian eugenics during the 1940s. Eugen Fischer, for example, visited Transylvania in early November 1941 and gave a lecture to the University of Cluj-Sibiu on 'Race as a Force in History' (*Anuarul Universității Regele Ferdinand I, Cluj-Sibiu* 1941–1942: 189). Guido Landra had been to Romania before the war as a participant at the 17th Congress of Anthropology and Prehistoric Archaeology. Two years later, and again in 1941, he published articles about Romanian eugenics in *La Difesa della Razza*, in the travel magazine, *Le Vie del Mondo*, and in Telesio Interlandi's *Il Tevere*. In 1942, he collected and refined these articles into a book, *Il Problema della razza in Romania* (*The Question of Race in Romania*), published by the short-lived Romanian-Italian Institute of Racial and Demographic Studies (*Istituto Italo-Romeno di Studi Demografici e Razziali*) in Bucharest. He also travelled throughout Romania in 1941, before settling down in Bucharest in 1942. Landra was intimately familiar with the anthropological, serological, biopolitical, and eugenic work produced by his Romanian colleagues (Landra 1942b: 28–35); he cultivated a particularly close relationship with Iordache Făcăoaru, whose writings Landra discussed in detail (Landra 1942b: 41–6).

Landra associated 'racial hygiene' with countries 'where the Nordic racial element prevailed': that is, in Germany, the United States, and Scandinavia. In contrast, the 'Latin countries demonstrated an aversion towards any measure to protect and strengthen the race according to racial hygiene.' However, Romanian eugenicists were making progress, in Landra's view. The Romanians who followed the German model of racial hygiene were not hesitant to 'engage boldly with eugenics' (Landra 1942b: 46–8). They were 'at least theoretically', in favour of eugenic laws informed by racial hygiene, as demonstrated by the 1936 Romanian Penal Code, the debate on voluntary and compulsory sterilization carried out at the Congress of Anthropology in 1937, and Banu's 1939 book.

According to Landra, Romania faced two racial enemies: the Roma and the Jews (Landra 1942b: 51–6 and 137–85). Once these 'problems' were 'solved', Romania could overcome the widespread 'biological crisis' characterizing modern European civilization, and make full use of its 'almost pristine racial patrimony' (Landra 1942b: 191). These views echoed the arguments put forward by Romanian eugenicists themselves at this time. Debates in Romania on authoritarian projects for national survival, especially after the territorial losses of 1940, were closely associated with the eugenic quest for comprehensive answers to social questions.[1] In the highly unsettling wartime years, these concerns took a distinctively racial turn: focus now turned to the alleged source of national degeneration posed by ethnic minorities (Turda 2011: 336–48).

As in Fascist Italy and Nazi Germany, various forms of radical biopolitics emerged in Romania during the Second World War that endorsed the totalitarian state as the quintessential expression of Romanian ethnic supremacy. The fascist sociologist Traian Herseni, for instance, believed that 'With the help of eugenics, a nation controls its destiny. It can systematically improve its qualities and can reach the highest stages of accomplishment and human creativity' (Herseni 1940: 2). In 1941, Herseni suggested the introduction of biopolitical laws, such as segregation and deportation, as the basis for national regeneration. 'The racial purification of the Romanian nation', he claimed, was 'a matter of life and death. It cannot be neglected, postponed or half-solved.' Like contemporaneous racial hygienists across Europe, Herseni attributed the alleged degeneration of the Romanian nation to 'the infiltration in our ethnic group by inferior racial elements; to the contamination of the ancient, Dacian-Roman blood by Phanariot and Gypsy blood, and recently by Jewish blood' (Herseni 1941: 1).

The Roma (Gypsies), however, were singled out for their perceived otherness and the 'dysgenic' danger they posed to the Romanian majority.[2] Outlining the 'racial problem in Romania' up to 1940, Sabin Manuilă similarly accused both Jews and Roma of existing outside of, and in opposition to, the Romanian national body. This was identified almost exclusively through a racial representation of each group's social and ethnic functions. The Jews were 'the most important social problem, the most sensitive political problem and most serious economic problem of Romania'; however, they did 'not constitute a racial problem as racial mixing between Romanians and Jews occurs very rarely' (Manuilă 1940a: 5).

The Roma, however, represented 'the most important, sensitive and serious racial problem for Romania' (Manuilă 1940a: 5). This ethnic minority had mixed with Romanians in villages and urban slums, thus creating a new racial hybrid which, in turn, infiltrated all spheres of Romanian social life. Unsurprisingly, Manuilă's assessment was meant to indicate a racially textured national drama: 'The mixing of Gypsy with Romanian blood is the most dysgenic occurrence affecting our race' (Manuilă 1940a: 5). A year later, Manuilă determined that eugenic sterilization was necessary to carry out his anti-miscegenation mission: 'Obstructing dysgenics, the unwanted, should be pursued until their complete sterilization' (Manuilă 1941: 2).

Other authors agreed, at least insofar as the Roma were concerned. The Orthodox theologian Liviu Stan thus complained that, contrary to their racial philosophies, 'neither National Socialism nor Fascism' had introduced a 'racial policy towards the Gypsies', erroneously assuming that in Germany and Italy the Roma people, the 'centre of infection and degeneration represented by the Gypsies', were non-existent (Stan 1941: 1). However, Stan believed that such a policy was imperative in Romania, where 'racial promiscuity between Gypsies and Romanians', especially in the southern regions, was leading to the moral and biological degeneration of the dominant race. Like Manuilă, Stan perceived that the Roma had caused more 'biological damage' to the Romanian racial body than had Jews, suggesting 'prophylactic measures' such as their 'segregation that included the prohibition of marriage between Gypsies and Romanians' (Stan 1941: 2). This 'racial policy towards the Gypsies' was intended to serve both moral and biological purposes, and Stan unhesitatingly presented it as part of the glorious destiny that God had planned for the Romanians.

In response to such ‘racial fears’, Gheorghe Făcăoaru suggested that sterilization be used as a means to ethnically cleanse Romania of the Roma:

> Nomadic and semi-nomadic Gypsies are to be interned in forced labour camps. There their clothes will be changed; they will be shaved, receive a haircut and sterilized. To cover the costs of their maintenance, they should be put into forced labour. We will be rid of them from the first generation. Their place will be taken by national elements, capable of disciplined and creative work. Sedentary Gypsies will be sterilized at home, so that within a generation the place will be cleansed of them. (Făcăoaru 1941: 17)

In addition to this programme of racial purification, sterilizing the Roma was presented as a cost-saving solution in a period of economic depression: ‘The state spends almost a third of its budget on the maintenance of hospitals and various institutions of social assistance and vice squads, yet the social dirt increases daily. There is an explanation and easy solution to this: evil must be cut at the roots and not cultivated’ (Făcăoaru 1941: 18).

These examples indicate how eugenic sterilization became intertwined with the ethnic nationalism at the centre of the biopolitical programme envisioned by Marshal Ion Antonescu’s regime. Traian Herseni made this connection clear: ‘Dysgenics must not be allowed to reproduce; inferior races should be completely isolated from the [Romanian] ethnic group. The sterilization of certain categories of individuals must not be conceived stupidly as a violation of human dignity but as a tribute to beauty, morality, and perfection’ (Herseni 1941: 7).

The support given to sterilization by prominent sociologists and statisticians like Herseni and Manuilă was consonant with the new ideological objectives of Romanian eugenics emerging after 1940, such as the introduction of premarital examination (Cupcea 1941: 105–26). Once set against this background, the biopolitical measures envisioned by Marshal Ion Antonescu’s regime during the War appeared to reflect a broader consensus among the political elite and the eugenicists (Ioanid 2000; Achim 2004; Solonari 2010). In 1941, Mihai Antonescu, the deputy prime minister and Minister of Foreign Affairs of Romania, thus spoke on the desirability of the ‘ethnic and political purification’ of Bessarabia and Bukovina. The Greater Romanian nation, Antonescu explained, ‘was in its most favourable historical moment to achieve its complete ethnic freedom, [and] for our people to be purified of all those elements foreign to its soul’. Describing the Jews as a ‘race’ alien to the Romanian nation, Antonescu was unambiguous in stating the aims of the ‘policy of ethnic purification’, namely, 'elimination or isolation in work-camps of all Jews' (Antonescu 1991: 139). This further attests to the intimate connection between eugenic discourses on the protection of national health and various policies of ethnic homogeneity promoted by the Romanian government in the 1940s.

To this effect, on 27 September 1943, with the approval of Marshal Ion Antonescu, the Presidency of Council of Ministers (*Preşedinţia Consiliului de Miniştri*) established a Commission for the Promotion and Protection of the Biological Capital of the Nation (*Comisia pentru promovarea şi ocrotirea capitalului biologic al naţiunii*). Iuliu Moldovan was president; other members included Sabin Manuilă, Iordache Făcăoaru (see Figure 7.5), Petru Râmneanţu, and Gheorghe Banu.

The Commission was to prepare a comprehensive account of the health of the population, to be submitted to Antonescu together with concrete quantitative and qualitative eugenic proposals. Făcăoaru also proposed the creation of an Institute of Ethnoracial Biology (*Institut de Biologie Etnorasială*), composed of five sections: human genetics; bioanthropology; biopolitics and euthenics; and an office dealing with ethnic talents. These five sections, in turn, were divided into over twenty sub-sections, including heredo-pathology, serology, biotypology, demography, negative eugenics, migration, and so on.

Figure 7.5 Iordache Făcăoaru (standing up) and Sabin Manuilă (on Făcăoaru's left)
Source: Iordache Făcăoaru's Personal File, Consiliul Naţional pentru Studierea Arhivelor Securităţii, Bucharest.

A month later, Făcăoaru presented the full description of the Institute at the next meeting of the Commission. This time, he also gave it a name: the 'Marshal Ion Antonescu' Institute of Ethnoracial Biology. The Institute was described as the pinnacle of more than twenty years of eugenic and biopolitical work in Romania, initiated in Cluj (Transylvania) by Iuliu Moldovan and his disciples. According to Făcăoaru, the Institute's research would concentrate on the 'racial and biological-hereditary foundations of the Romanian nation' (Fond Personal Caranfil, Nr. 384). In its final form, the Romanian Institute of Ethnoracial Biology resembled both Martial's Institute of Anthroposociology and Carrel's Foundation for the Study of Human Problems, while at the same time integrating some research agenda proposed by Pende's Institute of Orthogenesis and Racial Improvement. The main difference, of course, was that the Romanian institute never materialized.

In 1943, the gynaecologist Constantin I. Andronescu (see Chapter 4) sternly remarked that, in contrast to other European states and the United States, eugenic sterilization had still not been introduced in Romania – an impediment towards the racial improvements of the Romanians he hoped would soon be rectified (Andronescu 1943: 46). Indeed, in January 1944, the Commission proposed a 'Law for the Protection of the Family' (*Decret Lege pentru Ocrotirea familiei*). Similar to the French law of December 1942, the Romanian bill proclaimed 'the family as the life foundation of the Romanian nation and state'. It reinstated the importance of the prenuptial medical certification (which had been introduced in October 1943), and requested additional eugenic measures to protect the family. Yet, the Romanian proposal went further than its French counterpart: it also stipulated the introduction of compulsory sterilization for those with mental and physical hereditary diseases (Fond Personal Caranfil, files 35–7).

The embrace of negative eugenics by the Romanian government and political elite in the last two years of the war was partly the result of the constant need to assert Romanian racial supremacy within the confines of a rump country, alongside Romania's 'civilizational mission' in the East. The anthropological research carried out in Romania's eastern provinces, particularly in Transnistria, a territory administered by the Romanian army between 19 August 1941 and 29 January 1944, was of particular importance in this context. This research gave new scientific authority to the Romanian state's attempts to control its diverse ethnic groups and legitimize its power over them.

This Romanian 'Ostforschung' ('Research in the East') was organized and supervised by the Central Institute of Statistics, together with the Romanian Social Institute and the Civil Government of Transnistria. The intense politicization and total subordination of these institutions to the Romanian government illustrates the dynamic and symbiotic relationship between science and politics during the Second World War. To legitimate Romania's policy in the occupied eastern territories, researchers were required to produce social, economic, and cultural evaluations of the local Romanian population. For the first time, these scientists faced the challenges of applying their methodologies outside their country and research institutes, as well as in a war environment. Yet they actively complied; at long last, their expertise was required in the service of the nation. The ultimate goal was, according to Vladimir Solonari, 'to restore Greater Romania, to extend its borders, and thus to guarantee the country an important place in a new Europe and a new world dominated by Nazi Germany' (Solonari 2010: 150).

Iordache Făcăoaru, the director of the Demographic, Anthropological and Eugenics Department of the Central Institute of Statistics, conducted this research in Transnistria. Făcăoaru completed his doctorate in anthropology at the Ludwig-Maximilian University in Munich in 1931, under Theodor Mollison, and published his dissertation as *Soziale Auslese: Ihre Biologischen und Psychologischen Grundlagen* (Social Selection: Its Biological and Psychological Foundations) in 1933. As mentioned in Chapters 4 and 6, Făcăoaru was an assiduous researcher and assembled a long list of publications devoted to Romania's biological problems (Făcăoaru 1935, 1936, and 1940).

As Landra pointed out, by the 1940s Făcăoaru was an internationally recognized authority on anthropology and eugenics, concerned as much with mapping the racial structure of the Romanian national body as with protecting it through applied eugenics. His scientific views gravitated around the dominant myths of Romanian nationalism, and this commitment was in evidence as well in Făcăoaru's research in Transnistria, which he described in 1943. Here, Făcăoaru outlined the importance of these racial-biological investigations for defining Romania's territorial expansion and its policies of ethnic cleansing in the eastern territories: 'Racial research about our co-nationals living outside the borders of the country has both a scientific and biopolitical importance' (Făcăoaru 1943a: 1).

It was essential to establish the racial composition of the local Romanian population, Făcăoaru maintained, so that racial scientists could determine who could or could not be resettled in Romania. Having lived for centuries next to the Russians, it was assumed that many Romanians were of 'mixed origins'. Considered to be predominantly of Asiatic origin, the Russians were deemed Romania's 'greatest racial danger', with Hungarians coming second (Făcăoaru

1943a: 3). Through this racial screening in Transnistria, Făcăoaru hoped to identify those Romanians 'contaminated with Asian blood', thus preventing their eventual resettlement and ultimately further racial mixing. The methods employed were anthropometric (height, cephalic, facial and nasal indexes, eye and hair colour), with emphasis on both 'the ethnic community in general' and 'each race in particular' (Făcăoaru 1943a: 5). The aim was to prove the high proportion of 'European racial elements' within the population's biological structure and, more broadly, to understand how this population fit into Romania's racial history.

Făcăoaru's anthropological research in Transnistria served as the basis for two further studies (Făcăoaru 1943b and 1944). The first study concentrated on the 'bio-racial value' of each of the territories constituting Greater Romania; the second put forward an interpretation of the autochthonous Romanian race. His racial evaluation was to some degree consistent with previous anthropological commentaries about Romania's ethnic diversity written during the 1920s and 1930s; yet Făcăoaru used this ethnic analysis to suggest an internal racial hierarchy within Romania itself. Făcăoaru's theory, in short, was that whatever their racial origins, the various waves of racial mixing Romanian territories had experienced since ancient times had been swept into the dominant autochthonous race. Based on his belief in the existence of superior (Nordic-European) and inferior (Asiatic) races, Făcăoaru envisioned a superior racial type within the Romanian nation, which he then located in Romania's 'Western provinces'. Undoubtedly, Făcăoaru was ideologically predisposed to racial theories proposed by Nazi scientists, but his Nordicism did not restrain him from pronouncing the racial superiority of the Romanian nation. He did so by subsuming the racial types found in Romania within the 'Nordic-European races', not unlike the 'Aryanization' of the 'Italian race' decreed by Mussolini.

In his 1944 article, Făcăoaru argued that this group of people, 'the Carpathian race', was the archetypal model of the autochthonous Romanian race: 'the most beautiful and biologically endowed' of all races (Făcăoaru 1943b and 1944: 7). The 'Carpathian race' were supposedly 'a race of tall, brachycephalic people' (Făcăoaru 1944: 6), predominant in Romania's mountainous regions. Făcăoaru did not hesitate to build a historical genealogy for the 'Carpathian race' going back to the visual representation of the Dacian prisoners of war immortalized on Emperor Trajan's column in Rome, commissioned after he conquered Dacia in 106 CE.

At a time when ethnic distinctions between the Romanian majority and various ethnic minorities were being reinforced within the borders of the

Greater Romanian state, anthropology entered the final struggle to articulate a meaningful Romanian national identity. A more racial interpretation became entrenched amongst Romanian anthropologists and eugenicists. As Făcăoaru confessed, his naming the autochthonous Romanian race 'Carpathian' also had 'a nationalist-sentimental reason: the name shows reverence to our mountains, which are joined with the body of the nation. For millennia, these mountains were the cradle of Romanianism' (Făcăoaru 1944: 6). As the end of the war approached, such statements became more numerous, combined with the Romanian state's determination to gain back the Northern Transylvania from Hungary.

***

During the late 1930s and into the Second World War, many eugenicists in the Latin countries adopted a narrative of social and biological improvement influenced by German racial hygiene. The range of debates about the nation's eugenic improvement, as we have outlined in this chapter, powerfully illustrates how conceptually versatile Latin eugenics had become after 1940.

In general, a theory of race provided eugenicists with a unique opportunity to redefine the history of their nation; in particular, it tended to legitimize policies of exclusion and anti-Semitism. Encouraged by the example of Nazi Germany, eugenicists in Italy, Portugal, Spain, France, and Romania helped forge a narrative of national renewal that they had heretofore generally avoided.

The allegiance of many eugenicists to the nationalist ideas of racial survival promoted by European regimes during the war indicates the transformed nature of Latin eugenics. On one hand, many of the traditional elements of Latin eugenics, such as pronatalism and family protection, remained; on the other hand, the dominance of Nazi Germany over the European continent encouraged the appropriation of racialized nationalisms, as illustrated by the Manifesto of Racial Scientists in Italy or Romanian eugenic research in Transnistria. Quite often, eugenicists in France, Italy, and Romania endorsed official racial politics during the war – a stance that many of them would later come to regret.

At that time, the national disruptions unleashed by the territorial changes experienced by the Latin European states led Latin eugenicists to promote a new mythology of national belonging thoroughly suffused with ideas of historical continuity and racial distinctiveness. In France, the exigencies of German domination added layers of pro-Germanism and anti-Semitism to the pre-existing Latin eugenics stratum. In Italy, the scientific establishment largely resisted Mussolini's dramatic re-orientation of Italian eugenics towards

Aryanism and anti-Semitism; thus, the Duce had to rely on more marginal allies from the scientific community to keep up the pretence of a new racial consciousness in Italy. In Spain, devastated by civil war yet still free from German domination, eugenicists attempted to resurrect a mythical 'conquistador' race. Similar colonial projects were voiced in Portugal. In Romania, however, attempts to introduce practical eugenic measures such as the Law for the Protection of the Family failed, due less to the lack of political support than to the military dynamics of the War.

As this chapter demonstrates, after 1940, eugenicists in Italy, France, Spain, Portugal, and Romania affirmed the need to embark on the quest for the biological protection of the nation. Their embrace of scientific racism became progressively more trenchant and some of them openly promoted programmes of racial engineering, such as France's and Romania's attempts to institutionalize the protection of the nation and race along biopolitical lines. Amidst difficult wartime circumstances, the ideal of a healthy nation voiced by the Latin eugenicists since the early twentieth century had finally become encoded in their respective states' official politics. Their ideas of social and biological improvement were put into practice, although not without changes reflecting the political exigencies of the time.

Conclusion

As we have shown, by the mid-1930s Latin eugenics had developed into a coherent set of social, biological, and cultural ideas centred on a unique definition of the individual and the national community. Neo-Lamarckism, puériculture, biotypology, and homiculture provided the intellectual foundation for Latin eugenics. Unlike other versions of eugenics, often described as Anglo-Saxon and Nordic, Latin eugenics sought the biological betterment of the individual and the collective by means of preventive medicine, social hygiene, demographic studies, and public health, rather than genetic engineering, racial selection, and compulsory sterilization.

From the mid-nineteenth to the mid-twentieth centuries, many nations in Latin Europe and Latin America sought to use eugenics as a tool to advance programmes of modernization and national renewal. Given that it remained fluid, Latin eugenics could be invoked to provide the social and biological improvement of modern societies that were in many ways quite distinct. However, all had significant obstacles to overcome: France suffered from population stagnation; Italy, Spain, Portugal, Romania, and the countries of Latin America, from various levels of under-development, along with their attendant medical and social problems. In many Latin countries, moreover, traditional cultural and religious values mediated the adaptation to an increasingly modern, urban, secular world. Educational progress was relatively slow, labour efficiency mediocre. Ethnic and racial divisions remained or even intensified. Fears of stagnation, if not degeneration, abounded.

However, as a modern, scientifically credentialed panacea, eugenics appeared to be capable of solving these many problems simultaneously. Eugenics developed from a combination of evolutionary theories; advances in medicine and biology; and scientists' confident assertions that they could effectively manage the nation's social and biological improvement. Latin eugenicists claimed they had finally acquired the knowledge that would fulfil science's promise to the modern world.

As with all such synoptic remedies, eugenicists made some promises on which they could not deliver. Evolutionary theories still lacked the mathematical precision that would only come with the modern Darwinian synthesis. Biological heredity, though conceptually attributed to 'genes', was poorly understood. Human variation undoubtedly existed, but which characteristics arose from heredity and which arose from the environment could not yet be determined with any scientific credibility. Nevertheless, eugenicists promised that they would solve the nation's predicaments with the beneficent power of science, if only political leaders left population management to them and implemented their prescriptions. As Alexis Carrel put it confidently in 1935:

> Science, which has transformed the material world, gives man the power of transforming himself. It has unveiled some of the secret mechanisms of his life. It has shown him how to alter their motion, how to mould his body and his soul on patterns born of his wishes. For the first time in history, humanity, helped by science, has become master of its destiny. (Carrel 1935: 241)

The variety of diachronical influences on Latin eugenic thinking since the late nineteenth century is one of the most important findings of this book. As argued here, the union of Latin culture and the aspirations of the modernizing Latin nations gave rise to a unique form of eugenics. This provides scholars with an opportunity to revisit current historiographical models and to take seriously the crucial role played by Latin eugenics in shaping modern ideas of social and biological improvement.

Of course, the power of Latin eugenicists to control human improvement was limited, regardless of their desire that it be otherwise. After the First World War, in some Latin countries eugenics was confronted with new ideological claimants to the role of secular saviours: nationalism and fascism. Nationalists and fascists also declared that they had the capability to fundamentally empower the nation through the application of eugenic measures, including the management of the population, the protection of mothers and infants, and various schemes of social assistance and public health. Given their similarities, it is hardly surprising that nationalism, fascism and Latin eugenics made for such comfortable allies. Indeed, in one form or another, nationalist and fascist eugenicists became prominent in many Latin societies in the 1920s through the 1940s, with Italy and Romania as the most notable examples. However, Latin eugenics was restrained to some degree by the continued political influence of Catholicism and Orthodoxy; thus, Latin eugenicists in Europe did not enjoy the same level of state support (apart from brief periods during the Second World War), or the ability to employ the most radical eugenic measures, as did eugenicists in Nazi Germany.

In Latin America, eugenicists believed that the racial vitality of their national communities could be revivified, unified, enriched, and empowered through eugenics, while also retaining their distinctive Latin identities. To do so, it was necessary to genetically improve their populations, and create homogeneous national communities based upon Latin eugenic principles. By the time Latin eugenicists created their own international organization, in the mid-1930s, they were united in regarding their version of eugenics as more 'humane' than those prevailing in Nazi Germany, the USA, and Scandinavia – all of which promoted compulsory eugenic sterilization, a measure rejected by most Latin eugenicists.

During the Second World War, when many feared that Europe would fall under complete and permanent submission to National Socialism, Latin eugenicists once again placed their theories of social and biological improvement at the service of the state. This time they were successful. The relationship between the future of the nation and the health of the population was already an established eugenic trope, but the War transformed it into a national obsession.

However, the willingness of German scientists during the Second World War to devote their energies to eugenic-inspired genocide deepened the growing suspicions of many people that biological utopias of racial improvement were only dangerous fantasies. After the end of the War, the rhetoric of anti-fascism and inclusive democracy (though not necessarily its practice) made it difficult for any country to advocate the benefits of a racially homogenous national community cleansed of the most vulnerable members of society. International eugenic movements of any variety could hardly survive in such a radically changed world.

But we should be careful not to downplay the elements of continuity between the pre- and post-war periods for several reasons. By the end of the war, the main tenets of Latin eugenics – as broadly defined in this book – were no longer the exclusive property of the eugenicists, but were fully integrated into their countries' system of social welfare and public health. As they relied less on race, the foundations of Latin eugenics (biotypology, puériculture, and homiculture) could be more readily adapted to accommodate the changes in political cultures after 1945. This is a point where Latin eugenics differed significantly from Anglo-Saxon and Nordic eugenics: it needed few fundamental changes to survive political and academic condemnation after the war. In this crucial respect, Latin eugenics finally prevailed over other eugenic movements. And in this sense, at least, Latin eugenicists had achieved one of their most important goals.

During the late nineteenth and the first half of the twentieth centuries, Latin eugenics embodied the promise of a healthier and a more productive life put forward by theories of social and biological improvement. It also legitimated demands for demographic growth, the protection of mothers and children, and validated attempts to resist, as well as impose, negative eugenic measures such as compulsory sterilization or euthanasia. It is partly for this reason that Latin eugenicists in Europe and Latin America continued to exert scientific influence for several decades after the war. By then, however, Latin eugenics dissipated into a myriad of national projects, having achieved a significant portion of its ambitious programme; tellingly, not 'in the name of eugenics', but in the name of one of the most prominent twentieth-century biomedical projects: the welfare state.

All eugenicists shared Alexis Carrel's belief in the omnipotence of science during the twentieth century – so expressively nuanced in the above quotation. They also shared the eugenic vision of a society guided by scientists and scientific institutions. As Carrel indicated, eugenics aimed to change the *nature* of the human body, both in *quantity* and in *quality*, according to a set of principles based on the laws of heredity, as well as knowledge of the social and biological environment. But eugenics provided not only the biological and social foundations upon which the improvement of the individual could be achieved; it also served the dream of human and racial perfectibility, so brutally instrumentalized during the Second World War by Nazi Germany.

After 1945, eugenics would be referred to continuously in its association with the Holocaust. However, since Latin eugenics was less closely associated with Nazism than was German racial hygiene, eugenicists in Latin countries found it easier to adapt to post-Second World War realities. As well as bringing the nation and the state together (indeed synthesizing them in many cases) these eugenicists argued that they had always promoted a theory of human improvement as a philosophy of social and biological regeneration across the political spectrum without favouring one race over another. Not surprisingly, then, during the 1950s and 1960s, Latin eugenicists continued to campaign for the nation's health according to principles they had outlined during the 1930s and 1940s. The ultimate goal, they claimed, was always to achieve a healthy nation by establishing a modern system of health care that was able to detect recurrent social and biological problems. As Jacob Tanner pointed out, eugenic programmes 'were integrated into the fabric of everyday life and corresponded with widely shared opinions about the foundations of heredity with regard to private and public health' (Tanner 2012: 476). To some extent, what was left of Latin eugenics, and what assured its post-1945 survival, was what made it different from coercive racial hygiene in the first place: its focus on demographic growth, biotypology, puériculture, neonatal and maternal care, preventive medicine, and social hygiene.

In conclusion, Latin eugenics comprised a wide range of views: from Catholicism to anarchism and from fascism to communism; and was geographically diverse. To understand this complex biopolitical discourse, therefore, this book has paid greater attention to a wide range of scientific arguments about social and biological improvement, as well as to the political cultures within which such arguments circulated as well. Generally, the accomplishments of Latin eugenics derived from the creative rivalry between differing social and biological visions of human improvement. We should certainly neither exaggerate nor privilege Latin eugenics, but try to offer an appropriate historical explanation for its difference from, and similarity to, eugenic movements elsewhere. Ultimately, understanding Latin eugenics serves as a good example of our responsibility to historically contextualize the past.

Epilogue: Latin Eugenics after 1945

On 1 December 1944, a short article entitled 'To Rehabilitate a Science: Eugenics' ('Pentru reabilitarea unei ştiinţe: eugenia') was published in the official newspaper of the Romanian Communist Party, *Scânteia* (The Spark). Its author, the writer Octav Mărgărit, was familiar with the history of eugenics, which he presented succinctly, focusing especially on Francis Galton's ideas. He also described eugenics in neo-Lamarckian terms, insisting on its social role and the improvement of the general health of the population. More importantly, Mărgărit noted the 'tragic destiny' that befell Galton's 'innovative and progressive ideas'. Eugenics was 'abducted by Nazi ideologues' to serve as a 'scientific basis for the German racist theory and the justification of the crimes committed in its name for 12 years'. The Nazis, Mărgărit continued, 'distorted eugenics', using it to promote the improvement of 'the Nordic race', while simultaneously practising 'sterilization' and the 'extermination of peoples declared inferior' – all with the intended purpose of 'universal domination'.

Galton would have disapproved of the abuse of eugenics by the Nazis, Mărgărit believed. As the defeat of Nazi Germany was foreseeable, it was essential that eugenics be restored to its 'original meaning', as envisioned by Galton at the end of nineteenth century. Eugenics, Mărgărit concluded, needed to 'reflect [Romania's] new social and political conditions', and thus 'concentrate on the physical and mental improvement of the individual, aiming to restore his true human condition' (Mărgărit 1944: 2).

This article is important in two respects: first, it indicates that the association between eugenics and Nazism occurred already in Romania by the end of 1944; second – and in anticipation of the scientific and political purges of eugenics, following the end of the war – it endeavours to clarify the 'true' meaning of eugenics and its importance in ensuring the health of the Romanian nation. Romanian eugenicists themselves were aware of the changes the defeat of Nazi Germany and the Allied victory would bring to their position and discipline. To prepare the implementation of new interpretations of human improvement that were politically acceptable, in 1944, the oldest eugenic journal in Romania, *Buletin eugenic și biopolitic*, began to publish articles on Soviet medicine, public health, and social hygiene (Stoichiță 1944: 213–41).

One such article was authored by one of Iuliu Moldovan's closest collaborators, Salvator Cupcea. It dealt with 'theoretical and practical biology' in the Soviet Union ('Biologia teoretica şi aplicată în URSS'), particularly with neo-Darwinism, neo-Lamarckism, animal and plant genetics, and Michurinism (Lysenkoism). Cupcea also reflected on Soviet research on 'human biology', with a special reference to the work carried out at the Maxim Gorky Medico-Biological Research Institute in Moscow. He commended Soviet biology, for having seriously studied both 'heredity and the environment' (Cupcea 1944: 316), as well as the Soviet policies for the protection of the family.

Tellingly, Cupcea concluded his analysis by referring to Soviet eugenics during the 1920s. He thus noted how the term 'eugenics, was considered tainted and abandoned' in the Soviet Union, due to the fact that 'German biologists had tried to bring eugenics closer to racism'. Yet, by promoting all 'individual values' equally, and by 'eliminating counter-selective social and economic factors', the scientific policies promoted by the Soviet Union were, Cupcea believed, 'so entirely *eugenic* that there was no need to maintain eugenics as a separate discipline, as it had infiltrated diffusely all [Soviet] biological schools' (Cupcea 1944: 317. Italics in the original).

Immediately after the War, however, eugenics remained central to research on psychology (Mărgineanu 1944), bio-sociology (Herseni 1947: 184–96), education and anthropology (Preda 1947; Rădulescu-Motru and Nestor 1948), and heredo-pathology (Sulică 1944: 265–74). Moreover, the main Romanian eugenicists like Banu, Moldovan, and Râmneanţu continued to publish on hygiene, biopolitics, public health, and preventive medicine (Moldovan 1946: 1–7; Banu 1947; Moldovan, Stoichiţă, and Râmneanţu 1947).

The increased Soviet presence in Romania after 1944, however, had a profound impact on the history of eugenics in this country. Between 1945 and 1950, Romania – like other Eastern European countries – underwent the troubled transition from an independent country to Soviet occupation and ultimately to the transformation into a communist satellite state. Most Romanian eugenicists (I. Moldovan, G. Banu, I. Făcăoaru, G. K. Constantinescu, and so on) were gradually imprisoned and professionally marginalized; university chairs and departments were dissolved, and 'bourgeois' eugenics was deemed 'incompatible' with the new scientific ideologies imported from the Soviet Union.

Some of them, however, most notably C. I. Parhon, the last president of the International Federation of Latin Eugenic Societies, were politically more accommodating. After 1946, Parhon became a member of the new Romanian National Assembly, and for a short period of time even its president, thus in effect Romania's head of state. Due to his political influence, endocrinology continued to dominate as a scientific discipline in Romania, assuring the survival of pre-war eugenic theories of biological betterment, with a special emphasis on glandular rejuvenation, organotherapy, and longevity. During the 1950s, Parhon and his disciples provided a concrete and compelling attempt to maintain the relationship between human constitution, endocrinal health, and chronic degenerative diseases. In this area of applied biology the interwar relationship between eugenics, chemical embryology, and endocrinology (cultivated also by Marañon and Pende) survived in communist Romania. When the Institute of Gerontology and Geriatrics (*Institutul de Gerontologie și Geriatrie*) was established in 1952 its directorship was given to one of Parhon's students, Ana Aslan, a former head of department at the Institute of Endocrinology. During the 1960s and 1970s, Aslan's rejuvenation therapy became Romania's most-celebrated scientific export.

With the establishment of the communist regime in Romania in 1947, the Latin eugenic tradition was officially terminated but not forgotten. Romanian scientists were not deterred in their attempts to synchronize interwar eugenic narratives about the health of the population with communist nationalist principles. Medicine, sociology, and anthropology were also disciplines that perpetuated eugenic themes in Romania in the 1950s and 1960s. For instance, ideas of national biology involving notions of racial differentiation, cycles of growth and decay, genetic genealogies, and the interconnectedness of nurture and nature, were abundantly present in the first collective anthropological investigations published in communist Romania. The fact that Traian Herseni was a contributor to both publications is illustrative, as he provides an exemplary case of a Romanian eugenicist's post-war professional and theoretical adjustment. Although in his studies, Herseni generally reflected on genetic genealogies, his main argument focused on the importance of ethnic biology in connecting forms of the nation's micro- and macro-physical development over time (Herseni 1958: 47–65 and 1961: 57–71).

Finally, this was the very period in which a new narrative on national identity emerged in Romania, allowing social scientists to reposition autochthonous ideas within their discipline. The eugenic codes of the interwar period were brought back in a nuanced form (Râmneanțu 1975: 25–30). A further, most dramatic eugenic project occurred in Romania with the ascension of Nicolae Ceaușescu to power in 1965. During his regime strict pronatalism was introduced in Romania, and the state aimed at the complete control of reproduction (Kligman 1998). Population control and political demography, coupled with the ideological prioritization of family and mothers (all themes revered by Latin eugenicists since the beginning of the twentieth century) became political tools in communist Romania through which the regime hoped to create a healthy national community, numerically and physically strong.

A different political context shaped the post-1945 evolution of Latin eugenics in the other Latin countries, situated in Western Europe and Latin America. In France, for example, protecting the eugenic movement from its association with Nazism was rather unproblematic. The French Eugenics Society did not re-form after the war, but the Society of Biotypology continued to exist, not least due to the efforts of Henri Laugier and Eugène Schreider. On 13 May 1966, however, it changed its name to the Society of Human Biometry (*Société de Biométrie Humaine*); Schreider was now the director of the Laboratory of Human Biometry at the French National Centre for Scientific Research (*Centre national de la recherche scientifique*).

In 1945, the Foundation for the Study of Human Problems was transformed into the National Institute for Demographic Studies (*Institut National d'Études Démographiques*, INED), under the leadership of Alfred Sauvy (Drouard 1992b: 1453–66). Most of its personnel remained untouched by accusations of collaborationism, including the anthropologist Robert Gessain; moreover, the Foundation provided the much-needed scientific expertise to assist France's 'demographic renewal' (Reggiani 2007: 166; Rosental 2012: 547–51). A 1950 book by Jean Sutter, a physician and one of the Foundation's members, demonstrates that INED did not abandon eugenics (Rosental 2012: 553–4). Entitled *L'eugénique: problèmes, méthodes, résultats* (*Eugenics: Problems, Methods, Results*), the book positioned eugenics at the confluence of hygiene, health, social policy, medicine, and genetics, reviving pre-war themes of Latin eugenics such as the importance of social environment and the relationship between biological, social, and cultural factors in determining human improvement (Sutter 1950).

In 1945 the Consultative Committee of Population (*Haut comité de la population*) became the Consultative Committee of Population and Family (*Haut comité de la population et de la famille*), demonstrating that the new regime was also concerned with the protection of the family, the decline of birth rates and various measures to counteract it (Drouard 1999: 171–97). Important eugenicists, demographers, and natalist campaigners of the interwar period formed the core membership of the Consultative Committee on Population and the Family, including Adolphe Landry and Alfred Sauvy. The fact that the Committee adopted the racial demographer Georges Mauco's immigration strategy on its second session, in May 1945, is another indication of the continuities between the Vichy and the post-war period, particularly in the fields of demography, family planning, and social assistance (Burgess 2011: 167–7; Rosental 2012: 559).

In Portugal, both biotypology and eugenics remained on the scientific agenda, as illustrated by work carried out by the anthropologist Leopoldina Ferreira Paulo for the National Institute of Statistics (*Instituto Nacional de Estatística*) in Lisbon (Paulo 1945: 115–38), as well as by the articles published in the late 1940s by one of the country's most respected experts of legal medicine, L. A. Duarte Santos and the physician Emílio Aparício Pereira. These articles were published in the Oporto-based *Jornal do Médico* (*Physician's Journal*), edited by Mário Cardia, who in 1948 also founded *Acta Endocrinológica e Ginecológica Hispano-Luisitana* (*Hispano-Luisitan Journal of Endocrinology and Gynecology*). *Jornal do Médico*, in particular, promoted 'social medicine, with strong Catholic sentiments', and – as Richard Cleminson has argued – 'eugenics continued to have a presence in the journal well into the 1960s' (Cleminson 2014: 157).

In Italy, both Nicola Pende and Corrado Gini continued to promote their eugenic ideas after the war. At various medical conferences during the 1950s, Pende advocated the application of biotypological principles and 'Christian marriage' as post-war biomedical strategies of renewal and protection. However, while Pende's international status as endocrinologist and biotypologist remained unchallenged, Gini's position as the leader of the Italian eugenic movement came under criticism. As a means of proving his 'long-standing opposition' to fascism at his post-Second World War anti-fascist trial, Gini reiterated his own tribulations trying to keep Latin eugenics alive in the late 1930s. He explained that he led the separation of the Italian Eugenics and Genetics Society from the International Federation of Eugenic Organizations in the early 1930s because of his fight against 'racial discrimination in the civilian population, as well as against measures harmful to personal liberty and integrity (sterilization, inspections, or certifications for marriage), adopted in many of the states of the USA and then even more severely in Germany'. To clearly distance himself from such 'racist tendencies that hid themselves in pseudo-scientific dress in the heart of the International Federation of the Eugenics Society [sic!]', he created the International Federation of Latin Eugenic Societies in 1935.

Gini stubbornly refused to accept the demise of the eugenic movement after the war. In 1948, he led the Italian delegation to the Eighth International Congress of Genetics in Stockholm, but his leadership came under attack from younger Italian geneticists (Cassata 2013: 220). In the same year, Gini sent a letter to all members of the Italian Society of Genetics and Eugenics, informing them of his intention to 'reanimate the Society, which in the inauspicious wartime and post-war period was forcedly inactive' (quoted in Cassata 2011: 288). But Gini's hopes to reactive the Italian Society for Eugenics and Genetics were unsuccessful, not least due to the opposition of a younger generation of Italian geneticists led by Adriano Buzzati-Traverso and Claudio Barigozzi. They insisted that, although the society initially included the words 'eugenics' and 'genetics', it was now an anachronism to keep the two together. These younger scientists sought to distance themselves from Gini's fascist eugenics and his increasingly awkward endorsement of scientific racism. Gini's subsequent affiliation with the International Association for the Advancement of Ethnology and Eugenics, and its journal, *The Mankind Quarterly*, further deepened his alienation from the mainstream post-war Italian discussions on social welfare and medical eugenics.

Sociological demography and genetic demography were two other areas of research that Gini cultivated after the war. In 1949, he returned to the helm of the Italian Statistical Society, and continued to lecture on sociology at the University of Rome (Giorgi 2011: 12). His views remained unchanged, however (Gini 1967: 261–75). Equally important, in 1950, Gini became president of the International Institute of Sociology (established in Paris in 1893 by the sociologist René Worms), a position he kept until 1963. During this period, the Institute brought together many of the pre-war Latin demographers and eugenicists such as Severino Aznar and Sabin Manuilă, and some of former members the Italian Society for Genetics and Eugenics such as Franco Savorgnan. Manuilă, for instance, led the Institute's Committee for the Study of Sociological Consequences of Displacements of Populations (Cassata 2006: 203–5).

Considering the progress of genetics in Italy during and immediately after the War (Cassata 2011: 285–6), as exemplified by the work of Buzzati-Traverso, Barigozzi, Giuseppe Montalenti, Corlo Jucci, Luigi Gedda, Luisa Gianferrari, and others, it was not surprising that dissenting voices with the Italian scientific community demanded a clean beginning, detached from Gini's eugenics and racialism. Ultimately, the secessionist geneticists succeeded in separating genetics from eugenics, and in establishing their own society, the Italian Genetics Association (*Associazione Genetica Italiana*) in 1953 (Cassata 2011: 309).

The conflicts within the Italian Society for Eugenics and Genetics did not prevent the continuity of eugenic ideas in post-war Italy. In 1946, the first genetic counselling centre was established at the University of Milan, followed in 1948 by the 'first municipal eugenic counselling' at the Milan Policlinic (Cassata 2011: 309–10). By then the medical community was again divided over the introduction of the premarital eugenic examination, debated at a succession of conferences during the late 1940s, including the National Conference on Social Assistance (*Convegno per gli studi di Assistenza Sociale*); at the International Congress for Treatment of Medical and Social Problems of Premarital Prophylaxis (*Convegno internazionale per la trattazione dei problemi medico-socialie di profilassi pre-matrimoniale*) and the 4th International Congress of Catholic Physicians (*IV Congresso internationale dei medici cattolici*) (Cassata 2011: 312–4). While distancing themselves from negative eugenics, Italian eugenicists and medical geneticists remained adamant in their support for eugenic counselling and education. Eugenics took a new role in shaping the rhetoric and substance of the emerging post-war welfare state, one that even the Catholic Church, through Pope Pius XII, reaffirmed to participants of the Ninth International Congress of Genetics held in Bellagio (Italy) in August 1953.

The situation was less dramatic in Latin America. During the 1950s and 1960s, several Congresses of Social Prophylaxis maintained the relevancy of the importance of eugenic marriage, prenuptial medical certificates, the fight against venereal disease, and the association of immigrants with disease (Vallejo and Miranda 2005: 145–92). State funding continued, coupled with attempts to centralize and unify biomedical research. In Mexico, for instance, it was the Ministry of Health (*Secretaría de Salud y Asistencia Pública*) and the Ministry of Public Education (*Secretaría de Educación Pública*) that provided a new institutional framework not only for public health and educational policies but also for research on eugenics, biotypology, and medical genetics (Suárez-Diaz and Barahona 2013: 105–8).

The creation in 1945 of the Argentinian Society of Integral Eugenics (*Sociedad Argentina de Eugenesia Integral*), by the physician Carlos Bernaldo de Quirós, certainly contributed to a wider dissemination of the 'eugenic ideal' during the 1950s (Miranda 2005: 197; Cecchetto 2008: 42). The new eugenic society and the Argentinian Association of Biotypology, Eugenics and Social Medicine remained influential until the early 1970s. As elsewhere in Latin America, many eugenic ideas in Argentina survived through biotypology into the 1950s and 1960s, not least due to their embrace by Peronist discourses on public health and social assistance. The National Institute of Biotypology was financially supported and politically endorsed by government, particularly by the neurosurgeon Ramón Carrillo, who served as the Public Health Secretary between 1946 and 1949, and then as the Minister of Health until 1954. Carrillo also hoped to transform the Institute into a sort of 'Argentinian Institute of Man', resembling Carrel's Foundation for the Study of Human Problems (Haidar 2011: 318).

In this and other respects, Mexican and Argentinian eugenicists joined their European counterparts, who after 1945 remained committed to the core component of the Latin eugenic discourse and policy: the ideal of a healthy and numerous nation. This was the core of the Latin eugenic programme of social and biological improvement revealed in this book, and by exploring it across various historical periods and in many countries we have hopefully redefined the existing understanding of the history of eugenics in the late nineteenth and twentieth centuries.

Notes

Introduction

1 Latin identity was referred to by the term *latinité* in French, *latinità* in Italian, *latinidad* in Spanish, *latinitate* in Romanian, and so on. This book will use these terms interchangeably.

Chapter 1

1 Unless otherwise indicated ('quoted in') all translations in this book are those of the authors.

Chapter 3

1 Italian eugenicists, particularly Gini, must have known of these conferences. Not only did he attend the Fifth Meeting of the International Federation of Eugenic Organizations, which met in Paris in July 1926, but the topic of prenuptial certificates was on the meeting's agenda, next to consanguine marriages and immigration from the point of view of eugenics. The Argentinian eugenicist, Victor Delfino, also attended the meeting ('Report of the Meeting of the International Federation of Eugenic Organizations', 1926: 1–3).

Chapter 4

1 Namely Canada, the Swiss Canton of Vaud, 'The Free City of Danzig', Denmark, Estonia, Finland, Germany, Vera Cruz (Mexico), Norway, Sweden, and the United States.

Chapter 5

1 The meeting reportedly occurred about one year before the date of this letter.

2 The proposed institute was apparently never established. Nevertheless, in 1943 Ramos would be given an honorary degree from the University of Miami.

Chapter 6

1 Like Gini in Italy, Manuilă occupied the position between eugenicists and demographers in Romania. Manuilă remained at the helm of the Central Institute of Statistics until 1947, providing the Romanian government with the much-needed expertise and statistics during the Second World War.

2 The other important Romanian eugenicist from Transylvania, Petru Râmneanțu, presented papers only on anthropology and was thus in a different section.

Chapter 7

1 In 1940, Romania lost Bessarabia and northern Bukovina to the USSR; northern Transylvania to Hungary, and southern Dobrudja to Bulgaria.

2 According to the 1930 census there were 262,501 'Gypsies' (Roma) in Romania. Some 221,726 (84.5 per cent) lived in rural areas and 40,775 (15.5 per cent) lived in urban areas (Manuilă 1940b: 34–7).

Bibliography

Abbreviations

ACS	Archivio Centrale dello Stato (Rome)
ANR	Arhivele Naționale ale României (Bucharest)
APS	American Philosophical Society (Philadelphia, PN)
BMG	Biblioteca Museo Galileo (Florence)
CIWA	Carnegie Institute of Washington Archives (Washington, DC)
CNSAS	Consiliul Național pentru Studierea Arhivelor Securității (Bucharest)
CSHLA	Cold Spring Harbor Laboratory Archives (Cold Spring Harbor, NY)
NARA	National Archives and Records Administration (Adelphi, MD)
SIAA	Smithsonian Institution Anthropological Archives (Washington, DC)
TSUL	Truman State University Library (Kirksville, MO)
USHMM	United States Holocaust Memorial Museum (Washington, DC)
WL	Wellcome Library (London)

*

(1908), *The Eugenics Education Society: The Annual Report, 1908*, London: The Eugenics Education Society.

(1908), 'Restriction du marriage', *Archives d'anthropologie criminelle, de médicine légale et de psychologie normale et pathologique*, 23: 196.

(1911), 'Inauguration de l'Institute de Puériculture', *Paris Médical*, 4: 81.

(1913), 'Atti del comitato italiano per gli studi di eugenica', *Rivista di Antropologia*, 18: 543–4.

(1913), 'Eugenics and Public Health', *The Eugenics Review*, 5: 157–65.

(1913), 'Fondation d'une Société Française d'Eugénique', *La Presse Médicale*, 4: 44.

(1913), 'Scientific Alliance of the Latin Races', *The Lancet*, 181: 1081.

(1914), 'Italians Take up Eugenics', *The Journal of Heredity*, 5: 138.

(1917), 'Vers la fédération: Possibilité et conditions d'une federation latine', *Revue des Nations Latines*, 5 (1 September): 60–4.

(1919), 'Revue des Congrès', *Annales d'hygiène publique et de médicine légale*, 4: 50–7.

(1920), 'Société Belge d'Eugénique', *Eugenical News*, 5: 54.

(1920), 'Belgian Society of Eugenics', *Eugenical News*, 5: 63.

(1920), 'The Eugenization of America', *Eugenical News*, 5: 14.

(1920), 'The Italian Society of Genetics and Eugenics', *American Journal of Sociology*, 25: 638.

(1921), 'Foreign Notes', *Eugenical News*, 6: 72.
(1921), 'Insanity in the Balkan States', *The Lancet*, 197 (8 January): 94.
(1921), *Memoria del primer congreso Mexicano del niño*, Mexico City: El Universal.
(1921), 'Revue d'Eugénique', *Eugenical News*, 6: 43.
(1922), 'Belgian Eugenics Society', *Eugenical News*, 7: 14.
(1922), 'Dr. Albert Govaerts of Belgium', *Eugenical News*, 7: 64.
(1922), 'International Commission of Eugenics', *Eugenical News*, 7: 117.
(1922), 'National Office of Eugenics in Belgium', *Eugenical News*, 7: 120–1.
(1922), 'Social Welfare in Italy', *Eugenical News*, 7: 62.
(1922), 'The New Belgian Eugenics Office', *Eugenical News*, 7: 92.
(1924), *Conférence Internationale de l'Émigration et de l'Immigration, Rome, 15–31 mai 1924*, Rome: Commisariat Généra Italien de l'Émigration.
(1925), 'Eugenical Activities in Different Countries', *Eugenical News*, 10: 51.
(1926), 'Fascist Asks Latin Opposition to German, Slavs, Moslems', *New York Times* (21 May): 25.
(1927), *Atti del Primo Congresso italiano di Eugenetica Sociale, Milano, 20–23 settembre 1924 Roma*, Rome: Stabilimento Poligrafico per l'Amministrazione dello Stato.
(1927), *Proyecto de Codigo Panamericano de Eugenesia y Homicultura, Project of Panamerican Code on Eugenics and Homiculture*, Havana: Imp, Montalvo y Cardenas.
(1928), *Actas de la primera conferencia Pan Americana de eugenesia y homicultura de las repúblicas Americanas*, Havana: El Gobierno de la Republica de Cuba.
(1928), 'International Population Union', *Eugenical News*, 13: 131–2.
(1928), 'Pan-American Conference on Eugenics and Homiculture', *Eugenical News*, 13: 17–19.
(1928), 'The Second Emigration Conference: Havana', *Eugenical News*, 13: 73–5.
(1929), *Actas e Trabalhos. Primeiro Congresso Brasileiro de Eugenia*, vol. 1, Rio de Janeiro: Academia National de Medicina.
(1929), 'Second Italian Congress of Genetics and Eugenics', *Eugenical News*, 14: 157 and 159–61.
(1929), 'Society for Preventive Medicine and Eugenics', *Eugenical News*, 14: 173.
(1929), 'The Meeting of the International Federation of Eugenic Organizations', *Eugenical News*, 14: 153–7.
(1930), 'Membership and Organization of the International Federation of Eugenic Organizations', *Eugenical News*, 15: 11–13.
(1931), 'On Catholicism', *The Eugenics Review*, 23: 41–5.
(1932), *Atti del Secondo Congresso Italiano di Genetica ed Eugenica, Roma, 30 settembre–2 ottobre 1929*, Rome: Tipografia Failli.
(1932), 'L'activité de la Société de Biotypologie', *Biotypologie*, 1: 35–9.
(1932), 'Stérilisation des inadaptés sociaux', *Revue Anthropologique*, 42: 228–44.
(1934), *Acta de la Segunda Conferencia Pan Americana de eugenesia y homicultura de las repúblicas Americanas*, Buenos Aires: Imprenta Frascoli y Bindi.

(1934), *Actas Generales de la Novena Conferencia Sanitaria Panamericana celebrada en Buenos Aires del 12 al 22 de Noviembre de 1934*, Washington, DC: Pan American Sanitary Bureau.

(1934), *A Decade of Progress in Eugenics. Scientific Papers of the Third International Congress of Eugenics*, Baltimore: Williams and Wilkins.

(1934), 'An Investigation of Some Indian Tribes in Mexico', *Eugenical News*, 19: 114–15.

(1934), 'Eugenics in Belgium', *Eugenical News*, 19: 148.

(1934), 'Eugenics in France', *Eugenical News*, 19: 8–9.

(1934), 'Eugenics in Mexico', *Eugenical News*, 19: 144–5.

(1934), 'Mociones y resolutiones', *Actas de la Segunda Conferencia Pan Americana de eugenesia y homicultura*, 275–84

(1934), 'Notes et informations', *Biotypologie*, 2: 140–1.

(1934), *Segunda Conferencia Panamericana de Eugenesia y Homicultura*. Actas. Buenos Aires: Frascoli y Hindi.

(1935), *Bericht über die 11. Versammlung der Internationalen Föderation Eugenischer Organisationen, Konferenzsitzungen vom 18. bis 21. Juli 1934 im Waldhaus Dolder, Zürich, Schweiz*, Zürich: Orell Füssli.

(1935), 'Ha sido inaugurado nuestro Instituto de Biotipología', *Anales de Biotipología, Eugenesia y Medicina Social*, 3: 2 and 12–6.

(1935), 'Primera reunion de la Sección Española de la Asociación Argentina de Biotipología, Eugenesia y Medicina Social', *Anales de Biotipología, Eugenesia y Medicina Social*, 3: 3.

(1936), '4me Congrès de la Presse Médicale Latine', *Revue Médicale Française*, 17: 480.

(1936), 'Ergebnisse und Beschlüsse', *St. Lukas*, 4: 173–7.

(1936), 'Eugenics in Mexico', *Eugenical News*, 21: 114.

(1936), 'Eugenics in Romania', *Eugenical News*, 21: 84–5.

(1936), 'Informations', *Revue Anthropologique*, 46: 190–1.

(1936), *Programme: 4me Congrès de la Presse Médicale Latine*, Venice: Arti Grafiche Fantoni.

(1937), 'Comptes rendus de la Réunion Internationale de Biotypologie', *Biotypologie*, 5: 73–5.

(1937), 'Decret', *Diari Oficial de la Generalitat de Catalunya*, 5: 114–15.

(1937), *Memoria del VII Congreso Panamericano del Niño: Reunido en la Ciudad de México del 12 al 19 de octubre de 1935*, vol. 1, Mexico City: Departamento de Salubrida Pública.

(1937), *Premier Congrès Latin d'Eugénique*, Paris: Masson.

(1938), 'Acţiunea Italiei pentru apărarea rasei', *Buletin Eugenic şi Biopolitic*, 9: 254–5.

(1938), 'Congrès des Sociétés Latines d'Eugénique', *Revue Anthropologique*, 48: 86.

(1938), 'Première Réunion de la Fédération des Sociétés Latines d'Eugénique', *Revue Anthropologique*, 48: 83–5.

(1939), 'Al II-lea Congres Internaţional al Federaţiunii Societăţilor Latine de Eugenie', *Revista de Ştiinţe Medicale*, 28: 247–8.

(1939), 'Atti della Terza Riunione della Società Italiana di Genetica ed Eugenica. Bologna 5–7 settembre 1938', *Genus*, 3: 1–371.

(1939), *Tercera Conferencia Panamericana de Eugenesia y Homicultura, Bogotá, septiembre 4–14, 1938*, Washington, DC: Oficina Sanitaria Panamericana.

(1940), *Primera Jornada Peruana de Eugenesia, Lima 3–5 de mayo de 1939*, Lima: Imprenta La Gotera.

(1941), *Primer Congreso de la Población: 26 a 31 de Octubre de 1940*, Buenos Aires: Museo Social Argentino.

(1943), 'Loi 941 du 16 décembre 1942 relative á la protection de la maternité et de la première enfance', *La Presse Médicale*, (9 January): 5–6.

(1943), *Segunda Jornada Peruana de Eugenesia. Lima, 25 a 29 Mayo de 1943*, Lima: Imprenta La Gotera.

(1945), *În amintirea profesorului Fr. J. Rainer (1874–1944)*, Bucharest: Monitorul Oficial si Imprimeriile Statului.

(2000), *Hitler's Table Talk, 1941–1944: His Private Conversations*, New York: Enigma Books.

(2007), 'Casti Connubii, Encyclical of Pope Pius XI on Christian Marriage to the Venerable Brethren, Patriarchs, Primates, Archbishops, Bishops, and Other Local Ordinaries Enjoying Peace and Communion with the Apostolic See (December 1930)', in Mazumdar, P. M. H. (ed.), *The Eugenics Movement: An International Perspective*, vol. 4, New York: Routledge, 69–100.

Accampo, E. A. (2003), 'The Gendered Nature of Contraception in France, Neo-Malthusianism, 1900–1920', *Journal of Interdisciplinary History*, 34: 235–62.

Achim, V. (2004), *Documente privind deportarea ţiganilor în Transnistria*, Bucharest: Enciclopedică.

ACS. 16.49 Ministero della Pubblica Istruzione, Direzione Generale Istruzione Superiore, Professori Universitari epurati (1944–1946), b. 16, 'Gini, Corrado.'

ACS. Corrado Gini Papers, BD6, Fasc. 1947.

ACS. Ministero della Pubblica Istruzione, Commissione centrale per l'epurzione del personale universitario, Roma, No. 557: Gini, Corrado, 27 November 1944.

ACS. SPD (CO) 1922–1943, Busta 1959, Fasci. 532283, T. Interlandi.

ACS. SPD (CO) 509.058 Pende, Sen. Nicola. 'Prof. Nicola Pende di Genova'; Telespresso 225771/604; 1 August 1930.

ACS. SPD (CO) 509.058, Pende, Sen. Nicola. n.d. Prof. Nicola Pende di Genova', *c.* February 1931.

ACS. SPD (CO) 509.058, Pende, Sen. Nicola. Tagliano, Telespresso, No. 241317, Min. degli Affari Est. Dir. Gen. A.A.A. Uff. II, to R. Min, Ed. Naz. Gab. Ministero, 17 December 1930. Ogg: 'Vista del Prof. Pende in Argentina'.

ACS. SPD (CO) 509.058, Pende to Mussolini, 18 October 1938; Pende, Sen. Nicola. n.d. SEG Part di SE il Cap del Gov, Roma, 20 October 1938.

ACS. SPD, Carteggio riservato (1922–1943), fasc. 242/R, Farinacci avvocaro Roberto.

Aguirre, R. C. (1934), 'Discurso', *Actas de la Segunda Conferencia Pan Americana de eugenesia y homicultura*, 22–4.

Albarrán, E. J. (2008), *Children of the Revolution: Constructing the Mexican Citizen, 1920–1940*, Ph.D. dissertation, The University of Arizona.

Alexandresco, V. (1907), 'Enseignement official et particular de la puériculture et vulgarization de l'hygiene infantile en Roumanie', *Annales de medicine et de chirurgie infantile*, 11: 474–9.

Aliano, D. (2008), *Identity in Transatlantic Play: Il Duce's National Project in Argentina*, Ph.D. dissertation, City University of New York.

Almaça, C. (2000), 'Neo-Lamarckism in Portugal', *Asclepio*, 52: 85–98.

Alonso, A. F. (1921), 'La herencia eugénica y el future de México', in *Memoria del Primer Congreso Mexicano del Niño*, Mexico: El Universal, 34–42.

Amador, N. (1914), 'La eugénica y sus relacciones con la sociología y economía politica. El factor biológico', *Estudio*, 8: 167–99.

Anderson, B. (1983), *Imagined Communities: Reflections on the Origin and Spread of Nationalism*, London: Verso.

Andronescu, C. I. (1912), 'Eugenia', *Higiena*, 1: 4.

Andronescu, C. (1936), *Le diner de printempts de l'Union Medicale Latine (UMFIA) 1935*, Bucharest: Imp. 'Indépendence'.

——. (1943), *Pentru ce ne îmbolnăvim? Noțiuni de patologie socială*, Bucharest: Tip. 'Cartea Românească'.

Antonescu, M. (1941), 'Principiile nouei reforme a statului român', *Societate de mâine*, 19: 85–7.

——. (1991), 'Directive și îndrumări date inspectorilor administrativi și pretorilor trimiși în Basarabia și Bucovina (1941)', in Lya Benjamin (ed) *Martiriul evreilor din România, 1940–1944. Documente și mărturii*, Bucharest: Hasefer, 139.

ANR. Fond Personal Nicolae Caranfil.

ANR. Fond Personal Gheorghe Marinescu.

Anuarul Universității Regele Ferdinand I, Cluj-Sibiu 1941–42: 189

Apert, E. (1919), *L'hérédité morbide*, Paris: Flammarion.

——. (1922a), 'Les Journées Eugéniques Internationales de Bruxelles et la fondation de la Ligue Nationale Belge contre le Péril Vénérien', *Paris Médical*, 46: 322–33.

——. (1922b), 'Eugénique et santé nationale', in Apert, E. et al. (eds), *Eugénique et selection*, Paris: Félix Alcan, 59–75.

——. (1923), 'Immigration et métissage: Leur influence sur la santé de la nation', *La Presse Médicale*, 75 (19 September): 1565–9.

——. (1924), 'Le problème des races et de l'immigration en France', *Eugénique*, 3: 149–67.

——. (1937), 'L'importance sociale des etudes eugéniques', in *Premier Congrès Latin d'Eugénique*, Paris: Masson, 7–12.

APS. Frederick Osborn Papers.

——. B: P 312 Raymond Pearl Papers.

——. B: D 27 CB Davenport Papers. Ramos Files. File 1. 1922–1923: Ramos to Davenport, 27 June 1922.

——. B: D 27 CB Davenport Papers. Ramos, D.F. Folder 1, 1922–1923. 'Programme for the Constitution of the Pan American Association of Eugenics and Homiculture, Santiago de Chile, March 1923'.

——. B: D 27 CB Davenport Papers. First Panamerican Conference on Eugenics and Homiculture 1926: Cuba. Folder 1. 'Race Crossing' C. B. Davenport, Havana.

——. B: D 27 CB Davenport Papers. First Panamerican Conference on Eugenics and Homoculture [sic] 1926: Cuba. Folder 2. 'The Eugenical Principle in Immigration' given at Pan American Conference at Havana, 22 December 1927.

——. B: D 27 CB Davenport Papers. First Panamerican Conference on Eugenics and Homiculture 1926: Cuba. Folder 3. 'Primera session de la Pimera Conferencia Panamericana de Eugenesia y Homocultura—Jueves 22 de Diciembre de 1927.

——. B: D 27 CB Davenport Papers n.d. Panamerican Conference on Eugenics and Homiculture First, 1926: Cuba. Folder 4. Domingo F. Ramos' Closing Speech.

——. B: D 27 CB Davenport Papers. Gini to Davenport, 29 March 1929; Gini to Davenport, 10 December 1929; Gini to Davenport, 9 June 1930; Gini to Davenport, June 1931; Gini to Davenport, 16 June 1932.

——. B: D 27 CB Davenport Papers. International Federation of Eugenic Organizations. C.B.S. Hodson. Hodson to Davenport, 9 August 1930.

——. B: D 27 CB Davenport Papers. International Federation of Eugenic Organizations. C.B.S. Hodson. Folder 2. Hodson to Davenport, 17 June 1932.

——. B: D 27 CB Davenport Papers. International Federation of Eugenic Organizations. C.B.S. Hodson. Folder 2. Hodson to Davenport, 28 July 1932.

——. B: D 27 CB Davenport Papers. Pestalozza, Ernesto, Pestalozza to Davenport, 10 June 1931; Davenport to Pestalozza, 6 July 1931; Gini to Davenport, June 1931; Gini to Davenport, 16 June 1932; Davenport to Gini, 7 July 1932; Gini to Davenport, 20 August 1932.

Aragon, H. O. (1933), 'La biotypologie a l'université de Mexico', *Biotypologie*, 1: 136–8.

Arnould, J. (1930), 'L'examen médical prenuptial', in *L'église et l'eugénisme: La famille a la croisée des chemins*. Paris: Éditions Mariage et Familie, 105–26.

Artom, C. (1926), *La teoria dell'evoluzione e gli studi di genetica*, Milan: Istituto editoriale scientifico.

Arvey, S. R. (2012), 'Sex and the Ordinary Cuban: Cuban Physicians, Eugenics, and Marital Sexulity, 1933–1958', *Journal of the History of Sexuality*, 21: 93–120.

Ayala-Carcedo, F. J. and Driever, S. L. (eds) (1998), *Lucas Mallada-La futura revolución Española y otros escritos Regenerationistas*, Madrid: Editorial Biblioteca Nueva.

Ayarragaray, L. (1904), *La anarquía argentina y el caudillismo. Estudio psicológico de los origins nacionales, hasta el año XXIX*, Buenos Aires: F. Lajouane.

Azevedo, Ayres de. (1940a), 'População e Império', in *Congressos do Mundo Português. Congresso Nacional de Ciências da População*, vol. XVII, tom 1, section 2, Lisbon: Comissão Executiva does Centenários, 61–75.

——. (1940b), 'A pureza bioquímica do Povo Português', in *Congressos do Mundo Português. Congresso Nacional de Ciências da População*, vol. XVII, tom 1, section 2, Lisbon: Comissão Executiva does Centenários, 551–64.

Aznar, S. (1943), 'El regimen de subsidios familares, la fraternidad Cristiana y las consignas del Nuevo Estado', *Revista Internacional de Sociología*, 2–3: 97–110.

Bagdasar, D. (1923), 'Sterilizarea', *Adevărul*, 1.

Baldwin, P. (2005), *Contagion and the State in Europe, 1890–1930*, Cambridge: Cambridge University Press, 1999.

Balzagette, L. (1903), *Le probleme de l'avenir latin*, Paris: Librairie Fischbacher.

Banu, G. (1927), 'Biologia satelor', *Arhiva pentru știință și reformă socială*, 7: 87–121.

———. (1930), *L'hygiene sociale de l'enfance*, vol. 1, Paris: Masson.

———. (1933), 'Legea sterilizării în Germania', *Revista de Igienă Socială*, 3: 550–4.

———. (1936), 'Principes d'un programme d'hygiène raciale', *Bulletin de l'Académie de Médicine de Roumanie*, 2: 835–69.

———. (1937), 'Les facteurs dysgéniques en Roumanie: Principes d'un programme pratique d'eugénique', *Premier Congrès Latin d'Eugénique*, 296–319.

———. (1939a), 'Dysgénies de nature toxique dans le milieu rural', in *XVIIe Congrès International d'Anthropologie et d'Archéologie Préhistorique*, Bucharest: Imprimerie Socec, 615–20.

———. (1939b) 'Hérédité multiple de malformations de la bouche', *XVIIe Congrès International d'Anthropologie*, 620–6.

———. (1939c), 'L'orientation moderne de l'éugenique', *XVIIe Congrès International d'Anthropologie*, 635–40.

———. (1939d), *L'hygiène de la race: étude de biologie héréditaire et de normalisation de la race*, Paris: Masson.

———. (1943), *Eugenia*, Bucharest: Tip. Presa D. Andreescu.

———. (1944), *Tratat de medicină socială. Vol. 1: Medicina socială ca știință. Eugenia. Demografia, Vol. 2: Maternitatea. Prima vârstă. Medicina școlară, Vol. 3: Asistența socială, Vol. 4: Tuberculoza. Bolile venerice*, Bucharest: Casa Școalelor.

———, together with Şapira, B. and Dinu, I. (1947), *Recherches sur la biologie du people romain*, Bucharest: Bucovina.

Barbero, M. I. and Felder, S. (1988), 'El rolo de los italianos en el nacimiento y desarrollo de las asociaciones empresarias en la Argentina (1880–1930)', in Devoto, F. J. and Rosoli, G. (eds), *L'Italia nella societa* Argentina, Rome: Centro Studi Emigrazione, 146–53.

Barrachina, M.-A. (2004), 'Maternidad, Feminidad, Sexualidad. Algunos Aspectos de las Primeras Jornadas Eugénicas Españolas (Madrid, 1928–Madrid, 1933)', *Hispania*, 64: 1003–26.

Bashford, A. and Levine, P. (eds) (2010), *The Oxford Handbook of the History of Eugenics*, New York: Oxford University Press.

Beck, M. (1938), 'Unabhängigkeit der geistigen Kultur von der Rasse', in *Congrès International de la Population. Paris 1937*, vol. 8, (Problèmes qualitatifs de la population), Paris: Hermann, 106–16.

Bellessort, A. et al. (1941), *France 1941: La révolution et un programme*, Paris: Éditions Alsatia.

Benavente, A. (1940), 'Colaboración al plan eugenésico que requiere el Perú', in *Primera Jornada Peruana de Eugenesia, Lima 3–5 de mayo de 1939*, Lima: Imprenta La Gotera, 26–32.

Bertillon, J. (1897), 'Le problème de la depopulation: Le programme de l'Alliance nationale pour l'accroissement de la population française', *Revue politique et parlementaire*, 4: 538–9.

Bertrand, L. (1918), *Les Pays méditerranéens et la guerre*, Paris: La Renaissance du Livre.

Beruti, J. A. (1934), 'Eugenesia y maternidad', *Actas de la Segunda Conferencia Pan Americana de eugenesia y homicultura*, 164–70.

Bethell, L. (1998), *Latin America: Economy and Society since 1930*, Cambridge: Cambridge University Press.

Betta, E. (2008), 'Note sull'eugenica cattolica', in Bredi, D. et al. (eds), *Scritti in onore di Biancamaria Scarcia Amoreti*, vol. 1, Rome: Edizioni Q, 133–44.

———. (2012), *L'altra genesi, Storia della fecondazione artificiale*, Rome: Carocci editore.

Bettiol, G. (1934), 'Sterilizzazione e diritto penale in Germania', *Rivista Italiana di Diritto Penale*, 6: 754–61.

Biernat, C. and Ramacciotti, K. (2008), 'La tutela estatal de la madre y el niño en la Argentina: Estructuras administrativas, legislación y cuadros técnicos (1936–1955)', *História, Ciências, Saúde - Manguinhos*, 15: 331–51.

Birn, A.-E. (2005), 'No More Surprising than a Broken Picher? Maternal and Child Health in the Early Years of the Pan American Sanitary Bureau', in Warsh, K., Lynn, C. and Strong-Boag, V. J. (eds), *Children's Health Issues in Historical Perspective*, Waterloo, ON: Wilfrid Laurier University Press, 72–100.

———. (2008), 'Doctors on Record: Uruguay's Infant Mortality Stagnation and its Remedies, 1895–1945', *Bulletin of the History of Medicine*, 82: 311–54.

Blum, A. S. (2009), *Domestic Economies: Family, Work, and Welfare in Mexico City, 1884–1943*, Lincoln: University of Nebraska Press.

BMG, 'Documenti inerenti l'Esposizione universale di Roma E 42'. 24 June 1938/Mss. 41/Sez. III.5 Mostra dell'ortogenesi, 1938.

Boalick, A. R. (2013), *Revitalizing Youth in the Body Politics of Contemporary Spanish Culture*, Ph.D. Dissertation, University of Michigan.

Boas, F. (1938), 'Heredity and Environment', in *Congrès International de la Population. Paris 1937*, vol. 8, (Problèmes qualitatifs de la population), Paris: Hermann, 83–92.

Bocci, M. (2003), *Agostino Gemelli rettore e francescano. Chiesa, regime e democrazia*, Brescia: Morcelliana.

Bogdan-Duică, G. (1924), 'Eduard Herriot', *Societate de mâine*, 1: 155–6.

Boldrini, M. (1928), 'Il movimento eugenico in alcuni paesi', *Difesa sociale*, 7: 51–4.

———. (1937), 'Constitution et eugénique', *Premier Congrès Latin d'Eugénique*, 228–31.

Bolea, R. C. (2013), 'Marañón y la Identidad Sexual: Biología, Sexualidad y Género en la España de la Década de 1920', *Arbor: Ciencia, Pensamiento y Cultura*, 189: 1–9.

Bonfante, P. (1915), 'Verso la confederazione europea', *Scientia: Rivista Internazionale di Sintesi Scientifica*, 18: 325–42.

Borges, D. (1993), '"Puffy, Ugly, Slothful and Inert": Degeneration in Brazilian Social Thought, 1880–1940', *Journal of Latin American Studies*, 25: 235–56.

Bowler, P. (2003), *Evolution: The History of an Idea* (3rd ed.), Berkeley: University of California Press.

Brenier, H. (1930), 'La limiation des naissances et le problème économique', *L'église et l'eugénisme*, 19–30.

Briand, H. (1938a), 'L'eugénique et la conservation des qualities de la race', *Revue Anthropologique*, 48: 55–68.

———. (1938b), 'Les progress de l'eugénique et de la génétique en France au cours des derniéres années', *Revue Anthropologique*, 48: 307–14.

———. (1941a), 'Une conference du professor Eugen Fischer in Paris', *Revue Anthropologique*, 51: 195–6.

———. (1941b), 'Education physique et sauvegarde de la race', *Revue Anthropologique*, 51: 8–15.

Broberg, G. and Roll-Hansen, N. (eds) (1997), *Eugenics and the Welfare State. Sterilization Policy in Denmark, Sweden, Norway, and Finland*, East Lansing: Michigan State University Press.

Bronfman, A. (2004), *Measures of Equality: Social Science, Citizenship, and Race in Cuba, 1902–1940*, Chapel Hill: University of North Carolina Press.

Brousseau, A. (1937), 'De la viabilité et de la fécondité des insuffisants intellectuels', *Premier Congrès Latin d'Eugénique*, 187–95.

Bucur, M. (2002), *Eugenics and Modernization in Interwar Romania*, Pittsburgh: University of Pittsburgh Press.

Burga, C. (1940), 'La eugenesia desde el punto de vista católico', *Primera Jornada Peruana de Eugenesia*, 33–8.

Burgess, G. (2011), 'The Demographers' Moment: Georges Mauco, Immigration and Racial Selection in Liberation France, 1945–1946', *French History and Civilization*, 4: 167–77.

Burns, E. B. (1983), *The Poverty of Progress: Latin America in the Nineteenth Century*, Berkeley: University of California Press.

Businco, L. (1939), 'Gente d'italia in terra francese', *La Difesa della Razza*, 2 (5 March): 15–17.

Caja, F. (2009), *La raza catalana. El núcleo doctrinal del catalanismo*, Madrid: Ediciones Encuentro.

Camavitto, D. (1937), 'Premiers résultats d'une recherché anthropologique sur les Zambos de la Costa Chica', *Premier Congrès d'Eugénique*, 40–59.

Camiscioli, E. (2001), 'Reproducing Citizens: Reproducing the "French Race": Immigration, Demography and Pronatalism in Early Twentieth Century France', *Gender and History*, 13: 593–621.

———. (2009), *Reproducing the French Race: Immigration, Intimacy, and Embodiment in Early Twentieth Century*, Durham, NC: Duke University Press.

Canali, S. (2001), 'Il consiglio nazionale delle richerche e la medicina', *Medicina nei secoli*, 13: 143–67.

Capuano, C. (2009), *Vichy et la Famille. Réalités et faux-semblants d'une politique publique*, Rennes: Presses Universitaires de Rennes.

Carelli, A. (1928), 'Valore della sterilizzazione eugenica nel miglioramento della razza umana', *Difesa sociale*, 7: 341–5.

Carneiro, L. (1929), 'Educação e Eugenia', *Actas e Trabalhos do Primeiro Congresso Brasileiro de Eugenia*, vol. 1, 112–21.

Carol, A. (1995), *Histoire de l'eugénisme en France: Les medicines et la procreation XIXe–XXe siècle*, Paris: Éditions du Seuil.

Carrel, A. (1935), *Man, the Unknown*, New York: Harper.

Casagrandi, O. (1937), 'Tentatives microscopiques et biologiques en vue de l'identification de certaines tares organiques séminales, héréditaires et acquises', *Premier Congrès Latin d'Eugénique*, 147–9.

Cassata, F. (2006), *Il fascismo razionale: Corrado Gini fra scienza e politica*, Rome: Carocci.

——. (2008), '*La Difesa della razza*': *Politica, ideologia e immagine del razzismo fascista*, Turin: Giulio Einaudi.

——. (2011), *Building the New Man. Eugenics, Racial Science and Genetics in Twentieth-Century Italy*, Budapest: Central European University Press.

——. (2013), 'The Struggle for Authority over Italian Genetics: The Ninth International Congress of Genetics in Bellagio, 1948–1953', in Gausemeier, B. et al. (eds), *Human Heredity in the Twentieth Century*, London: Pickering and Chatto, 217–28.

Castanheira, J. P. (2010), *Um Cientista Português no Coração da Alemanhã Nazi*, Coimbra: Tenacitas.

Castejón, F. (1934), 'Eugénesia y criminalidad', in Noguera, E. and Huerta, L. (eds), *Genética, eugenesia y pedagogía sexual: Libro de las Primeras Jornadas Eugénicas Españolas*, vol. 2, Madrid: Javier Morata, 181–201.

Castrilli, V. (1937), 'La nuptialité et la fécondité des diplomés de l'enseignement secondaire en Norvège', *Premier Congrès d'Eugénique Latine*, 110–19.

Caulfield, S. (2000), *In Defense of Honor: Sexual Morality, Modernity and Nation in Early-Twentieth-Century Brazil*, Durham: Duke University Press.

Cecchetto, S. (2008), *La biología contra la democratia: Eugenesia, herencia, y perjuicio de Argentina, 1880–1940*, Mar del Plata: EUDEM.

Chadwick, H. M. (1945), *The Nationalities of Europe: The Growth of National Ideologies*, Cambridge: The University Press.

Chamberlain, H. S. (1912), *The Foundations of the Nineteenth Century*, London: J. Lane.

Champy, Ch. (1937), 'L'importance des variations raciales de sensibilité aux hormones dans l'appréciation de la valeur sexuelle de l'individu', *Premier Congrès Latin d'Eugénique*, 150–3.

Cházaro, L. (2005), 'Reproduccíon y muerte de la población Mexicana: Cálculos estadísticos y preceptos higiénicos a fines del siglo diecinueve', in Agostoni, C. and Speckman, E. (eds), *Discursos, prácticas y sanciones, Ensayos de historia social en una perspectiva comparada (México, Brasil y Argentina, 1850–1950)*, Mexico City: Universidad Nacional Autónoma de México, 55–81.

Childers, K. S. (2003), *Fathers, Families, and the State in France, 1914–1945*, Ithaca, NY: Cornell University Press, 2003.

Ciceri, M. (2009), *Origini controllate: La prima eugenetica in Italia (1900–1924)*, Rome: Prospettiva Editrice.

Ciocarlan, I. (1939), 'Distribution géographique et raciale des naissances multiples en Roumanie', *XVIIe Congrès International d'Anthropologie*, 668–78.

CIWA File: Eugenics Record Office. n.d. Harry Laughlin to Walter Gilbert, 6 September 1934.

Cleminson, R. (2000), *Anarchism, Science, and Sex: Eugenics in Eastern Spain, 1900–1937*, Oxford: Peter Lang.

——. (2006), '"A Century of Civilization under the Influence of Eugenics": Dr. Enrique Diego Madrazo, Socialism and Scientific Progress', *Dynamis*, 26: 221–51.

——. (2008), 'Eugenics without the State: Anarchism in Catalonia, 1900–1937', *Studies in History and Philosophy of Biological and Biomedical Sciences*, 39: 232–9.

——. (2014), *Catholicism, Race and Empire: Eugenics in Portugal, 1900–1950*, Budapest: Central European University Press.

Closson, C. C. (1897), 'The Hierarchy of European Races', *American Journal of Sociology*, 3: 314–27.

CNSAS. Fond Iordache Făcăoaru.

Colajanni, N. (1890), *Ire e sponpositi di Cesare Lombroso*, Catania: Tropea.

——. (1905), *Latins et Anglo-Saxons, Races supérieures et races inférieures*, trans. by Julien Dubois, Paris: Félix Alcan.

——. (1906), *Latini e anglo-sassoni: Razza inferiore e razza superiore*, Rome: Rivista Popolare.

Cole, J. H. (1996), 'There Are Only Good Mothers: The Ideological Work of Women's Fertility in France before World War I', *French Historical Studies*, 19: 639–72.

Colombo, D. (2003), 'Psychoanalysis and the Catholic Church in Italy: The Role of Father Agostino Gemelli, 1925–1953', *Journal of the History of the Behavioral Sciences*, 39: 333–48.

Comfort, N. (2012), *The Science of Human Perfection: How Genes became the Heart of American Medicine*, New Haven and London: Yale University Press.

Conklin, A. L. (2013), *In the Museum of Man. Race, Anthropology, and Empire in France, 1850–1950*, Ithaca, NY: Cornell University Press.

Conry, I. (1974), *L'introduction du darwinisme en France au XIXe siècle*, Paris: Vrin.

Consiglio, P. (1914), 'Problemi di eugenica', *Rivista Italiana di Sociologia*, 18: 444–66.

'Constitution y Nomina de la Oficina Central Panamericana de Eugenesia y Homicultura', *Actas de la Segunda Conferencia Pan Americana de eugenesia y homicultura*, 284–5.

Conversi, D. (1997), *The Basques, the Catalans, and Spain, Alternative Routes to Nationalist Mobilisation*, London: Hurst.

Cosmacini, G. (1985), *Gemelli*, Milan: Rizzoli.

Cova, A. (1997), *Maternité et droits des femmes en France, XIXe–XXe siècles*, Paris: Anthropos, coll, Historiques.

CSHLA. File: Davenport, Charles B.

Cuénot, L. (1922), 'Génétique et adaptation', in Apert, E. et al. (eds), *Eugénique et selection*, 201–47.

——. (1923), 'Génétique et adaptation', in *Eugenics, Genetics and the Family, vol. 1. Scientific Papers of the Second International Congress of Eugenics*, Baltimore: Williams & Wilkins, 29–58.

Cueto, M., Jorge, L. and Carol, P. (eds) (2009), *El rastro de la salud en el Perú*, Lima: Universidad Peruana Cayetano Heredia.

Cupcea, S. P. (1941), 'Examenul prenupţial', *Buletin eugenic şi biopolitic*, 12: 105–26.

——. (1944), 'Biologia teoretica şi aplicată în URSS', *Buletin eugenic şi biopolitic*, 15: 299–318.

D'Agostino, P. (2002), 'Craniums, Criminals, and the "Cursed Race": Italian Anthropology in American Racial Thought, 1861–1924', *Comparative Studies in Society and History*, 44: 319–43.

Dalsace, J. (1936), 'À propos de la stérilisation', in Harmsen, H. and Lohse, F. (eds), *Bevölkerungsfragen: Bericht des Internationalen Kongresses für Bevölkswissenschaft, Berlin 26 August–1 September 1935*, Berlin: J. F. Lehmanns Verlag, 706–12.

D'Aroma, N. (1957), *Vent'anni insieme: Vittorio Emanuele e Mussolini*, Bologna: Capelli.

Dartigues, L. (ed.) (1926), *Livre d'or et Annuaire de L'Union Médicale Franco-Ibéro-Américaine*, Paris: Laboratoires Darrase.

Darwin, L. (1922), 'L'eugénique pratique', in Apert, E. et al. (eds), *Eugénique et selection*, 189–99.

Dávila, J. (2003), *Diploma of Whiteness: Race and Social Policy in Brasil, 1917–1945*, Durham and London: Duke University Press.

De Bont, R. (2001), 'Meten en verzoenen: Louis Vervaeck en de Belgische criminele antropologie, circa 1900–1940', *Journal of Belgian History*, 9: 63–104.

De Castro y Serrano, J. (1871), 'De cómo principió y cómo ha concluido: El año de 1870', *La Ilustración Española y Americana*, 1 (5 January): 3–6.

De Donno, F. (2006), 'La Razza Ario-Mediterranea', *Interventions*, 8: 394–412.

De Felice, R. (1966), *Mussolini il fascista, vol. 1, La conquista del potere, 1921–1925*, Turin: Giulio Einaudi.

De Grazia, V. (1992), *How Fascism Ruled Women, Italy, 1922–1945*, Berkeley: University of California Press.

——. (2000), 'Die Radikalisierung der Bevölkerungspolitik im faschistischen Italien: Mussolinis "Rassenstaat"', *Geschichte und Gesellschaft*, 26: 219–54.

De Gubernatis, A. (1898), *L'Argentina: Ricordi e Letture*, Florence: Bernardo Seeber.

De Lacerda, J. B. (1911), 'The *Metis*, or Half-Breeds, of Brasil', in Spiller, G. (ed.), *Papers on Inter-Racial Problems Communicated to the First Universal Races Congress, London, July 26–29, 1911*. London: P. S. King and Son, 377–82.

De La Fuente, A. (2001), *A Nation for All: Race, Inequality, and Politics in Twentieth-Century Cuba*, Chapel Hill: University of North Carolina Press.

Delfino, V. (1907), *El alcoholismo y sus efectos en el individuo, la familia y la sociedad*, Barcelona, Buenos Aires: Biblioteca Ambos Mundos.

——. (1925), 'Eugenics in Argentina', *Eugenical News*, 10: 123–4.

——. (1927), 'El problema de la raza', *Atti del Primo Congresso Italiano di Eugenetica Sociale*: 445–8.

De Luca, V. B. (2005), 'Restoring the Notion of Family in France: Pronatalist and Pro-family Propaganda in Schools and Army Barracks (1920–1940)', *Population*, 60: 11–35.

——. (2008), *Les familles nombreuses. Une question démographique, un enjeu politique, France (1880–1940)*, Rennes: Presses universitaires de Rennes.

De Matos, P. F. (2010), 'Aperfeiçoar a "raça", salvar a nação: Eugenia, teorias nacionalistas e situação colonial em Portugal', *Trabalhos de Antropologia e Etnologia*, 50: 89–111.

——. (2012), *Mendes Correia e a Escola de Antropologia do Porto: Contribuição para o estudo das relações entre antropologia, nacionalismo e colonialismo (de finais do século XIX aos finais da década de 50 do século XX)*, Ph.D. Dissertation, University of Lisbon.

——. (2013), *The Colours of Empire: Racialized Representations during the Portuguese Colonialism*, New York: Berghahn.

Dendle, B. J. (1970), 'The Racial Theories of Emilia Pardo Bazán', *Hispanic Review*, 38: 17–31.

De Plauzoles, S. and Weisweiller, E. (eds) (1919–1920), *Congrès Interallié d'Hygiène Sociale pour la Reconstitution des Régions Dévastées par la Guerre*, vol. 3, Paris: Éditions Ernest Leroux.

De Raes, W. (1989), 'Eugenetika in de belgische medische wereld tijdens het interbellum', *Journal of Belgian History*, 20: 399–464.

Dermine, J. (1930), 'Les lois du marriage et les devoirs des époux', *L'église et l'eugénisme*: 174–90.

De Sandoval, I. E. (1932), 'L'Eugénésie en Amerique', *Atti del Secondo Congresso italiano di Genetica ed Eugenica*, 201–5.

De Souza, M. L. and Boarini, M. L. (2008), 'A deficiência mental na concepção da liga brasileira de higiene mental', *Revista Brasileira de Educação Especial*, 14: 273–92.

De Souza, V. S. (2006), 'Em nome da raça: A propaganda eugênica e as idéias de Renato Kehl nos anos 1910 e 1920', *Revista de História Regional*, 11: 29–70.

De Souza, V. S. et al. (2009), 'The National Museum's Physical Anthropology Archives: Sources on the History of Eugenics in Brazil', *História, Ciências, Saúde – Manguinhos*, 16: 763–77.

Destrée, J. (1929), 'L'examen prénuptial doit-il être obligatoire?', *Revue Belge*, 6 (1 July): 52–3.

De Vernejoul, R. (1930), 'La limitation des naissances et le problème moral', *L'église et l'eugénisme*: 161–73.

Devers, G. (2008), 'La médicine sous Vichy, étude de legislation, 1940–1944', *Droit, Déontologie & Soin*, 8: 134–60.

Dietrich, D. J. (1992), 'Catholic Eugenics in Germany, 1920–1945: Hermann Muckermann, S. J. and Joseph Mayer', *Journal of Church and State*, 3: 575–601.

Di Liscia, S. M. (2005), 'Relaciones peligrosas: Sobre bocio, cretinismo e inferioridad (Argentina, 1870–1920)', in Speckman, E. and Agostoni, C. (eds), *De normas y transgressiones: Enfermedad y crimen en América Latina, 1850–1950*, Mexico City: Universidad Nacional Autónoma de México, 21–54.

Doizy, H. and De Plauzoles, S. (1920), 'Projet d'un Institut interallié d'Hygiene publique à créer à Paris par la Société des Nations', in De Plauzoles, S. and Weisweiller, E. (eds), *Congrès Interallié d'Hygiène Sociale*, vol. 3: 9–11.

Domingo, A. (2012), 'Immigració i política demogràfica en l'obra de Josep Antoni Vandellós', *Treballs de la Societat Catalana de Geografia*, 73: 9–26.

Drouard, A. (1992a), 'Aux origins de l'eugénisme en France: Le néo-malthusianisme (1896–1914)', *Population*, 6: 435–59.

———. (1992b), 'La création de l'INED', *Population*, 6: 1453–66.

———. (1992c), *Une inconnue des science sociale: La Fondation Alexis Carrel, 1941–1945*, Paris: Fondation Maison des sciences de l'homme.

———. (1999), 'Le Haut comité de la population et la politique de population de la France (1939–1966)', *Annales de démographie historique*, 2: 171–97.

Dubourg, M. L. (1930), 'Le veritable eugénisme', *L'église et l'eugénisme*: 224–7.

Dunn, P. M. (2006), 'Adolphe Pinard (1844–1934) of Paris and Intrauterine Paediatric Care', *Archives of Disease in Childhood: Fetal and Neonatal*, 91: 231–2.

Enăchescu, S. D. (1936), 'Sterilizarea eugenică', *Revista de Medicină Legală*, 1: 273–9.

Eraso, Y. (2013), *Representing Argentinian Mothers: Medicine, Ideas and Culture in the Modern Era, 1900–1946*, Amsterdam: Rodopi.

Făcăoaru, Gh. (1941), *Familia și statul biopolitic*, Bucharest: Bucovina.

Făcăoaru, I. (1934), 'Legiurile recente pentru sterilizarea eugenică', *Buletin Eugenic și Biopolitic*, 5: 231–9.

———. (1935), *Curs de eugenie*, Cluj: Institutul de Igienă și Igienă Socială.

———. (1936), *Criteriile pentru diagnoza rasială*, Cluj: Tip. Universală.

———. (1937), *Din problematica și metodologia cercetărilor eugenice și genetice în cadrul monografiei sociologice*, Bucharest: Institutul Social Român, 1937.

———. (1939a), 'Étude généalogique d'une famille tarée', *XVIIe Congrès International*: 626–34.

———. (1939b), 'Eugénique et biopolitique-Hérédité-Sélection sociale', *XVIIe Congrès International*, 718–99.

———. (1940), *Structura rasială a populației rurale din România*, Bucharest: F. Göbl.

———. (1943a), 'Cercetări antropologice în patru sate românești din Transnistria'. USHMM, Microfilm, Fond 2242, Opis no. 1, RG-31.004, Reel 4.

———. (1943b), 'Valoarea biorasială a națiunilor europene și a provinciilor românești', *Buletin eugenic și biopolitic*, 14: 278–310.

——. (1944), *Contribuţie la studiul compoziţiei morfologice a românilor din Republica Moldovenească*, Bucharest: Imprimeriile Institutului de Statistică.

Fallon, V. (1929), 'L'examen prénuptial doit-il être obligatoire?', *Revue Belge*, 6 (1 October): 52–4.

Federici, N. (1937), 'Mortalité, mortalité infantile et mortalité prénatale chez les familles nombreuses italiennes', *Premier Congrès Latin d'Eugénique*, 278–82.

Felix, I. (1897), *Geografia medicală a României*, Bucharest: I. V. Socec.

Fernández-Ruiz, C. (1939), *Eugenesia y esterilidad: Factores sociales de disminución de la natalidad*, San Sebastián: Editorial Española.

Ferracuti, S. (1996), 'Cesare Lombroso (1835–1907)', *The Journal of Forensic Psychiatry*, 7: 130–49.

Ferrándiz, A. and Lafuente, E. (1999), 'El pensamiento eugénico de Marañón', *Asclepio*, 51: 133–48.

Ferrero, G. (1897), *L'Europa giovane*, Milan: Treves.

Fischer, E. (1936), 'Einleiter Vortrag', in Harmsen, H. and Lohse, F. (eds), *Bevölkerungsfragen*, 42–60.

——. (1942), 'Le problème de la race et la legislation raciale allemand', *Cahiers de l'Institut Allemand*, 1: 81–110.

Flaubert, G. (1975), *Oeuvres completes, vol. 14: Correspondence, 1859–1871*, Paris: Club de l'Honnête Homme.

Fogarty, R. S. (2008), *Race and War in France: Colonial Subjects in the French Army, 1914–1918*, Baltimore, MD: The Johns Hopkins University Press.

Fortunati, P. (1937). 'Le métabolisme social d'après des recherches sur les etudiants de l'Université de Padoue', *Premier Congrès Latin d'Eugénique*, 79–90.

Foschi, R., Giannone, A. and Giuliani, A. (2013), 'Italian Psychology under Protection: Agostino Gemelli between Catholicism and Fascism', *History of Psychology*, 16: 130–44.

Frader, L. L. (2008), *Breadwinners and Citizens: Gender in the Making of the French Social Model*, Durham: Duke University Press.

Frassetto, F. (1934), 'La standardisation internationale des methods et leur synthèse en anthropologique et en eugénique', *Biotypologie*, 2: 40–2.

——. (1940), 'Per una sistematizione razionale della Biotipologia', in *Actas, memórias e comunicações apresentadas ao Congresso Nacional de Ciências da População*, vol. 17, tom. 1, Lisbon: Comissão Executiva dos Centenários, 512–34.

——. (1941), *Per una sistematizione razionale biotipologia, Rocca S.* Casciano: L. Cappelli.

Frétigné, J.-Y. (2002), *Biographie intellectuelle d'un protagoniste de l'Italie libérale: Napoleone Colajanni (1847–1921)*, Rome: École Française de Rome.

Frick, W. (1936), 'Eröffnungsansprache', in Harmsen, H. and Lohse, F. (eds), *Bevölkerungsfragen*, 6–20.

Fuchs, R. G. (1992), *Poor and Pregnant in Paris: Strategies for Survival in the Nineteenth Century*, New Brunswick, NJ: Rutgers University Press.

Garner, J. W. (1914), 'The Decreasing Population of France', *Popular Science Monthly*, 85: 247–59.

Garrett, A. (1938), *A puericultura na Itália*, Lisbon: Impr. Médica.

Gaudillière, J.-P. (2006), 'Le syndrome nataliste: Hérédité, médicine et eugénisme en France et en Grande-Bretagne', in Gayon, G. and Jacobi, D. (eds), *L'éternel retour de l'eugénisme*, Paris: PUF, 177–99.

Gausemeier, B. et al. (eds) (2013), *Human Heredity in the Twentieth Century*, London: Pickering & Chatto.

Gemelli, A. (1915), 'Si ereditano le qualità psichiche?', *Vita e pensiero*, 1: 273–83.

——. (1927), 'Religione ed eugenica', *Atti del Primo Congresso Italiano di Eugenetica Sociale*, 53–66.

——. (1928), 'Punti di vista diversi nel problema della diminuzione delle nascite', *Vita e pensiero*, 19 (November): 645–52.

——. (1931), 'Ancora sulla condanna eugenetica, Echi e critiche alla enciclica Casti Connubi', *Vita e Pensiero*, 22 (October): 606–14.

——. (1936), 'Lücken und Unsicherheiten der Eugenik als Begründung der Unzulässigkeit der Präventiv-Sterilisierung', *St. Lukas*, 4: 80–96.

Georgescu, D. C. (1936) 'Secția de Demografie, Antropologie și Eugenie a Institutului Social Român', *Sociologie românească*, 1: 56–7.

——. (1939), 'La fertilité différentielle en Roumanie', *XVIIe Congrès International*, 640–67.

Georgescu-Roegen, N. (1939), 'La corrélation entre l'âge maternel et le sexe de l'enfant dans les naissances simples et gémellaires', *XVIIe Congrès International*, 580–5.

Gervasoni, M. (1997), '"Cultura della degenerazione" tra socialismo e criminologia alla fine dell'Ottocento in Italia', *Studi Storici*, 38: 1087–119.

Gibson, M. (1998), 'Biology or Environment? Race and Southern "Deviancy" in the Writings of Italian Criminologists, 1990–1920', in Schneider, J. (ed.), *Italy's Southern Question: Orientalism in One Country*, Oxford: Berg, 99–116.

——. (2002), *Born to Crime: Cesare Lombroso and the Origins of Biological Criminality*, Westport, CT; London: Praeger.

Gillette, A. (2001), 'The Origins of the 'Manifesto of Racial Scientists', *Journal of Modern Italian Studies*, 6: 305–23.

——. (2002a), 'Guido Landra and the Office of Racial Studies in Fascist Italy', *Holocaust and Genocide Studies*, 16: 357–75.

——. (2002b), *Racial Theories in Fascist Italy*, London: Routledge.

——. (2007), *Eugenics and the Nature–Nurture Debate*, New York: Palgrave Macmillan.

Gini, C. (1912a), *I fattori demografici dell'evoluzione delle nazioni*, Torin: Bocca.

——. (1912b), 'The Contributions of Demography to Eugenics', in *Problems in Eugenics. Papers communicated to the First International Eugenics Congress*, University of London, 24–30 July1912, London: The Eugenics Education Society, 294–334.

——. (1923), 'The War from the Eugenic Point of View', *Eugenics in Race and State: vol. 2, Scientific Papers of the Second International Congress of Eugenics*, 430–1.

——. (1926), 'Decline in the Birth-Rate and the "Fecundability" of Woman', *The Eugenics Review*, 17: 258–74.

——. (1927a), 'Le relazioni dell'eugenica con le altre scienze biologiche e sociali', *Atti del Primo Congresso Italiano di Eugenetica Sociale*, 3–26.

——. (1927b), 'The Scientific Basis of Fascism', *Political Science Quarterly*, 42: 99–115.

——. (1930), *Nascita, evoluzione e morte delle nazioni: La teoria ciclica della popplazione e I vari sistemi di politica demografica*, Rome: Libreria del littorio.

——. (1931), *Le basi scientifiche della politica della popolazione*, Catania: Studio editoriale modern.

——. (1932), 'Discorso inaugurale del Presidente della S.I.G.E', *Atti del Secondo Congresso Italiano di Genetica ed Eugenica*, 17–27.

——. (1934a), *Atti del Congresso Internationale per gli Studi sulla Popolazione*, Rome: Istituto Poligrafico dello Stato, 10 vols.

——. (1934b), 'Response to the Presidential Address', in *A Decade of Progress in Eugenics. Scientific Papers of the Third International Congress of Eugenics, held at American Museum of Natural History*, New York, 21–23 August 1932, Baltimore: The Williams & Wilkins Company, 25–8.

——. (1936), 'Parole inaugurali del Prof. Corrado Gini: Presidente della Società Italiana di Genetica ed Eugenica e della Federazione Latina fra le Società di Eugenica: lette alla Riunione delle Societàdi Eugenica dell'America Latina tenutasi a Messico il 12 Ottobre 1935', *Genus*, 2: 75–81.

——. (1937a), 'Allocution', *Premier Congrès Latin*: 5–6.

——. (1937b), 'Biotypologie et eugénique', *Premier Congrès Latin*: 200–11.

——. (1937c), 'De quelques recherches récentes sur les variations que présenteraient certains caractères suivant le nombre d'enfants de la famille', *Premier Congrès Latin*: 262–7.

——. (1940), 'Les facteurs de la décroissance de la natalité à l'époque contemporaine', *Actas, memórias e comunicações apresentadas ao Congresso Nacional de Ciências da População*, vol. 17, tom. 1, 76–123.

——. (1941), 'Le Rôle du Facteur Démographique dans la Politique Internationale', *Affairs Danubiennes: Revue de l'Europe Centrale et du Sud-Est*, 4: 67–94.

——. (1967), 'Race and Sociology', in Kuttner, R. E. (ed.), *Race and Modern Science*, New York: Social Science Press, 261–75.

Giorgi, G. M. (2011), 'Corrado Gini: The Man and the Scientist', *Metron-International Journal of Statistics*, 69: 1–28.

——. (2012), 'Corrado Gini', in Clericuzio, A. and Ricci, S. (eds), *Il contributo italiano alla storia del pensiero*. Ottava appendice. Economia, Rome: Istituto della Enciclopedia Italiana, 620–6.

Giuffrida-Ruggeri, V. (1912), 'The So-called Laws of Inheritance in Man', *Problems in Eugenics*: 38–47.

Glass, D. V. (1935), 'The Berlin Population Congress and Recent Population Movements in Germany', *Eugenics Review*, 27: 207–12.

———. (1936), 'Italian Attempts to Encourage Population Growth', *The Review of Economic Studies*, 3: 106–19.

Glick, T. F. (2001), 'The Reception of Darwinism in Uruguay', in Puig-Samper, M. A., Ruiz, R. and Glick, T. F. (eds), *The Reception of Darwinism in the Iberian World: Spain, Spanish America and Brazil*, Dordrecht: Kluwer Academic Publishers, 28–52.

———. (2005), 'Marañón, Intersexuality and the Biological Construction of Gender in 1920s Spain', *Cronos*, 8: 121–38.

Goemaere, P. (1929), 'L'examen prénuptial doit-il être obligatoire?', *Revue Belge*, 6 (1 June): 408–10.

Gomes, A. C. V. (2012), 'A emergência da biotipologia no Brasil: Medir e classificar a morfologia, a fisiologia e o temperamento do brasileiro na década de 1930', *Boletim do Museu Paraense Emílio Goeldi*, 7: 705–19.

Gómez, M. J. B. (2005), 'Eugenesia y pediatría: Hygiene infantil y "degeneración de la raza" en España a principios del siglo XX', in Miranda, M. and Vallejo, G. (eds), *Darwinismo social y eugenesia en el mundo latino*, Buenos Aires: Siglo XXI, 641–64.

González, A. (2007), *Las trampas del poder: Sanidad, eugenesia y migración, Cuba y Estados Unidos (1900–1940)*, Madrid: Consejo Superior de Investigaciones Científicas.

González, A. G. (2001), 'Darwinism, Eugenics and Mendelianism in Cuban Biological Education: 1900–1959', in Puig-Samper, M. A., Ruiz, R. and Glick, T. F. (eds), *The Reception of Darwinism in the Iberian World*, 153–70.

González, A. G. and Peláez, R. Á. (1999), *En busca de la raza perfecta, Eugenesia e higiene en Cuba (1898–1958)*, Madrid: Colección Estudios sobre Ciencia.

Goode, J. (2009), *Impurity of Blood: Defining Race in Spain, 1870–1930*, Baton Rouge: Louisiana State University Press.

Gori, G. (2000), 'Model of Masculinity: Mussolini, the "New Italian" of the Fascist Era', in Mangan, J. A. (ed.), *Superman Supreme: Fascist Body as Political Icon-Global Fascism*, London: Routledge, 27–61.

Govaerts, A. (1927), 'La pratique de l'examen médical avant le mariage en Belgique', in Sand, R. et al. (eds), *L'examen médical en vue du mariage*, Paris: Ernest Flammarion, 177–84.

Grădinescu, A. (1936), 'Dr Nicola Pende', *Endocrinologie, Gynecologie, Obstetrică*, 1: 196–202.

Graham, R. (ed.) (1990), *The Idea of Race in Latin America, 1870–1940*, Austin: University of Texas Press.

Griffing, J. B. (1940), 'Natural Eugenics in Brazil', *The Journal of Heredity*, 31: 13–16.

Grimoult, C. (2012), 'Les anthropologues et la Révolution nationale (1940–1944)', *Parlement[s]*, 18: 55–66.

Guerra, L. (2002), *From Revolution to Involution in the Early Republic: Conflicts over Race, Class, and Nation, 1902–1906*, Chapel Hill: University of North Carolina Press.

Guy, D. J. (1990), *Sex and Danger in Buenos Aires: Prostitution, Family and Nation in Argentina*, Lincoln: University of Nebraska Press.

Haidar, V. (2011), '"Every Man in his due Place": The Biotypological Solution to the Conflict between Productivity and Health (Argentina, 1930–1955)', *Salud Colectiva*, 7: 317–32.

Hansen, R. and King, D. (2013), *Sterilized by the State: Eugenics, Race, and the Population Scare in Twentieth-Century North America*, New York: Cambridge University Press.

Harmsen, H. and Lohse, F. (eds) (1936), *Bevölkerungsfragen: Bericht des Internationalen Kongresses für Bevölkswissenschaft, Berlin 26 August–1 September 1935*, Berlin: J. F. Lehmanns Verlag.

Haro, F. (1932), *Eugenesia y Matrimonio, El Certificado Médico Prenupcial*, Madrid: Javier Morata.

Harris, R. (1989), *Murders and Madness: Medicine, Law, and Society in the Fin-de-Siecle*, Oxford: Clarendon Press.

——. (1993), 'The "Child of the Barbarian": Race and Nationalism in France during the First World War', *Past & Present*, 141: 170–206.

Harris, R. G. (1922), 'Eugenics in South America', *Eugenical News*, 7: 17–42.

Harsin, J. (1985), *Policing Prostitution in Nineteenth-Century Paris*, Princeton: Princeton University Press.

Haškovec, L. (1920), 'Sur le certificat médical avant le mariage', in De Plauzoles, S. and Weisweiller, E. (eds), *Congrès Interallié d'Hygiène Sociale*, vol. 3, 152–65.

Hațieganu, I. (1936), 'Senatorul Nicolae Pende', *Endocrinologie, Gynecologie, Obstetrică*, 1: 193–5.

Hațieganu, I. and Voina, A. (1925), 'ASTRA medicală şi biopolitică', *Societatea de mâine*, 2: 813–4.

Hawthorn, É. (1930), 'La doctrine et le mouvement néo-malthusiens', *L'église et l'eugénisme*, 3–18.

Haycraft, J. B. (1895), *Darwinism and Race Progress*, London: Sonnenschein.

Hedrick, T. (2003), *Mestizo Modernism: Race, Nation, and Identity in Latin American Culture, 1900–1940*, New Brunswick, NJ: Rutgers University Press.

Helg, A. (1990), 'Race in Argentina and Cuba, 1880–1930: Theory, Policies, and Popular Reaction', in Graham, R. (ed.), *The Idea of Race in Latin America*, 37–69.

Henriques, B. M. C. (2012), *A ressurreição da raça portuguesa no pensamento de Mendes Correia. História, Antropologia, Eugenia (1911–1960)*, Porto: Universidade do Porto, Faculdade de Letras.

Hèritier, Ph. (1937), 'Mise en evidence de l'existence de differences d'ordre biométrique et physiologique, entre diverses souches de drosophilia melanogaster, meig. Identiques morphologiquement', *Premier Congrès d'Eugénique Latin*, 101–7.

Hernández, E. and and Ramos, D. F. (1911), *Homicultura*, Havana: Secretaria de Sanidad y Beneficencia.

Herseni, T. (1940), 'Mitul sângelui', *Cuvântul*, 17: 2.

——. (1941), 'Rasă şi destin naţional', *Cuvântul*, 18: 1 and 7.

——. (1947), 'Ancheta bio-sociologică', *Buletin eugenic şi biopolitic*, 18: 184–96.

——. (1958), 'Studiul genealogic al populaţiei', in Milcu, Şt. M. and Dumitrescu, H. (eds), *Cercetări antropologice în Ţara Haţegului. Clopotiva*, Bucharest: Editura R. P. R., 47–65.

——. (1961), 'Familia şi grupurile genealogice', in Milcu, Şt. M. and Dumitrescu, H. (eds), *Cercetări antropologice în Ţara Pădurenilor. Satul Bătrîna*, Bucharest: Editura R. P. R., 57–71.

Hesse, P.-J.and Le Crom, J.-P. (2001), *La Protection sociale sous le régime de Vichy*, Rennes: Presses Universitaires de Rennes, 2001.

Heuyer, G. and and Courthial, A. (1937), Constitution et eugénique', *Premier Congrès Latin d'Eugénique*, 232–8.

Hochman, G., Lima, N. T. and Maio, M. C. (2010), 'The Paths of Eugenics in Brazil: Dilemmas of Miscegenation', in Bashford, A. and Levine, L. (eds), *The Oxford Handbook of the History of Eugenics*, 493–510.

Hoffmann, G. (1913), *Rassenhygiene in den Vereinigten Staaten von Nordamerika*, Munich: J. F. Lehmanns Verlag.

Horn, D. G. (1994), *Social Bodies: Science, Reproduction, and Italian Modernity*, Princeton: Princeton University Press.

Houssay, F. (1912), 'Eugenics Selection and the Origin of Defects', *Problems in Eugenics*: 158–61.

——. (1922), 'Eugénique et régimes alimentaires', in Apert, E. et al. (eds), *Eugénique et selection*, 21–32.

Huerta, L. (1918), *Eugénica, Maternología y Puericultura, Ensayo de un estudio sobre el cultivo de la especie humana por las leyes biológicas*, Madrid: Imprenta de Fontanet.

——. (1934), 'Hacia una nueva cultura del amor', in Noguera, E. and Huerta, L. (eds), *Genética, eugenesia y pedagogía sexual*, 1: 7–11.

Hunt, N. R. (1999), *A Colonial Lexicon: Of Birth Ritual, Medicalization, and Mobility in the Congo*, Durham and London: Duke University Press.

Huss, M.-M. (1990), 'Pronatalism in the Inter-war Period in France', *Journal of Contemporary History*, 25: 39–68.

Husson, R. (1937), 'Mesure des accroissements quantitatifs et qualitatifs d'une population', *Premier Congrès Latin d'Eugénique*, 72.

——. (1938), 'Théorie du classement des individus en biométrie individuelle sur la base de combinaisons linéaires de mesures', *Congrès International de la Population*, 8: 1–2.

Ichok, G. (1938), 'La classification des professions et les methods de biotypologie', *Congrès International de la Population*, 8: 33–7.

Ide, M. (1929), 'L'examen prénuptial doit-il être obligatoire?', *Revue Belge*, 6 (1 July): 50–2.

Illanes, M. A. (2007), *Cuerpo y sangre de la política: La construcción histórica de las visitadoras sociales, Chile, 1887–1940*, Santiago de Chile: Lom Ediciones.

Interlandi, T. (1939), 'Carta razziale Italiana', *La Difesa della Razza*, 2: 7.

Ioanid, R. (2000), *The Holocaust in Romania: The Destruction of Jews and Gypsies under the Antonescu Regime, 1940–1944*, Chicago: Ivan R. Dee.

Ionescu-Muscel, P. and Lombroso, C. (1929), *Opera sa 'L'uomo delinquente'*, Craiova: Scrisul Romanesc.

Iorga, N. (1936), *Pentru Italia şi ce ne leagă de Italia*, Bucharest: Tip. 'Universul'.

Ipsen, C. (1996), *Dictating Demography: The Problem of Population in Fascist Italy*, Cambridge: Cambridge University Press.

Israel, G. (2007), 'Il documento *Il fascismo e i problemi della Razza* del luglio 1938', *La Rassegna Mensile di Israel*, 73: 103–18.

———. and Nastasi, P. (1998), *Scienza e razza nell'Italia fascista*, Bologna: Il Mulino.

Izquierdo, J. J. (1923), 'Breve reseña genealogica de la familia Izquierdo', *Eugenics, Genetics and the Family*, 348–73.

Jackson, J. (2001), *France: The Dark Years, 1940–1944*, Oxford: Oxford University Press.

Jarnot, S. (2000), 'Une relation récurrente. Science et racisme', *L'exemple de L'Ethnie Française'*, *Les Cahiers du Cériem*, 5: 17–34.

Jennings, É. (2002), 'Discours corporatiste, propagande nataliste, et contrôle social sous Vichy', *Revue d'histoire moderne et contemporaine*, 4: 101–31.

Jonas, R. (2011), *The Battle of Adwa: African Victory in the Age of Empire*, Cambridge, MA: Belknap Press.

Jordan, É. (1930), 'Préface', in *L'église et l'eugénisme*, v–xi.

———. (1931), *Eugénisme et morale*, Paris: Libraire Bloud & Gay.

Kehl, R. (1920), *Eugenia e medicina social – problemas da vida*, Rio de Janeiro: Livraria Francisco Alves.

———. (1921), 'Eugenics in Brazil', *Eugenical News*, 6: 18.

———. (1929a), *Lições de Eugenia*, Rio de Janeiro: Francisco Alves.

———. (1929b), *A eugenia no Brasil: Esboço historico e bibliographico*, Rio de Janeiro: Editora Sodré.

———. (1930), *Pedagogía sexual: Lecciones de eugenesia*, Madrid: Javier Morata.

———. (1937a), *Por que sou eugenista: 20 anos de campanha eugenica 1917–1937*, Rio de Janeiro: Livraria Francisco Alves.

———. (1937b), 'Valeur comparée de l'accroissement qualitative et quantitative d'une population', *Premier Congrès Latin d'Eugénique*, 73–8.

Kevles, D. J. (1985), *In the Name of Eugenics: Genetics and the Uses of Human Heredity*, New York: Alfred A. Knopf.

Kligman, G. (1998), *The Politics of Duplicity: Controlling Reproduction in Ceausescu's Romania*, Berkeley: University of California Press.

Knight, A. (1990), 'Racism, Revolution, and *Indigenismo*: Mexico, 1910–1940', in Graham, R. (ed.), *The Idea of Race in Latin America*, 71–113.

Knudsen, E. R. (2010), 'Recasting the Italian: The Influence of Fascist Racism. Enforcing the Racial Ideal, 1938–1940', Paper for ISS Colloquium 2010, Yale University, http://iss.yale.edu/node/30/attachment (accessed 5 April 2014).

Korherr, R. (1928), *Regresso delle nascite: Morte dei popoli*, Rome: Libreria del Littorio.

Kühl, S. (1994), *The Nazi Connection: Eugenics, American Racism, and German National Socialism*, New York: Oxford University Press.

——. (2013), *For the Betterment of the Race: The Rise and Fall of the International Movement for Eugenics and Racial Hygiene*, New York: Palgrave Macmillan.

Lackerstein, D. (2012), *National Regeneration in Vichy France: Ideas and Policies, 1930–1944*, Farnham: Ashgate.

Lamarck, J. B. (1914), *Zoological Philosophy: An Exposition with Regard to the Natural History of Animals*, London: Macmillan.

Lamas, C. S. (1934), 'Discurso', *Actas de la Segunda Conferencia Pan Americana de eugenesia y homicultura*, 18–20.

Landau, E. (1918), 'Notes eugéniques', *The Eugenics Review*, 10: 81–4.

Landra, G. (1938), 'Per una carta della razza italiana in Francia', *La Difesa della Razza*, 2 (January): 8–10.

——. (1942a), 'Razzismo biologico e scientismo', *La Difesa della Razza*, 6 (November): 9–11.

——. (1942b), *Il problema della razza in Romania*, Bucharest: Istituto Italo-Romeno di Studi Demografici e Razziali.

Landry, A. (1934), *La révolution démographique: Études et essais sur les problèmes de la population*, Paris: Librairie du Recueil Sirey.

Langeron, A. (1942), *Recherche d'un Français nouveau*, Paris: Editions Baudiniere.

Larbiou, B. (2005), 'René Martial, 1873–1955. De l'hygiénisme à la raciology, une trajectorie possible', *Genèses*, 60: 98–120.

Laughlin, H. (1927), 'O privire asupra sterilizării eugenice în America', *Buletin Eugenic și Biopolitic*, 1: 253–7.

Laugier, H. (1933), 'La biotypologie au Mexique', *Biotypologie*, 1: 145–9.

Lavrin, A. (1995), *Women, Feminism & Social Change in Argentina, Chile & Uruguay, 1890–1940*, Lincoln: University of Nebraska Press.

Lázaro, L. M. (2009), 'Luis Huerta: Eugenesia, medicina y pedagogía en España', *Historia de la educacíon*, 28: 61–88.

Leclercq, J. (1929), 'L'examen prénuptial doit-il être obligatoire?', *Revue Belge*, 6 (1 July): 53–5.

Le Dantec, F. (1899), *Lamarckiens et darwiniens: Discussion de quelques théories sur la formation des espèces*, Paris: Félix Alcan.

Lefaucheur, N. (1992), 'La puériculture d'Adolphe Pinard', in Tort, P. (ed.), *Darwinisme et société*, Paris: Presses Universitaires de France, 413–35.

Le Gall, P. (2007), *A History of Econometrics in France: From Nature to Models*, London and New York: Routledge.

Leon, S. M. (2013), *An Image of God: The Catholic Struggle with Eugenics*, Chicago and London: The University of Chicago Press.

Léonard, J. (1983a), 'Eugénisme et darwinisme, Espoirs et perplexities chez des medicines francais du XIXe siècle et du début du XXe siècle', in Conry, Y. (ed.), *De Darwin au Darwinisme: Science et idéologie*, Paris: Vrin, 187–207.

——. (1983b), 'Le premier Congrès international d'eugénique (Londres, 1912) et ses consequences françaises', *Société Française d'Histoire de la Médicine*, 17: 141–6.

Leonida, I. (1935), 'Ce poate realiza practic eugenia la noi', *Mişcarea Medicală Română*, 8: 366–71.

Lepicard, É. (1998), 'Eugenics and Roman Catholicism: An Encyclical Letter in Context: *Casti Connubii*, 31 December 1930', *Science in Context*, 11, 3–4: 527–44.

Leriche, R. (1941), 'Medicine', in Bellessort, A. et al. (eds), *France 1941: la Révolution Nationale Constructive. Un bilan et un programme*, Paris: Éditions Alsatia, 146–86.

Lesné, E. (1937), 'Influence des régimes carencés et déséquilibrés, suralimentation et sous-alimentation, sur la natalité, la mortinatalité et la mortalité des petits rats', *Premier Congrès Latin d'Eugénique*, 144–6.

Letard, E. (1937), 'Les leçons de l'expérimentation animale dans le problème du métissage', *Premier Congrès d'Eugénique*, 61–9.

Levi della Vida, G. (1937), 'Le métabolisme social comme facteur de dégénération dans la société', *Premier Congrès d'Eugénique*, 120–31.

Lévy-Solal, E. (1919), *La puériculture*, Paris: Leroux.

Lima, N. T. (2007), 'Public Health and Social Ideas in Modern Brazil', *American Journal of Public Health*, 97: 1168–77.

Livingstone, M. A. (2014), *The Fascists and the Jews of Italy: Mussolini's Race Laws, 1938–1943*, New York: Cambridge University Press.

Lolli, E. (1888), *La confederazione latina: Utopia di Liello Lamenue*, Rome: Tip. Romana.

Lombroso, C. (2004), *Criminal Woman, the Prostitute, and the Normal Woman*, Durham, NC: Duke University Press.

——. (2006), *Criminal Man*, Durham, NC: Duke University Press.

López-Cordón, M. V. (1975), *El pensamiento politico-internacional del federalismo español (1868–1874)*, Barcelona: Editorial planeta.

López-Guazo, L. S. (2001), 'The Mexican Eugenics Society: Racial Selection and Improvement', in Puig-Samper, M. A., Ruiz, R. and Glick, T. F. (eds), *The Reception of Darwinism in the Iberian World*, 143–52.

——. (2005), *Eugenesia y racismo en México*, Mexico City: Universidad Nacional Autónoma de México.

López, O. V. (1932), 'Los nuevos fundamentos biológicos de la eugenesia', *Annals de l'Acadèmia de Medicina de Barcelona*, 14: 141–50.

——. (1935), 'Cristianismo y eugenesia', *Anales de Biotipología, Eugenesia y Medicina Social*, 3: 6–7.

Loria, A. (1912), 'The Psycho-physical Elite and the Economic Elite', *Problems in Eugenics*, 179–83.

Löscher, M. (2009), '*... der gesunden Vernunft nicht zuwider...*'? *Katholische Eugenik in Österreich vor 1938*, Vienna: Studien Verlag.

Loyo, G. (1934), *Las deficiencias cuantitativas de la población de México y una politica demográfica nacional*, Mexico City: Partido Nacional Revolucionario.

———. (1936), *La politica demográfica de México: Ley general de población*, Mexico City: Ministerio de Prensa y Propaganda.

Lucamante, S. (2008), *A Multitude of Women: The Challenges of the Contemporary Italian Novel*, Toronto: University of Toronto Press.

Lucassen, L. (2010), 'A Brave New World: The Left, Social Engineering, and Eugenics in Twentieth-Century Europe', *International Review of Social History*, 55: 265–96.

Luchaire, J. (1965), *Confession d'un française moyen, vol. 2, 1914–1950*, Florence: Leo S. Olschki.

Maciel, M. E. (1991), 'A Eugenia No Brasil', *Anos*, 90, 11: 121–43.

Madrazo, E. D. (1903), *¿El pueblo español ha muerto? Impresiones sobre el estado actual de la sociedad española*, Santander: Blanchard y Arce.

Manias, C. (2009), 'The *Race prussienne* Controversy: Scientific Internationalism and the Nation', *Isis*, 100: 733–59.

Manicatide, M. (1937), 'Comment fait-on la statistique de la mortalité infantile', *Premier Congrès Latin d'Eugénique*, 292–5.

Manliu, I. (1921), *Crâmpeie de eugenie și igienă socială*, Bucharest: Tip. 'Jockey-Club'.

———. (1931), 'Sterilizarea degeneraților', *Revista de Igienă Socială*, 1: 374–85.

———. (1933), 'Sterilizarea degeneraților', *Progresul Social*, 2: 763–73.

Mantovani, C. (2003), 'Rigenerare la stirpe: Il movimento eugenico italiano e la Grande Guerra 1915–1924', *Ricerche di storia politica*, 2: 203–24.

———. (2004), *Rigenerare la società: L'eugenetica in Italia dalle origini ottocentesche agli anni trenta*, Rubbettino: Soveria Mannelli.

Manuilă, S. (1936a), 'Autorizarea de funcționare, actul constitutiv și statutele Societății Regale Române pentru Eugenie și Studiul Eredității', *Revista de Igienă Socială*, 6: 271–8.

———. (1936b), 'Societatea regală română de eugenie și studiul eredității', *Sociologie Românească*, 1: 31–2.

———. (1940a), 'Problema rassială a României', *România nouă*, 7: 5.

———. (1940b), *Studiu etnografic asupra populației României*, Bucharest: Editura Institutului Central de Statistică.

———. (1941), 'Acțiunea eugenică ca factor de politică de populație', *Buletin eugenic și biopolitic*, 12: 1–4.

Manuilă, V. and Domilescu, M. (1939), 'Eugenischer und demographischer Einfluss der unehelichen Verbindungen auf die Progenitur', *XVIIe Congrès International*, 683–94.

Marañon, G. (1966), *Obras completas*, vol. 1, Madrid: Espasa Calpe.

March, L. (1912), 'The Fertility of Marriages According to Profession and Social Position', *Problems in Eugenics*, 208–20.
——. (1913), 'Depopulation and Eugenics' (part 1), *The Eugenics Review*, 5: 234–51.
——. (1914), 'Depopulation and Eugenics' (part 2), *The Eugenics Review*, 5: 343–51.
——. (1919), 'Some Attempts towards Race Hygiene in France during the War', *Eugenics Review*, 10: 195–212.
——. (1921), 'The Consequences of War and the Birth Rate in France', *The Scientific Monthly*, 1 (July–December): 399–420.
——. (1922), 'Natalité et eugénique', in Apert, E. et al. (eds), *Eugénique et selection*, 77–129.
——. (1923), 'The Consequence of War and the Birth Rate in France', *Eugenics, Genetics and the Family*, 243–65.
——. (1927), 'Conclusions générales', in Sand, R. et al. (eds), *L'examen médical en vue du mariage*, 239–46.
Mărgărit, O. (1944), 'Pentru reabilitarea unei ştiinţe: Eugenia', *Scânteia*, 14: 2.
Mărgineanu, N. (1944), *Psihologia persoanei* (2nd ed.), Bucharest: Editura Universităţii din Cluj la Sibiu.
Marín, R. C. and Huertas, R. (2001), 'The Theory of Degeneration in Spain (1886–1920),' in Puig-Samper, M.A., Ruiz, R. and Glick, T. F. (eds), *The Reception of Darwinism in the Iberian World*, 171–87.
Marinescu, G. (1906), 'Progresele şi tendinţele medicinei moderne', *Academia Română. Discursuri de recepţie*, 28: 3–34.
——. (1934), 'Constituţia umană după lucrările şcoalei italiene', *Academia Română. Memoriile Secţiunii Ştiinţifice*, 9: 1–34.
——. (1936), 'Despre hereditatea normalăşi patologicăşi raporturile ei cu eugenia', *Academia Română. Memoriile Secţiunii Ştiinţifice*, 11: 1–85.
——. (1937a), 'Allocation', *Premier Congrès d'Eugénique*: 13–14.
——. (1937b), 'Les relations du gigantisme et de l'acromégalie considérées a la lumière des données histologiques et génétiques', *Premier Congrès d'Eugénique*: 351–68.
Marinesco, M. and Stroesco, G. (1939), 'L'influence des facteurs héréditaires sur le mécanisme de non-disjonction des chromosomes sexuels chez Drosophila melanogaster', *XVIIe Congrès International*: 575–80.
Marques, V. R. B. (1994), *A medicalização da raça. Médicos, educadores e discurso eugênico*, Campinas: Editora da Unicamp.
Marro, A. (1912), 'The Influence of the Age of the Parents upon the Psycho-physical Characters of the Children', *Problems in Eugenics*: 118–36.
Martial, R. (1931), *Traité de l'immigration et de la greffe inter-raciale*, Paris: Larose.
——. (1933), 'L'immigration et le pouvoir de résorption de la France', *Revue Anthropologique*, 43: 352–69 and 447–67.
——. (1934), *La Race française*, Paris: Mercure de France.
——. (1935a), 'Indésirables et refoulements', *Mercure de France*, 884: 83–92.

——. (1935b), 'Politique de l'immigration', *Mercure de France*, 884: 267–94.

——. (1937), 'Mètissage et immigration', *Premier Congrès Latin d'Eugénique*: 16–35.

——. (1938), *Race, hérédité, folie. Étude d'anthroposociologie appliquée à l'immigration*, Paris: Mercure de France.

——. (1943), *Notre race et ses aïeux*, Paris: Secretariat Général de la Jeunesse.

Mayer, J. (1930), 'Eugenics in Roman Catholic Literature', *Eugenics*, 3: 43–51.

Mayr, E. (1982), *The Growth of Biological Thought: Diversity, Evolution, and Inheritance*, Cambridge, MA: Belknap Press.

McDougall, M. L. (1983), 'Protecting Infants: The French Campaign for Maternity Leaves, 1890s–1913', *French Historical Studies*, 13: 79–105.

Mckiernan-González, J. (2012), *Fevered Measures: Public Health and Race at the Texas-Mexico Border, 1848–1942*, Durham and London: Duke University Press.

McLaren, A. (1983), *Sexuality and Social Order: The Debate over the Fertility of Women and Workers in France, 1770–1920*, London: Holmes & Meier.

Mendes Correia, A. (1914), *Os Criminosos Portugueses* (2nd ed.), Coimbra: F. França Amado Editor.

——. (1919), *Raça e Nacionalidade*, Porto: Editores Renascença Portuguesa.

——. (1933), 'La distribution géographique des hommes supérieurs', *Revue Anthropologique*, 43: 142–8.

——. (1940), 'O mestiçamento nas Colónias Portuguesas', *Actas, memórias e comunicações apresentadas ao Congresso Nacional de Ciências da População*, vol. 14, tom. 1, 111–33.

Michaelis, M. (1998), 'Mussolini's Unofficial Mounthpiece: Telesio Interlandi – *Il Tavere* and the Evolutioin of Mussolini's Anti-Semitism', *Journal of Modern Italian Studies*, 3: 217–40.

Michels, R. (1912), 'Eugenics in Party Organisation', *Problems in Eugenics*, 232–6.

Ministero dell Cultura Popolare, Gabinetto, Ufficio Razza, Guido Landra, 'Appunto per S.E. il Ministero', Roma, 31 dicembre 1938/XVII, 'Prima riunione del comitato Italo-Tedesco per le questioni razziali, relazione: Berlino–Monaco, 13–21 dicembre 1938/ XVII'.

Minkowska, F. (1937), 'Eugénique et généalogie', *Premier Congrès Latin d'Eugénique*, 341–50.

Miranda, M. A. (2005), 'La biotipología en el pronatalismo argentino (1930–1983)', *Asclepio*, 57: 189–218.

Missa, J. N. and Susanne, C. (eds) (1999), *De l'eugénisme d'État à l'eugénisme privé*, Paris: De Boeck.

Mjöen, A. (1913), 'Legal Certificates of Health before Marriage', *Eugenics Review*, 4: 356–62.

Moldovan, I. (1923), 'Dare de seamă asupra activității Institutul de Igienă şi Igienă Socială în anul şcolar 1921–22', in *Anuarul Universității din Cluj pe anul şcolar 1921–22*, Cluj: Institutul de Arte Grafice 'Ardealul', 66–80.

——. (1925), *Igiena națiunii: Eugenia*, Cluj: Institutul de Igienă şi Igienă Socială.

——. (1926), *Biopolitica*, Cluj: Institutul de Igienă şi Igienă Socială.
——. (1927), 'Necesitatea unui minister al vigoarei naţionale', *Buletin eugenic şi biopolitic*, 1: 338–40.
——. (1943), 'Statul etnic', *Buletin eugenic şi biopolitic*, 14: 1–19.
——. (1946), 'Concepte din domeniul igienei', *Buletin eugenic şi biopolitic*, 17: 1–7.
Moldovan, I., Stoichiţă, I. and Râmneanţu, P. (1947), *Tratat de sănătate publică* (2nd ed.), Cluj: n.p.
Monacelli, M. (1937), 'La protezione igienica del bambino in Italia nei raporti con la lotta antivenerea', in *Deliberatione Congressus Dermatologorum Internationalis, IX*, vol. 3, Leipzig: Johann Ambrosius Barth, 386–90.
Moncorvo Filho, A. (1926), *Histôrico da protecção à infancia no Brasil*, Rio de Janeiro: Empresa Graphica Editora.
Morcillo, A. G. (2010), *The Seduction of Modern Spain: The Female Body and the Francoist Body Politic*, Cranbury, NJ: Rosemund Publishing.
Morselli, E. (1912), 'La psicologia etnica e la scienza eugenistica', *Problems in Eugenics*, 58–9.
——. (1923), *L'Uccisione Pietosa (L'Eutanasía) in rapporto alla Medicina, alla Morale ed all'Eugenica*, Turin: Fratelli Bocca.
Mosso, A. (1897), 'Le cagione dell'efeminatezza latina', *Nuova Antologia: Rivista di scienze, letter ed arti*, 22 (November): 249–65.
Mucchielli, L. (2000), 'Criminologie, hygiénisme et eugénisme en France (1870–1914): Débats médicaux sur l'élimination des criminels réputés "incorrigibles"', *Revue d'histoire des sciences humaines*, 3: 57–89.
Muel-Dreyfus, F. (2004), 'La rééducation de la sociologie sous le regime de Vichy', *Actes de la recherche en sciences sociales*, 153: 65–77.
Munareto, G. D. (2013), *Por uma nova raça: Pensamento médico eugênico no rio grande do sul (1920–1940)*, M.A. Dissertation, Rio Grande do Sul: História da Faculdade Filosofia e Ciências Humanas da Pontifícia Universidade Católica do Rio Grande do Sul.
Murard, L. and Zylberman, P. (1996), *L'Hygiène dans la République, 1870–1918*, Paris: Fayard.
Mussolini, B. (1916), 'Italia e Francia (20 November)', in *Opera Omnia di Benito Mussolini*, vol. 11 (1953), Florence: La Fenice, 199.
——. (1919a), 'Italy and France', *Review of the Foreign Press, Allied Press Supplement*, 5 (26 February): 364.
——. (1919b), 'Benito Mussolini, *Popolo d'Italia*', 15 May 1919, *Review of the Foreign Press, Allied Press Supplement*, 6 (28 May): 104–6.
——. (1928), 'Preface', in Korherr, R. (ed.), *Regresso delle nascite: morte dei popoli*, 7–23.
——. (1929), 'Relazione alle camera dei deputati sugli accordi del Laterano (14 May)', in *Opera Omnia di Benito Mussolini*, vol. 24 (1958), Florence: La Fenice, 43–90.
NARA File 540.1A1a. C[larence] Campbell to US Secretary of State [Cordell Hull], *c.* 6 October 1934.
NARA File 540.1A1a. Report of the American Delegation to the Department of State, 14 August 1935.

NARA File 540.1A1a. n.d. Hugh Smith Cumming, Bolivar Lloyd, Kendall Emerson, 15 August 1935.

Nari, M. A. M. (1999), 'La eugenesia en Argentina, 1890–1940', *Quipu: Revista Latinoamericana de Historia de las Ciencias y la Tecnología*, 12: 343–69.

Nash, M. (1985), 'Aproximación al movimiento eugénico español: El primer curso eúgenico y la aportación del Dr. Sebastián Recansesn', *Gimbernat: Revista Catalana d'historia de la medicina i de la ciencia*, 4: 195–202.

———. (1991), 'Pronatalism and Motherhood in Franco's Spain', in Thane, G. and Bock, P. (eds), *Maternity and Gender Policies: Women and the Rise of the European Welfare States, 1880s–1950s*, New York: Routledge, 160–77.

———. (1992), 'Social Eugenics and Nationalist Race Hygiene in Early 20th Century Spain', *History of European Ideas*, 15, 4–6: 741–8.

———. (1999), 'Un/Contested Identities: Motherhood, Sex Reform and the Modernization of Gender Identity in Early Twentieth-Century Spain', in Enders, V. E. and Radcliff, P. B. (eds), *Constructing Spanish Womanhood: Female Identity in Modern Spain*, Albany, NY: State University of New York Press, 25–50.

Nasso, I. (1935), 'Commenti bibliografici: Fasti della dottrina tedesca della ereditarietà e delle razze', *Pediatria*, 43: 720–2.

Niceforo, A. (1897), *La delinquenza in Sardegna: Note di sociologia criminale*, Palermo: Sandron.

———. (1898), *L'Italia barbara contemporanea*, Milan; Palermo: Remo Sandron.

———. (1901), *Italiani del nord e italiani del sud*, Torino: Fratelli Bocca.

———. (1912), 'The Cause of the Inferiority of Physical and Mental Characters in the Lower Social Classes', *Problems in Eugenics*, 184.

Nisot, M.-T. (1927), *La question eugénique dans les divers pays*, vol. 1, Bruxelles: Librairie Falk fils.

———. (1929a), *La question eugénique dans les divers pays*, vol. 2, Bruxelles: Librairie Falk fils.

———. (1929b), 'La sterilization des anormaux', *Mercure de France*, 40: 595–603.

Noguera, E. and Huerta, L. (eds) (1934), *Genética, eugenesia y pedagogía sexual: Libro de las Primeras Jornadas Eugénicas Españolas*, 2 vols, Madrid: Javier Morata.

Novoa, A. and Levine, A. (2010), *From Man to Ape: Darwinism in Argentina*, Chicago: University of Chicago Press.

Nye, R. A. (1984), *Crime, Madness, and Politics in Modern France*, Princeton: Princeton University Press.

Odobescu, G. (1936), *Eugenie pentru neamul românesc*, Bucharest: Monitorul Oficial şi Imprimeriile Statului.

Offen, K. (1984), 'Depopulation, Nationalism, and Feminism in Fin-de-Siècle France', *American Historical Review*, 89: 648–76.

Orbegoso, A. (2012), 'Eugenesia, tests mentales y degeneración racial en el Perú', *Revista de Psicología*, 14: 230–43.

Ortega y Gasset, J. (1975), *Meditaciones del Quijote*, Madrid: Aguilar [first published 1914].

Orzaes, C. C. (2009), 'La institucionalización de la maternología en España durante la Secunda República y el Franquismo', *Historia de la Educación*, 28: 161–83.

Osborn, F. (1938), 'The Application of Measures of Quality', *Congrès International de la Population*, 117–22.

Otaola, J. M. (1934), 'Concepción y anticoncepción', in Noguera, E. and Huerta, L. (eds), *Genética, Eugenesia y Pedagogía sexual*, vol. 1, 273–309.

Ottaviani, A. (2010), 'Dallo spirito latino allo spirito della razza: Il caso Gemelli', in Gentili, S. and Foà, S. (eds), *Cultura della razza e cultura letteraria nell'Italia del Novecento*, Rome: Carocci, 69–92.

Padovan, C. (1937), 'La disproportion entre le travail libre et le travail servile comme cause perturbatrice dans la composition de la population', *Premier Congrès Latin d'Eugénique*, 132–42.

Palma, H. A. and di Vincenzo, G. J. A. (2009), 'Biotipología, eugenesia y orden social en la Argentina de 1930 a 1943', *Eä – Revista de Humanidades Médicas & Estudios Sociales de la Ciencia y la Tecnología*, 1: 1–21.

Papillault, G. 'Conséquences psycho-sociales de la derniere guere au point de vue eugénique', in Apert, E. et al. (eds), *Eugénique et selection*, 131–57.

Pasteau, O. (1936), 'Staatliche Eugenik und Materialismus', *St. Lukas*, 4: 77–9.

Paté, H. (1919), 'Discours', in De Plauzoles, S. and Weisweiller, E. (eds), *Congrès Interallié d'Hygiène Sociale*, vol. 1, 41–5.

Paulo, L. F. (1945), 'Os Tipos Constitucionais nos Portugueses', *Revista do Centro de Estudos Demográficos*, 1: 115–38.

Pedersen, J. E. (2003), *Legislating the French Family: Feminism, Theater, and Republican Politics, 1870–1920*, New Brunswick: Rutgers University Press.

Péhu, M., Noël, R. and Brochier, A. (1939), 'Les érythroblastoses familiales et raciales dans l'enfance', *XVIIe Congrès International d'Anthropologie et d'Archéologie Préhistorique*, 591–600.

Peláez, R. Á. (1985a), 'Introducción al estudio de la eugenesia española (1900–1936)', *Quipu: Revista Latinoamericana de Historia de las Ciencias y la Tecnología*, 2: 95–122.

——. (1985b), 'La eugenesia en España', *Mundo Cientifico*, 5: 622–9.

——. (1988a), 'El Instituto de Medicina Social: Primeros intentos de institucionalizar la eugenesia (en España)', *Asclepio*, 40: 343–58.

——. (1988b), 'Eugenesia y fascismo en la España de los años treinta', in Huertas, R. and Ortiz, C. (eds), *Ciencia y fascism*, Madrid: Doce Calles, 77–95.

——. (2005), 'Eugenesia, ideología y dicurso del poder en España', in Miranda, M. and Vallejo, G. (eds), *Darwinismo social y eugenesia en el mundo latino*, 87–114.

Pende, N. (1922), *Le debolezze di costituzione, Introduzione alla patologia constituzionale*, Rome: Bardi.

——. (1928), *Constitutional Inadequacies: An Introduction to the Study of Abnormal Constitutions*, Philadelphia, PA: Lea & Febiger.

——. (1930), *Trattado sintetico di patologia e clinica medica*, vol. 3, Genoa: Messinma.

——. (1933a), *Bonifica umana razionale e biologia politica*, Bologna: Licinio Cappelli.

——. (1933b), 'Les variétés raciales humaines de la Ligurie', *Biotypologie*, 1: 113–34.

——. (1935a), 'Biologia de las razas y unidad espiritual mediterrnea', *Anales de Biotipología, Eugenesia y Medicina Social*, 3: 2–4.

——. (1935b), 'Biotipologia şi ameliorarea constituţiei umane', *România Medicală*, 13, 5: 67–8.

——. (1936), 'La mentalita' mediterranea nella nuova medicina dell'italia imperiale', *Il Policlinico, sezione pratica*, 45, 9 (9 November): 2004–6.

——. (1937), 'La méthode biotypologique d'exploration de l'individu humain', *Biotypologie*, 5: 76–84.

——. (1938), 'La terra, la donna e la razza', *Gerarchia*, 18 (October): 663–9.

——. (1939a), *Scienza dell'ortogenesi*, Bergamo: Istituto italiano d'arti grafiche.

——. (1939b), *Trattato di biotipologia umana individuale e sociale con applicazioni alla medicina preventiva, all politica biologica, alla sociologia*, Milan: Francesco Vallardi.

——. (1940), 'Il principio biotipologico unitario ("Moderna dottrina della razza")', *Gerarchia*, 11 (November): 569–72.

——. together with Gualco, S. and Landogna Cassone, F. (1933), 'Les variétés raciales humaines de la Ligurie, Premières recherches de biotypologie métrique et fonctionnelle des races blanches', *Biotypologie*, 1: 113–34.

Penrose, L. S. (1939), 'Maternal Age and Congenital Disease', *XVIIe Congrès International d'Anthropologie et d'Archéologie Préhistorique*, 600–5.

Pérez, L. A. (1998), *The War of 1898, The United States and Cuba and History and Historiography*, Chapell Hill: The University of North Carolina Press.

Perrier, E. (1922), 'Eugénique et biologie', in Apert, E. et al. (eds), *Eugénique et selection*, 1–20.

Persell, S. M. (1999), *Neo-Lamarckism and the Evolution Controversy in France, 1870–1920*, New York: Edwin Mellen Press.

Pestalozza, E. (1927), 'Le operazioni operatorie in rapporto all'eugenica', *Atti del Primo Congresso Italiano di Eugenetica Sociale*, 81–5.

Petrovici, E. (1939), 'Le facteur héréditaire dans l'apparition de la syphilis nerveuse', *XVIIe Congrès International d'Anthropologie et d'Archéologie Préhistorique*, 605–14.

Petruccelli, J. L. (1993), 'Influences françaises sur la pensée brésilienne: Races, peuple et population (1890–1930)', *Annales de Démographie Historique*, 251–62.

Piéri, J. (1930), 'La stérilisation', *L'église et l'eugénisme*, 71–88.

Pimentel, I. F. (1998), 'O aperfeiçoamento da raça', *História*, 3: 18–27.

——. (1999), 'A assistência social e familiar do Estado Novo nos anos 30 e 40', *Análise Social*, 34: 477–508.

Pinard, A. (1895), 'Notes pour servir à l'histoire de la puériculture intra-utérine', *Bulletin de l'Académie de medicine* (3rd series), 34: 593–7.

——. (1899), 'De la conservation et de l'amélioration de l'espèce', *Bulletin Medical*, 13: 141–6.

——. (1908), *De la puériculture*, Lyon: Imprimeries Réunies.

——. (1912), 'Considérations générales sur 'la puériculture' avant la procreation', *Problems in Eugenics*, 457–9.

——. and Richet, C. (1903), 'Rapport sur les causes physiologiques de la diminution de la natalité en France', *Annales de Gynécologie et d'Obstétrique*, 59: 15–24.

Piot, E. (1900), *La question de la dépopulation en France: Le mal, ses causes, ses remèdes*, Paris: P. Mouillot.

——. (1902), *La dépopulation: Enquête personnelle sur la dépopulation en France: documents, discours, et rapports*, Paris: P. Mouillot.

Pitt-Rivers, G. H. L. F. (ed.) (1932), *Problems of Population*, London: George Allen.

Pogliano, C. (1984), 'Scienza e stripe: Eugenica in Italia (1912–1939)', *Passato e Presente*, 5: 61–97.

——. (2005), *L'ossessione della razza. Antropologia e genetica nel XX secolo*, Pisa: Scuola normale superiore.

Poisson, G. (1916), 'La race germanique et sa prétendue supériorité', *Revue anthropologique*, 26: 25–43.

Pollard, M. (1998), *Reign of Virtue: Mobilizing Gender in Vichy France*, Chicago: University of Chicago Press.

Popovici, A. C. (1910), *Naționalism sau democrație*, Bucharest: 'Minerva'.

Popovici, G. (1928), 'Biopolitica, puericultura și schimbarea de sistem în conducerea statului', *Societatea de mâine*, 5: 443–4.

Pozzi, L. (2012), 'Augustino Gemelli: L'eugenetica fascista e l'enciclica Casti connubii', *Annali di storia dell'educazione e delle istituzioni scolastiche*, 19: 161–74.

Preda, V. (1947). *Tratat elementar de antropobiologie*, Sibiu: Editura 'Dacia Traiana'.

Puig i Sais, H. (1915), *El problema de la natalitat a Catalunya, Un peril gravíssim per la nostra pàtria*, Barcelona: Impremta de Viuda de Badia Cantenys.

Punnett, R. C. (1911), *Mendelism* (3rd ed.), New York: Macmillan.

Querton, L. (1912), 'The Practical Organisation of Eugenic Action', *Problems in Eugenics*, 146–50.

Quine, M. S. (1996), *Population Politics in Twentieth Century Europe: Fascist Dictatorships and Liberal Democracies*, London: Routledge.

——. (2002), *Italy's Social Revolution: Charity and Welfare from Liberalism to Fascism*, Basingstoke: Palgrave.

——. (2010), 'The First-Wave Eugenic Revolution in Southern Europe: Science sans Frontières', in Bashford, A. and Levine, P. (eds), *The Oxford Handbook of the History of Eugenics*, 377–97.

——. (2012), 'Racial "Sterility" and "Hyperfecundity" in Fascist Italy. Biological Politics of Sex and Reproduction', *Fascism*, 1: 92–144.

——. (2013), 'Making Italians: Aryanism and Anthropology in Italy during the Risorgimento', in Turda, M. (ed.), *Crafting Humans: From Genesis to Eugenics and Beyond*, Göttingen: V&R Unipress, 127–52.

Quinlan, S. M. (2007), *The Great Nation in Decline: Sex Modernity and Health Crises in Revolutionary France, c.1750–1850*, Aldershot: Ashgate.

Rădulescu-Motru, C. and Nestor, I.-M. (1948), *Cercetări experimentale asupra inteligenței la români*, Bucharest: Imprimeria Națională.

Radzinowicz, L. (1999), *Adventures in Criminology*, London: Routledge.

Ramacciotti, K. I. (2003), 'El Museo Social Argentino y el Primer Congreso de Población de 1940', *Sociohistórica*, 13–14: 231–6.

Râmneanțu, P. (1937), 'Din metodele biotipologiei', *Buletin eugenic și biopolitic*, 8: 257–61.

———. (1938), 'Biotype and Fertility of Women from the Province of Banat (Rumania)', *Congrès International de la Population*, 14–20.

———. (1939), *Elemente de biometrie medicală și statistică vitală*, Bucharest: Institutul Central de Statistică.

———. (1940), 'Măsuri de politică demografică și politica demografică totalitară', *Buletin eugenic și biopolitic*, 11: 29–52.

———. (1975), 'Biometrical Variations of Some Anthropological Parameters of Romanians in the Last Century', *Annuaire roumain d'anthropologie*, 12: 25–30.

Ramos, D. F. (1923), 'Homiculture in Its Relations to Eugenics in Cuba', *Eugenics in Race and State*, 432–4.

———. (1934), 'Control of Immigration', *A Decade of Progress in Eugenics*: 79–86.

———. (1940), 'Agriculture and Homiculture: In its relations with Biophysics, Biocosmics, and Biocracy', *Informes*, 3: 35–44.

———. (1942), 'Public Health as an Important Part of Pan American Defense', *American Journal of Public Health*, 32: 627–30.

Reggiani, A. H. (1996), 'Procreating France: The Politics of Demography, 1919–1945', *French Historical Studies*, 19: 725–54.

———. (2002), 'Alexis Carrel, the Unknown: Eugenics and Population Research under Vichy', *French Historical Studies*, 25: 331–56.

———. (2005), 'La ecología institucional de la eugenesia: Repensando las relaciones entre biomedicina y política en la Argentina de entreguerras', in Miranda, M. and Vallejo, G. (eds), *Darwinismo social y eugenesia en el mundo latino*, 273–309.

———. (2007), *God's Eugenicist: Alexis Carrel and the Sociobiology of Decline*, New York: Berghahn Books.

———. (2010), 'Dépopulation, fascisme et eugénisme 'latin' dans l'Argentine des années 1930', *Le Mouvement Social*, 230: 7–26.

Regianni, A. and Bollo, H. G. (2007), 'Dénatalité, "Crise de la race" et politiques démographiques en Argentine (1920–1940)', *Vingtième Siècle. Revue d'histoire*, 95: 29–44.

Reis, J. R. F. (1999), 'Raça, Imigração e Eugenia: O Projeto de "Regeneração Nacional" da Liga Brasileira de Higiene Mental', *Estudos Afro-Asiáticos*, 36: 29–55.

Rey, E. (1912), *La Renaissance de l'orueil Français*, Paris: Bernard Grasset.

Ricci, M. (1938), 'Ereditarià ed eugenica', *La Difesa della Razza*, 1: 29–31.

Richards, M. (2004), 'Spanish Psychiatry c. 1900-1945: Constitutional Theory, Eugenics and the Nation', *Bulletin of Spanish Studies*, 81, 6: 823–48.

——. (2012), 'Antonio Vallejo Nágera: Heritage, Psychiatry and War', in Quiroga, A. and del Arco, M. (eds), *Right-Wing Spain in the Civil War Era: Soldiers of God and Apostles of the Fatherland, 1914–45*, London: Continuum, 195–224.

Richet, C. (1919), *La sélection humaine*, Paris: Alcan.

——. (1922), 'La sélection humaine', Apert, E. et al. (eds), *Eugénique et selection*, 33–57.

Richter, I. (2001), *Katholizismus und Eugenik in der Weimarer Republik und im Dritten Reich: Zwischen Sittlichkeitsreform und Rassenhygiene*, Paderborn: Ferdinand Schöningh.

Rodriguez, J. (2006), *Civilizing Argentina: Science, Medicine and the Modern State*, Chapel Hill: University Press of North Carolina.

Rodríguez-Ocaña, E. (1987), *La Constitución de la Medicina Social como Disciplina en España (1882–1923)*, Madrid: Ministerio de Sanidad y Consumo.

——. (2007), 'Medicine as a Social Political Science: The Case of Spain c. 1920', *Hygiea Internationalis*, 6: 37–52.

Rodríguez, M. F. (2012), 'La légalisation de l'eugénisme en Espagne. Les théories de Jiménez de Asúa dans *Libertad de amar y derecho a morir* (1928),' *Babel*, 26: 133–45.

Roquette-Pinto, E. (1929), 'Notas Sobre os tipos antropológicos do Brasil', *Actas e Trabalhos do I Congresso Brasileiro de Eugenia*, vol. 1, 119–47.

Rosen, C. (2004), *Preaching Eugenics: Religious Leaders and the American Eugenics Movement*, Oxford: Oxford University Press.

Rosental, P.-A. (2012), 'Eugenics and Social Security in France Before and After the Vichy Regime', *Journal of Modern European History*, 10: 540–61.

Rossi, A. (1940), 'Instituto de Biotipología y Medicina del Trabajo', *Anales de Biotipología, Eugenesia y Medicina Social*, 25: 20–4.

——. (1941), 'Herencia, constitución, eugenesia y ortogenesis', *Anales de Biotipología, Eugenesia y Medicina Social*, 26: 1–24.

Roussy, G. and Huguenin, R. (1937), 'Vues sur le role de l'hérédité dans le cancer humain', *Premier Congrès Latin d'Eugénique*, 171–86.

Rowold, K. (2010), *The Educated Woman: Minds, Bodies, and Women's Higher Education in Britain, Germany, and Spain, 1865–1914*, New York: Routledge.

Ryan, D. F. (1996), *The Holocaust & the Jews of Marseilles: The Enforcement of Anti-Semitic Policies in Vichy's France*, Urbana, IL: University of Illinois Press.

Sánchez-Albornoz, N. (1974), *The Population of Latin America: A History*, Berekely: University of California Press.

Sand, R. et al. (eds) (1922), *L'examen médical en vue du mariage*, Paris: Ernest Flammarion,

——. (1927), 'Examen médical prénuptial aux Étas-Unis', in Sand, R. et al. (eds), *L'examen médical en vue du mariage*, 37–43

——. (1934), *L'économie humaine par la médicine sociale*, Paris: Les Éditions Rieder.

——. (1935), *Health and Human Progress, An Essay in Sociological Medicine*, London: Kegan Paul.

Sanger, M. (ed.) (1927), *Proceedings of the World Population Conference, held at the Salle Centrale, Geneva, August 29th to September 3rd, 1927*, London: Edward Arnold.

Sauvage, P. (1941), *La politique familiale de l'État français*. Paris: Éditions Spes.

Savarino, F. (2002), 'The Sentinel of the Bravo: Italian Fascism in Mexico, 1922–1935', in Sorensen, G. and Mallett, R. (eds), *International Fascism, 1919–1945*, London: Frank Cass, 97–120.

Savorgnan, F. (1917), *La Guerra e la popolazione*, Bologna: Zanichelli.

Schell, P. A. (2010), 'Eugenics Policy and Practice in Cuba, Puerto Rico, and Mexico', in Brashford, A. and Levine, P. (eds), *The Oxford Handbook of the History of Eugenics*, 477–92.

Schneider, W. H. (1982), 'Toward the Improvement of the Human Race: The History of Eugenics in France', *The Journal of Modern History*, 54: 268–91.

——. (1986), 'Puericulture, and the Style of French Eugenics', *History and Philosophy of the Life Sciences*, 8: 265–77.

——. (1990), *Quality and Quantity: The Quest for Biological Regeneration in Twentieth-Century France*, Cambridge: Cambridge University Press.

——. (1991), 'The Scientific Study of Labor in Interwar France', *French Historical Studies*, 17: 410–46.

——. (1994), 'Hérédité, sang, et opposition à l'immigration', *Ethnologie française*, 24: 104–17.

——. (ed.) (2002), *Rockefeller Philantrophy and Modern Biomedicine, International Initiatives from World War I to the Cold War*, Bloomington, IN: Indiana University Press.

Schraenen, W. (1929), 'L'examen prénuptial doit-il être obligatoire?', *Revue Belge*, 6 (1 October): 58–61.

Schreiber, G. (1913a), 'Considérations sur l'amélioration de la race humaine', *La Presse Médicale*, 21: 285–7.

——. (1913b), 'Considérations sur l'amélioration de la race humaine', *La Presse Médicale*, 34: 489–92.

——. (1913c), 'Considérations sur l'amélioration de la race humaine', *La Presse Médicale*, 38: 545–8.

——. (1913d), 'Considérations sur l'amélioration de la race humaine', *La Presse Médicale*, 54: 777–80.

——. (1913e), 'Considérations sur l'amélioration de la race humaine', *La Presse Médicale*, 62: 897–9.

——. (1913f), 'Considérations sur l'amélioration de la race humaine', *La Presse Médicale*, 71: 1009–11.

——. (1913g), 'Considérations sur l'amélioration de la race humaine', *La Presse Médicale*, 91: 1273–7.

——. (1922), 'Eugénique et mariage', in Apert, E. et al. (eds), *Eugénique et selection*, 159–88.

——. (1926), 'Examen médical prénuptial devant la société française d'eugénique', *Vie Médicale*, 7: 2411–14.

——. (1927), 'Examen médical prénuptial dans les différents pays', in Sand, R. et al. (eds), *L'examen médical en vue du mariage*, 9–29.

——. (1930), 'France', *Report of the Ninth Conference of the International Federation of Eugenic Organisations, Farnham, Dorset, September 11th to 15th, 1930*, London: IFEO, 81–2.

——. (1935a), 'Compte rendu de la XIe Assemblée de la Fédération Internationale des Organisations Eugéniques', *Revue Anthropologique*, 45: 78–82.

——. (1935b), 'La stérilisation eugénique aux États-Unis', *Revue Anthropologique*, 45: 82–4.

——. (1935c), 'La stérilisation eugénique en Allemagne une année d'application de la loi du 14 Julliet 1933 pour l'enraiement de l'hérédité morbide', *Revue Anthropologique*, 45: 84–91.

——. (1936), 'Actual Aspects of the Problem of Eugenical Sterilization in France', *Eugenical News*, 11: 104–5.

——. (1937), 'Allocations familiales et eugénique', *Premier Congrès Latin d'Eugénique*, 91–100.

Schreider, E. (1933), 'L'école biotypologique italienne', *Biotypologie*, 1: 67–97.

——. (1938), 'Une mission biotypologique au Mexique', *Congrès International de la Population*, 8–13.

Scott, J. C. (1998), *Seeing like a State: How Certain Schemes to Improve the Human Condition Have Failed*, New Haven: Yale University Press.

Sergi, G. (1889), *La decadenza delle nazioni latine*, Turin: Fratelli Bocca.

——. (1912), 'Variation and Heredity in Man', *Problems in Eugenics*, 16–22.

SIAA. Arles Hrdlicka Papers n.d. Box 60: Società Italiana di Genetica ed Eugenica, Corrado Gini to Arles Hrdlicka, August 1919.

Sinclair, A. (2007), *Sex and Society in Early Twentieth-Century Spain: Hildegart Rodríguez and the World League for Sexual Reform*, Cardiff: University of Wales Press.

Sippial, A. T. (2013), *Prostitution, Modernity, and the Making of the Cuban Republic, 1840–1920*, Chapel Hill: University of North Carolina.

Skidmore, T. (1990), 'Racial Ideas and Social Policy in Brazil, 1870–1940', in Graham, R. (ed.), *The Idea of Race in Latin America*, 7–36.

Smith, J. S. (1914), 'Marriage, Sterilization and Committment Laws Aimed at Decreasing Mental Deficiency', *Journal of the American Institute of Criminal Law and Criminology*, 5: 364–70.

Sobral, J. M. (2008), 'Race and Space in Interpretations of Portugal: The North–South Division and Representations of Portuguese National Identity in the Nineteenth and Twentieth Centuries', in Roseman, S. R. and Parkhurst, S. S. (eds), *Recasting Culture and Space in Iberian Contexts*, New York: State University of New York Press, 205–24.

Solano, S. (1940), 'La cartilla biotipológica obligatoria contribuye a la higiene de la raza', *Primera Jornada Peruana de Eugenesia*, 96–101.

Soldán, C. E. P. (1916), *La medicina social, Ensayo de sistematización*, Lima: Imprenta de los Sagrados Corazones.

———. (1919), 'Eugenização da América', *Saúde*, 4: 95–6.

———. (1934a), 'Discurso', *Actas de la Segunda Conferencia Pan Americana de eugenesia y homicultura*, 24–31.

———. (1934b), 'La eugenesia Americana y sus posibilidades', *Actas de la Segunda Conferencia Pan Americana de eugenesia y homicultura*, 198–208.

Solonari, V. (2010), *Purifying the Nation: Population Exchange and Ethnic Cleansing in Nazi-Allied Romania*, Washington: Woodrow Wilson Centre Press.

Somogyi, S. (1934), 'La conception fasciste de la politique démographique', *Economia*, 12: 1–13.

Sòrgoni, B. (2003), '"Defending the Race": The Italian Reinvention of the Hottentot Venus during Fascism', *Journal of Modern Italian Studies*, 8: 411–24.

Speckman-Guerra, E. (2005), 'Menores delincuentes en la ciudad de Mexico (1884–1910)', in Speckman, E. and Agostoni, C. (eds), *De Normas y Transgresiones*, 225–53.

Spektorowski, A.and Ireni-Saban, L. (2013), *Politics of Eugenics: Reproduction, Population, and National Welfare*, New York and London: Routledge.

Spencer, H. (1897), *The Principles of Sociology*, New York: Appleton.

Spengler, J. J. (1938), *France Faces Depopulation*, Durham, NC: Duke University Press.

Stan, L. (1941), 'Rasism față de țigani', *Cuvântul*, 18: 1–2.

Stepan, N. L. (1985), 'Eugenesia, genética y salud pública: El movimiento eugenésico brasileño y mundial', *Quipu: Revista Latinoamericana de Historia de las Ciencias y la Tecnología*, 2: 351–84.

———. (1990), 'Eugenics in Brazil, 1917–1940', in Adams, M. (ed.), *The Wellborn Science: Eugenics in Germany, France, Brazil, and Russia*, New York: Oxford University Press, 110–52.

———. (1991), *The Hour of Eugenics: Race, Gender, and Nation in Latin America*, Ithaca and London: Cornell University Press.

———. (1992), 'Race, Gender and Nation in Argentina: The influence of Italian Eugenics', *The History of European Ideas*, 15, 4–6: 749–56.

Sterian, E. (1910), 'Sterilizarea nebunilor și criminalilor', *Medicul Poporului*, 1 (20 March): 113–14.

Stern, A. M. (1999a), *Mestizophilia, Biotypology, and Eugenics in Post-Revolutionary Mexico: Towards a History of Science and the State, 1920–1960*, Working Papers Series (No. 4), Chicago, Center for Latin American Studies.

———. (1999b), 'Responsible Mothers and Normal Children: Eugenics, Nationalism, and Welfare in Post-Revolutionary Mexico, 1920–1940', *Journal of Historical Sociology*, 12: 369–97.

———. (2003), 'From Mestizophilia to Biotypology: Racialization and Science in Mexico, 1920–1960', in Appelbaum, N. P., Macpherson, A. S. and Rosemblatt, K. A. (eds), *Race & Nation in Modern Latin America*, Chapel Hill, NC: The University of North Carolina Press, 187–210.

——. (2011), '"The Hour of Eugenics" in Veracruz, México: Radical Politics, Public Health and Latin America's Only Sterilization Law', *Hispanic American Historical Review*, 91: 431–43.

Stoichiță, I. (1944), 'Organizația sanitară în Rusia Sovietică', *Buletin eugenic și biopolitic*, 15: 213–41.

Strauss, P. (1901), *Dépopulation et puériculture*, Paris: Charpentier.

Suárez-Diaz, E. and Barahona, A. (2013), 'Post-war and Post-revolution: Medical Genetics and Social Anthropology in Mexico, 1945–70', in Gausemeier, B. et al. (eds), *Human Heredity in the Twentieth Century*, 101–12.

Sulică, M. (1944), 'Eredologie și patologie', *Buletin eugenic și biopolitic*, 15: 265–74.

Sutter, J. (1950), *L'eugénique: Problèmes, méthodes, résultats*, Paris: INED.

Taguieff, P.-A. (1991), 'L'introduction de l'eugénisme en France: Du mot à l'idée', *Mots: Les langages du politique*, 26: 23–45.

——. (1995), 'Le débat sur l'immigration dans l'entre-deux-guerres', *Vingtieme Siècle*, 47: 103–31.

——. (2001), *The Force of Prejudice. On Racism and Its Doubles*, Minneapolis, MN: University of Minnesota Press.

Tamagnini, E. (1938), 'Gründung der Portugiesischen Gesellschaft für eugenische Studien', *Archiv für Rassen- und Gesellschaftsbiologie*, 32: 554–9.

Tanner, J. (2012), 'Eugenics before 1945', *Journal of Modern European History*, 10: 458–79.

Tauro, G. (1937a), 'La transmigration des classes sociales par l'education', *Premier Congrès Latin d'Eugénique*, 320–1.

——. (1937b), 'Eugénique et pédagogie', *Premier Congrès Latin d'Eugénique*, 379–80.

Tedesco, L. (2011), '"For a Healthy, Peace-Loving and Hardworking Race": Anthropology and Eugenics in the Writings of Giuseppe Sergi', *Modern Italy*, 16: 51–65.

Thooris, M. A. (1937), 'Considérations ethnologiques et démographiques sur la population française', *Premier Congrès Latin d'Eugénique*, 214–27.

Thorp, R. (1998), *Progress, Poverty and Exclusion: An Economic History of Latin America in the 20th Century*, Washington, DC: Inter-American Development Bank.

Timm, A. F. (2010), *The Politics of Fertility in Twentieth-Century Berlin*, New York: Cambridge University Press.

Torrubiano, J. (1934), 'El Cristianismo es el mejor auxiliar de la Ciencia Eugénica', in Noguera, E. and Huerta, L. (eds), *Genética, eugenesia y pedagogía sexual*, vol. 1, 60–84.

Toscano, M. (1996), 'Marcello Ricci: Una testimonianza sulle origini del razzismo fascista', *Storia Contemporanea*, 27: 879–97.

Traetta, L. (2013), 'Docimology Enters into Pscyhology: Dagmar Weinberg's Work in French Applied Psychology Laboratories', *International Review of Social Sciences and Humanities*, 4: 149–58.

Treves, A. (2001), *Le nascite e la politica nell'Italia de Novecento*, Milan: LED.

TSUL. Harry H. Laughlin Papers, Ramos to Harry Laughlin and Charles Davenport, *c.* 1932.

———. Harry H. Laughlin Papers, The International Federation of Eugenic Organizations, Membership list and governing rules, September 1929.

———. (1926), Harry H. Laughlin Papers, Manuscript Collection, L 1, 'C' Boxes, C-2-1: 2. 'Report of the Meeting of the International Federation of Eugenic Organizations': 1–3.

———. Harry H. Laughlin Papers, Cora Hodson to Davenport, 9 August 1930.

———. Harry Laughlin Papers. Letter, *c.*1935.

———. Harry Laughlin Papers. Laughlin to Dr. Kendall Emerson, Director, National Tuberculosis Association, 16 January 1935.

———. Harry Laughlin Papers. Ramos to Laughlin, 18 January 1935; Ramos to Clarence Campbell, 11 February 1935.

Tumblety, J. (2012), *Remaking the Male Body: Masculinity and the Uses of Physical Culture in Interwar and Vichy France*, Oxford: Oxford University Press.

Turda, M. (2004), *The Idea of National Superiority in Central Europe, 1880–1918*, New York: Edwin Mellen.

———. (2007), 'The Nation as Object: Race, Blood and Biopolitics in Interwar Romania', *Slavic Review*, 66: 413–41.

———. (2008), *Eugenism și antropologia rasială în România*, Bucharest: Ed. Cuvântul.

———. (2009), '"To End the Degeneration of a Nation": Debates on Eugenic Sterilization in Interwar Romania', *Medical History*, 53: 77–104.

———. (2011), 'Controlling the National Body: Ideas of Racial Purification in Romania, 1918–1944', in Promitzer, C. et al. (eds), *Health, Hygiene and Eugenics in Southeastern Europe to 1945*, Budapest: CEU Press, 325–50.

———. (2012), 'Gheorghe Banu's Theory of Rural Biology in the 1920s Romania', in Bărbulescu, C. and Ciupală, A. (eds), *Medicine, Hygiene and Society from the Eighteenth to the Twentieth Centuries*, Cluj-Napoca: Ed. Mega, 125–40.

———. (2013), 'Crafting a Healthy Nation: European Eugenics in Historical Context', in Turda, M. (ed.), *Crafting Humans*, 109–26.

———. (2014), *Eugenics and Nation in Early Twentieth Century Hungary*, Basingstoke: Palgrave Macmillan.

Turpin, R. and Caratzali, A. (1937a), 'Influence de l'age maternel sur la mortinatalité des jumeaux', *Premier Congrès Latin d'Eugénique*, 283–5.

———. (1937b), 'Sexe et mortinatalité des jumeaux', *Premier Congres Latin d'Eugénique*, 286–90.

Turpin, R., Caratzali, A. and Georgescu-Roegen, N. (1937c), 'Influence de l'age maternel, du rang de naissance, et de l'order des naissances sur la mortinatalité', *Premier Congrès Latin d'Eugénique*, 271–7.

Turpin, R., Caratzali, A.and Gorny, M. (1937d), 'Contribution a l'étude de l'influence de l'age et de l'état de santé des procréateurs, du rang et du nombre des naissances sur les caractéres de la progéniture', *Premier Congrès Latin d'Eugénique*, 240–61.

Turpin, R., Caratzali, A., Rogier, H. (1937e), 'Étude étiologique de 104 cas de mongolisme et considerations sur la pathogénie de cette maladie', *in Premier Congrès Latin d'Eugénique*, 154–63.

Vacher de Lapouge, G. (1896), *Les selections sociales: Cours libre de science politique*, Paris: A. Fontemoing.

Vaida-Voevod, A. (1927), 'Politica națională și capitalul biologic național', *Buletin eugenic și biopolitic*, 1: 199–211.

Vallejo, G. (2004), 'El ojo del ponder en el espactio del saber: Los Institutos de Biotipología', *Asclepio*, 56: 219–44.

——. (2005), 'Las formas del organicismo social en la eugenesia latina', in Miranda, M. and Vallejo, G. (eds), *Darwinismo social y eugenesia en el mundo latino*, 231–72.

Vallejo, G. and Miranda, M. (2004), 'Los saberes del poder: Eugenesia y biotipología en la Argentina del siglo XX', *Revista de Indias*, 64: 425–44.

——. (2005), 'La eugenesia y sus espacios institucionales en Argentina', in Miranda, M. and Vallejo, G. (eds), *Darwinismo social y eugenesia en el mundo latino*, 145–94.

——. (2011), '"Civilizar la libido": Estrategias ambientales de la eugenesia en la Argentina', *Iberoamericana*, 11: 57–75.

Vallejo-Nágera, A. (1934), *Higiene de la raza: La asexualización de los psicópatas*, Madrid: Ediciones Medicina.

——. (1937), *Eugenesia de la Hispanidad y regeneración de la raza*, Burgos: Editorial Española.

——. (1938), *Política racial del Nuevo Estado*, San Sebastián: Editorial Española.

Vandellós, J. A. (1935), *Catalunya, poble decadent*, Barcelona: Biblioteca Catalana d'Autors Independents.

Van der Aa, J. S. (ed.) (1937), *Proceedings of the XIth International Penal and Penitentiary Congress held in Berlin, August 1935*, Bern: Bureau of the International Penal and Penitentiary Commission.

Variot, G. (1908), *Projet d'un Institut de Puériculture aux Enfants-Assistés*, Paris: Imprimerie A. Davy.

Vasconcelos, J. (1979), *The Cosmic Race*, Baltimore, MD: Johns Hopkins University Press.

Vasilescu-Bucium, I. (1935), 'Eugenia și studiul eredității', *Mișcarea Medicală Română*, 8: 169–74.

——. (1939), 'Tendances eugéniques dans le nouveau code penal roumain Carol II', *XVIIe Congrès International d'Anthropologie et d'Archéologie Préhistorique*, 678–81.

Verano, A. F. (1929), *Las doctrinas eugénicas (ensayo de sistematización)*, Buenos Aires: Liga Argentina de Profilaxis Social.

Verdicchio, P. (1997), *Bound by Distance: Rethinking Nationalism through the Italian Diaspora*, Madison, NJ: Fairleigh Dickinson University Press.

Vervaeck, L. (1929), 'Une enquête de la Revue Belge, L'examen prenuptial doit-il être obligatoire? Exposé général par le Docteur Louis Vervaeck', *Revue Belge*, 6 (1 June): 410–41.

Vestemeanu, M. (1939), 'La fertilité des femmes dans les provinces Banat et Dobrogea et l'influence des marriages prématurés sur la natalité', *XVIIe Congrès International*, 587–90.

Vignes, H. (1927), 'Certificat de mariage ou vulgarisation des notions d'eugénique', in Sand, R. et al. (eds), *L'examen médical en vue du mariage*, 197–238.

———. (1932), 'Sterilization des inadaptés sociaux', *Revue Anthropologique*, 42: 228–44.

———. (1934a), 'Les lois sterilization eugénique', *La Presse Medicale: Chroniques varieties, informations*, 40: 825–7.

———. (1934b), *Les Lois de stérilisation eugénique*, Paris: Masson.

———. (1937), 'De l'influence de l'intoxication alcoolique des procréateurs sur leur progéniture', *in Premier Congrès Latin d'Eugénique*, 165–70.

Villejean, A. (1930), 'La protection de la maternité et de l'enfance en France', *Bulletin de l'Office International d'Hygiene Publique*, 22: 1165–87.

Viola, G. (1937), 'L'évaluation de la constitution humaine individuelle', *Biotypologie*, 5: 85–92.

Vivó, V. (1910), *La sanidad nacional: Eugenesia y biometría*, Barcelona: La Neotipia.

———. (1912), 'A Healthy Sane Family Showing Longevity in Catalonia', *Problems of Eugenics*: 399.

Vlossak, E. (2010), *Marianne or Germania?* Oxford: Oxford University Press.

Voina, A. (1924a), 'Aspecte demografice', *Societate de mâine*, 1: 266.

———. (1924b), 'Doi factori de progres: Igiena și eugenia', *Societatea de mâine*, 1: 183–4.

———. (1924c), *Eugenia și igiena națiunii*, Cluj: Tipografia Națională.

Vornica, Gh. (1941), 'Concepția de rasă în Italia fascistă', *Buletin eugenic și biopolitic*, 12: 179–84.

Vyverberg, H. (1989), *Human Nature, Cultural Diversity, and the French Enlightenment*, Oxford: Oxford University Press.

Wadsworth, J. E. (1999), 'Moncorvo Filho e o problema da infância: Modelos institucionais e ideológicos da assistência à infância no Brasil', *Revista Brasileira de História*, 19: 103–24.

Wanrooij, B. P. F. (2001), '"The Thorns of Love": Sexuality, Syphilis and Social Control in Modern Italy', in Davidson, R. and Hall, L. A. (eds), *Sex, Sin, and Suffering: Venereal Disease and European Society since 1870*, London: Routledge, 137–59.

Ward, L. F. (1891), *Neo-Darwinism and Neo-Lamarckism*, Washington: Gedney & Roberts.

Wattiaux, H. (1981), 'Eugénisme et Morale Catholique', *La Nouvelle Revue Théologique*, 10: 801–17.

Weber, E. (1976), *Peasants into Frenchmen: The Modernization of Rural France, 1870–1914*, Stanford, CA: Stanford University Press.

Wegner, R. and De Souza, V. S. (2013), 'Eugenia "negativa", psiquiatria e catolicismo: Embates em torno da esterilização eugênica no Brasil', *História, Ciências, Saúde-Manguinhos*, 20: 263–88.

Weinberg, D. (1938), 'La biométrie différentielle et la biotypologie comme methodes pour la classification des individus et des groupes', *Congrès International de la Population*, 237–48.

Weismann, A. (1893), *The Germ-Plasm: A Theory of Heredity*, New York: Scribner's.

Weiss, S. F. (1990), 'The Race Hygiene Movement in Germany, 1904–1945', in Adams, M. (ed.), *The Wellborn Science*, Oxford: Oxford University Press, 8–68.

WL. Eugenics Education Society, SA/EUG/C, Hodson, C. S. to Davenport, C., 22 October 1924.

WL. Eugenics Education Society, International Federation of Eugenic Organisations, SA/EUG/E.12: 1932–1940.

Wishnia, J. (1995), 'Natalisme et nationalisme pendant la premiere guerre mondiale', *Revue d'Histoire*, 45: 30–9.

Wolff, E. C. (2013), 'Biological Racism and Antisemitism as Intellectual Constructions in Italian Fascism: The Case of Telesio Interlandi and *La difesa della razza*', in Weiss-Wendt, A. and Yeomans, R. (eds), *Racial Science in Hitler's New Europe, 1938–1945*, Nebraska: University of Nebraska Press, 175–99.

Zimmermann, E. (1992), 'Racial Ideas and Social Reform: Argentina, 1890–1916', *The Hispanic American Historical Review*, 72: 23–46.

Zola, É. (1992), *Nana*, trans., with an introduction and notes by D. Parmée, Oxford: Oxford University Press.

Zollschan, I. (1938), 'Die Bedeutung des Rassenfaktors für die Kulturgenese', *Congrès International de la Population*, 8: 93–105.

Zulawski, A. (2007), *Unequal Cures: Public Health and Political Change in Bolivia, 1900–1950*, Durham, NC: Duke University Press.

Zylberman, P. (2001), 'Hereditary Diseases and Environmental Factors in the "Mixed Economy" of Public Health: Rene Sand and the French Social Medicine, 1920–1934', in Gaudillière, J.-P. and Löwy, I. (eds), *Heredity and Infection: The History of Disease Transmission*, Chicago, IL: University of Chicago Press, 261–81.

Index

Accampo, E. A. 63
Achim, V. 230
Act of 31 July 1920 63
Aguirre, R. C. 135–6
Albarrán, E. J. 138
alcoholism 32–3
Alexandresco, V. 36
Aliano, D. 204
Almaça, C. 29–30
Alonso, A. F. 138
Amador, N. 55
Anales de Biotipología, Eugenesia y Medicina Social (Annals of Biotypology, Eugenics and Social Medicine) 133, 134
Anderson, B. 5
Andronescu, C. I. 104, 181, 232
  see also eugenic sterilization
Antonescu, M. 230–2
Apert, E. 29, 42, 47, 49, 69, 78, 81, 95, 107–8, 114–15, 185, 190, 201, 207
Aragon, H. O. 95
Argentina, eugenic developments in
  biotypology, promotion of 131–4
  eugenic achievements 130–1
  eugenic–medical–social welfare nexus 135
  eugenic prenuptial counselling 134
  eugenic sterilization 135
  homiculture 136
  institutions promoting eugenics 129–30
  prenuptial medical certification, introduction of 134
  puériculture 131
  see also Brazil, eugenic developments in; Cuba, eugenic developments in
Argentine Indo-Hispanic heritage 19
Argentinian Association of Biotypology, Eugenics and Social Medicine (*Asociación Argentina de Biotipología, Eugenesia y Medicina Social*) 132, 133, 174, 176, 248
Argentinian Eugenics Society (*Sociedad Eugenica Argentina*) 129, 130
Argentinian Society of Integral Eugenics (*Sociedad Argentina de Eugenesia Integral*) 248
Arnould, J. 82–3
Artom, C. 52, 88, 107
Arvey, S. R. 162
Avendaño, L. 32
Ayala-Carcedo, F. J. 27
Ayarragaray, L. 19
Azevedo, A. de 209–10, 212
Aznar, S. 169–70, 172, 206, 247

Bagdasar, D. 105
Baldwin, P. 5
Balzagette, L. 17
Banu, G. 32, 73–4, 124–5, 178, 189–90, 193, 225–8, 231, 243
Barahona, A. 108, 142, 248
Barbero, M. I. 19
Barrachina, M.-A. 75, 109
Barthou, L. 55
Bashford, A. 7
Battle of the Marne 57
Baudelocque Clinic 33
Bazán, E. P. 26
Beck, M. 184
Belgian eugenic programme 67–70
Belgian Society for Preventive Medicine and Eugenics (*Société Belge de Médicine Préventive et d'Eugénique*) 70
Belgian Society of Eugenics (*Société Belge d'Eugénique*) 67
Benavente, A. 161
Bertillon, J. 54
Bertrand, L. 59
Beruti, J. A. 133, 135, 174
Bethell, L. 21, 137
Betta, E. 113, 117
Bettiol, G. 124

Biernat, C. 133
biological determinism, Latin eugenics on 174–6
biological racism 8
biologization of national belonging 7
Biopolitical and Eugenics Section of the ASTRA Association (*Secția de eugenie și biopolitică a Asociațiunii ASTRA*) 72, 177
biopolitical purpose of eugenics 6–7
biotypological healthcare programme 93–5
  First International Meeting of Biotypology 183–5
biotypological identification card (*ficha biotipológica escolar*) 133
Birn, A.-E. 21, 29
Bismarck, O. von 65
Blum, A. S. 139
Boalick, A. R. 205
Boarini, M. L. 149
Boas, F. 184
Bocci, M. 117
Bogdan-Duica, G. 73
Boldrini, M. 88, 107, 145, 169, 188–9, 216
Bolea, R. C. 2, 76
Bollo, H. G. 131–2
Bombarda, M. 53
Bonfante, P. 59
Borges, D. 30
Bossi, L. M. 61
Boulenger, G. A. 68
Bowler, P. 11
Brazil, eugenic developments in
  government's active involvement in eugenic programmes 147–9
  prenuptial medical examinations and institutional segregation 145
  programmes of biological improvement 145
  public role of eugenics 143
  rural sanitation and preventive medicine 146
  in terms of childcare, maternal assistance, and family education 144
  see also Argentina, eugenic developments in; Cuba, eugenic developments in
Brenier, H. 114
Briand, H. 29, 108, 165, 170, 190–1, 217, 224
Broberg, G. 2
Bronfman, A. 37, 151, 153, 156
'brotherhood of Latin' nations 9
Brousseau, A. 187
Bucur, M. 70–1
Buletin Eugenic și Biopolitic (Eugenic and Biopolitical Bulletin) 104, 213, 242
Burga, C. 163
Burgess, G. 246
Burns, E. B. 20–1
Businco, L. 205

Caja, F. 55
Camavitto, D. 141, 186
Camiscioli, E. 31, 201, 204
Canali, S. 100
Cantón, E. 56
Capuano, C. 219
Caratzali, A. 187, 189, 193
Carelli, A. 106–7
Carneiro, L. 147–8
Carol, P. 163
Caron, A. 33
Carranza, V. 53
Carrel, A. 162, 221–4, 232, 238, 240, 248
Casagrandi, O. 187
Cassata, F. 2, 6, 13, 22, 44–5, 52–3, 59–61, 77, 88, 91, 96, 116, 118–20, 169–70, 200–1, 209, 211–13, 215–16, 246–8
Castanheira, J. P. 209
Castejón, F. 2, 109
Casti Connubii 11, 110, 115, 116, 119
castration 121, 126–8
Castrilli, V. 187
Catalonian Eugenics Society (*Societat Catalana d'Eugenèsia*) 110, 186
Catholic Church 30–1
  Catholic tradition, spread of 11
  role in social and reproductive functions 2
Caulfield, S. 5, 143
Cecchetto, S. 130, 248
Central Secretariat of the National Associations of Catholic Physicians (*Secrétariat Central des Sociétés Nationales de Médecins Catholiques*)
  Central Institute of Statistics 120
  Institutul Central de Statistică, Bucharest 179
  Istituto Nazionale di Statistica, Rome 91, 98

Chadwick, H. M. 39
Chamberlain, H. S. 16–17
Champy, Ch. 187
Cházaro, L. 138
Childers, K. S. 219
Children's Code (*Código del Niño*) 135
Chilean Institute of Puériculture (*Instituto de Puericultura*) 36
Christian eugenics 115, 117, 120, 121
Ciceri, M. 51
Cleminson, R. 2, 6, 24, 53, 75, 80, 84, 108–10, 124, 210, 246
Closson, C. C. 16
Code of Eugenics and Homiculture (*Códico Panamericana de Eugenesia y Homicultura*) 154, 155
Colajanni, N. 24–5
Cole, J. H. 31, 54
Colombo, D. 117
Comfort, N. 13
Comité France-Italie (French-Italian Committee) 57
Commission for the Promotion and Protection of the Biological Capital of the Nation (*Comisia pentru promovarea şi ocrotirea capitalului biologic al naţiunii*) 231
Commission for the Scientific Study of Population (*Comisión para los Estudios Científicos de la Población*) 197
Committee of Eugenic Studies (*Comitato Italiano per gli studi di Eugenica*) 51
  Gini's contribution 52
  proposed eugenic outline 52
Congrès International d'Anthropologie et d'Archéologie Préhistorique (Congress of Anthropology and Prehistoric Archaeology), Bucharest 107, 191
Coni, E. R. 52
Conklin, A. L. 207, 209
Conry, I. 29
Consiglio, P. 106
Conversi, D. 56
Correia, A. A. M. 17, 22, 29, 76, 84, 202–3
Cosmacini, G. 117
Coupeau, N. 15
Courthial, A. 188
Cuba, eugenic developments in
  'Code of Eugenics and Homiculture' 154–6
  Great Depression, effect of 158–9
  homiculture 150–2
  political situation and 158–61
  puériculture 150–2
  racial crossing and medical certificates 154–5
  under under Ramos's leadership 150–63
  rise of 'negative' eugenics 153
  see also Argentina, eugenic developments in; Brazil, eugenic developments in
Cuban Aryanism 27–8
Cuénot, L. 78–9
Cueto, M. 163
Cupcea, S. P. 230, 243
'cutting-edge' genetic research 191–2
Czech Eugenics Society 64

D'Agostino, P. 16, 24
D'Aroma, N. 213
Dalsace, J. 172–3
Dartigues, L. 57–8
Darwin, C. 28
Darwin, L. 88, 154, 166, 201
Davenport, C. B. 107, 155–6, 167–8
Dávila, J. 143
De Bandelac Pariente, A., 57
De Bont, R. 85
De Castro y Serrano, J. 16
De Donno, F. 11, 210
De Felice, R. 66
De Grazia, V. 97, 200
De Gubernatis, A. 20
De La Fuente, A. 28
De Lacerda, J. B. 143
De Luca, V. B. 5
De Matos, P. F. 2, 26, 30, 203–4
De Napoli, F. 77
De Plauzoles, S. 64
De Raes, W. 115, 124
De Sandoval, I. E. 158
De Souza, M. L. 149
De Souza, V. S. 2, 145–6, 148–9
De Vernejoul, R. 115
Decroly, O. 68, 74

defective genes 11
Delfino, V. 47, 88, 129–32, 135, 144–5, 151, 172, 250
Demographic, Anthropological and Eugenics Department (*Secția de Demografie, Antropologie și Eugenie*) 176, 179, 192
Dendle, B. J. 27
Dermine, J. 115
Destrée, J. 86
Di Liscia, S. M. 20
di Vincenzo, G. J. A. 133
Díaz, E. A. 19
Dietrich, D. J. 112
Doizy, H. 64
Domilescu, M. 193
Domingo, A. 110
Doumer, P. 47
Driever, S. L. 27
Drouard, A. 33, 223, 245
Dubourg, M. L. 114
Dunn, P. M. 33

early Latin eugenics
  First International Eugenics Congress, impact of 41–57
  First World War, impact of 57–66
  1910s 41
  premarital medical examination, debate on 77–87
  scientific interest 41
  see also Latin eugenics
education, importance of
  Brazil 144
  compulsory mass education 5
  eugenics-themed educational curricula 10
  Mexico 138–9
  Romania 177
Enachescu, S. D. 106
Ensch, N. 68
Eraso, Y. 2, 36, 56–7, 133–4
ethnic state 12
eugenic engineering of national communities 7
eugenic pastoralism 51, 105
eugenic regeneration 23
  aesthetic sensuality, development of 26
  biological engineering 25
  improvement of living conditions 30
  Sergi's measures to prevent degeneration 22, 24
  see also neo-Lamarckism
Eugenics Education Society 48, 68
eugenic organizations 2
Eugenics Record Office 68
Eugenics Section (*Sección de Eugénica*) 74
eugenic sterilization 43, 240
  in Argentina 135
  Catalonian eugenicists against 110
  Catholic Church on 119–22
  versus Christian morality 109–12
  condemnation of 78
  French eugenicists against 107–8
  impact of Nazi Sterilization Law of 1933 122–8
  Italian eugenicists against 106–7
  Latin eugenicists and 104–12
  and marriage restriction 53–4
  in Mexico 138–40
  Romanian eugenicists in favour of and against 105–6, 195–6
  Spanish eugenicists against 108–9
Eugénique 48
European political relations 6

Făcăoaru, Gh. 230
Făcăoaru, I. 106, 125, 192–5, 225, 228, 230–5, 243
Fallon, V. 70, 87, 119, 169
fascism and eugenics 87–101
Faure, F. 47
Federación International Latina de Sociedades de Eugenesia 174
Fédération de la Presse Médicale Latine (Federation of Latin Medical Press) 181
Fédération Internationale Latine des Sociétés d'Eugénique 9
Federațiunea Societatilor Latine de Eugenie 9
Federazione Latina fra le Società di Eugenica 9
Federici, N. 189, 216
Feinmann, E. 37
Felder, S. 19
Felix, I. 30
Fernández, J. S. 23, 28
Fernández-Ruiz, C. 206
Ferrándiz, A. 75

Ferrero, G. 17, 23, 25–6
Finlay Institute (*Instituto Finlay*) 154, 162
Fircks, A. F. von 18
First Congress of Latin Eugenics (*Ier Congrès Latin d'Eugénique*), discussions 185–96
  aetiology of 'mongolism' 187
  biotypology and eugenics, relationship between 188
  'constitutional type and eugenics,' relationship between 188
  debate on eugenic sterilization 192
  eugenics and demography, pathology and pedagogy, relationship 189
  heredity, eugenics, and selection 192
  'passive' and 'active eugenics' 188–9
  'qualitative and quantitative growth of population' 187
  relevance of eugenics 189–90
  social importance of eugenic studies 185–6
First International Eugenics Congress, discussions 41
  Catalonian eugenicists, impact on 55–6
  child welfare 42–3
  contribution to nation's population growth 47
  cyclical evolution of nations 44
  demography and fertility of social classes 44
  disagreements among delegates 42
  eugenic history of a family from Catalonia 47
  eugenic 'responsibility,' idea of 46–7
  eugenic sterilization 43
  French eugenicists, impact on 42, 47–51, 55
  impact of 41–57
  Italian eugenicists, impact on 51–2
  Latin eugenicists, impact on 52–3
  national efficiency 56
  national eugenics 46
  poverty and social misfortune 44
  practical strategy of biological improvement 43
  puériculture, discussion of 42
  racial specificity 45
  Romanian and Argentinean eugenicists, impact on 53
  Spanish eugenicists, impact on 55
  statistical investigations of differential fertility 45
  sterilization and marriage restriction 53–4
  universal eugenics 46
  variability index and mutability index 45
First International Meeting of Biotypology (*Réunion Internationale de Biotypologie*) 183
First Italian Congress of Social Eugenics (*Primo Congresso Italiano di Eugenetica Sociale*) 88
First National Congress of the Child (*Congreso Nacional del Niño*) 56
First World War
  negative effects 63
  positive effects of 61–2
  racial prejudice and stereotyping 58
  wartime rape and racial pollution 59–61
Fischer, E. 100, 167, 172–3, 183, 212–13, 217–18, 227
Flaubert, G. 15
Fogarty, R. S. 201
Fondation Française pour l'Étude des Problèmes Humains (French Foundation for the Study of Human Problems) 223
Fortunati, P. 187
Foschi, R. 117
France
  decline of geopolitical power of 17–18
  demographic decline 18
  French eugenics under the Vichy regime 217–24
  Latin eugenics in 12
  in mid-1850s 6
  see also French Eugenics Society (*Société Française d'Eugénique*)
Franco-Prussian War 15
Frassetto, F. 79, 197
French Anti-Alcohol Union 32
French Eugenics Society (*Société Française d'Eugénique*) 47
  on demographic decline in France 49–50
  on economic incentives for large families 50
  'Eugenics and Child Study' 49

focus of 48
internationalization of 50
programme of 48–9
rejuvenation of society, suggestions 50
relationship between eugenics and puériculture 49
statutes 48
French National Committee of Physical Education and Social Hygiene (*Comité National de l'Education Physique et de l'Hygiène Sociale*) 63
French rape victims, 'physiological impregnation' of 60
Frétigné, J.-Y. 24
Frick, W. 173
Fuchs, R. G. 35, 55

Galton, F. 94, 115, 171, 242
Galtonian eugenics 42
Garner, J. W. 18, 32
Garrett, A. 76
Gaudillière, J.-P. 2, 78
Gemelli, A. 100, 117–21, 170, 212, 214, 216
General Directorate of Demography and Race (*Direzione Generale per la Demografia e la Razza*) 211
Georgescu, D. C. 179
Georgescu-Roegen, N. 189, 193
Germanic (Teutonic) race, cultural supremacy of 17–18, 26
German racism, and militarism 11
German racial hygiene 207–17
germ-plasm theory 29
Giannone, A. 117
Gibson, M. 18, 22
Gillette, A. 11, 88, 99, 211–12
Gini, C. 10, 42, 44–5, 47, 52, 56, 62, 79–80, 88–91, 94–6, 100, 117, 141–2, 146–7, 151, 166–72, 174–6, 182–3, 185, 188–9, 191, 197, 201, 204, 212, 214–16, 246–7, 250–1
Giorgi, G. M. 91, 187, 247
Giuffrida-Ruggeri, V. 42, 45, 79, 80, 200
Giuliani, A. 117
Glass, D. V. 97–8
Glick, T. F. 76
Goemaere, P. 84
Gomes, A. C. V. 146
Gómez, M. J. B. 36–7
González, A. G. 2, 5, 27, 37, 151, 153–8, 161–2
Goode, J. 16
Gori, G. 100
Gorny, M. 189
Govaerts, A. 68–9, 79–82, 151, 166, 170
Gradinescu, A. 181
Griffing, J. B. 150
Grimoult, C. 217
Guardiola, M. 77
Guérin, A. 82
Guérin, G. 82–3
Guerra, L. 23
Guy, D. J. 5
Gynaecological Society (*Sociedad Ginecológica Española*) 74

Haidar, V. 133, 136, 248
Hallopeau, F. H. 47
Hamonic, P. 60
Hansen, R. 13
Harmsen, H. 112, 172
Haro, F. 109
Harris, R. G. 23, 60, 137
Harsin, J. 5
Haškovec, L. 64, 81
Haţieganu, I. 72, 181
Hawthorn, É. 114
Haycraft, J. B. 25
health institutions 67
Hedrick, T. 137
Helg, A. 19, 27
Henriques, B. M. C. 22, 53, 76, 109
Hèritier, Ph. 187
Hernández, E. 37, 150–1, 154
Herseni, T. 228–30, 243–4
Heuyer, G. 185, 188
Hochman, G. 143, 145–8
Hodson, C. B. S. 89, 168
Hoffmann, G. 77
homiculture 37, 38
Homo Europaeus 16
Homo Mediterranaeus 16
Horn, D. G. 96–7
Houssay, F. 29, 42–4, 47, 49, 78
Huerta, L. 74, 109
Huertas, R. 26
Huguenin, R. 187

Hunt, N. R. 70
Huss, M.-M. 62, 187–8
Husson, R. 187–8

Iberian states 4
Ichok, G. 185
Ide, M. 86
idealized national community 5
Il Problema della razza in Romania (The Question of Race in Romania) 227
Illanes, M. A. 36
Infante, C. D. 32
Ingenieros, J. 23
Institute of Anthroposociology (*Institut d'anthroposociologie*) 217
Institute of Ethnoracial Biology (*Institut de Biologie Etnorasială*) 231
Institute of Gerontology and Geriatrics (*Institutul de Gerontologie și Geriatrie*) 244
Institute of Hygiene and Social Hygiene (*Institutul de Igienă și Igienă Socială*) 71
Institute of Orthogenesis and Racial Improvement (*Istituto di Ortogenesi e Bonifica della Stirpe*) 215
Institute of Puériculture (*Institut de Puériculture*) 35
Institutos de Puericultura in Argentina 36
Interallied Congress of Social Hygiene (*Congrès Interallié d'Hygiène Sociale*) 63
Interallied Institute of Public Health (*Institut Interallié d'Hygiène Publique*) 64
Interlandi, T. 205, 212–15, 227
International Congress on Population (*Congrès International de la Population*) 183
international eugenic movement 9
International Federation of Eugenic Organizations (IFEO) 9
  Anglo-Saxon and Nordic eugenics, dominance of 166–7
  eugenic sterilization, debate on 167–9
  French eugenicists, influence of 166–7
  ideological differences between Latin and Nordic eugenics 166–73
  Italian eugenicists attitude 167
  Italian population policies and 170
International Institute of Sociology 247
international 'Latin sisterhood' 2
International Society for Racial Hygiene (*Internationale Gesellschaft für Rassenhygiene*) 48
intra-European dimension of Latin eugenics 2
intra-uterine puériculture 33–9, 64, 73
  care of individuals 37
  in Latin countries 36
  Pinard's description of 42
  pregnancy care 33–4
  social selection 34
Ioanid, R. 230
Ionescu-Muscel, P. 23
Iorga, N. 181
Ipsen, C. 54, 96, 170
Ireni-Saban, L. 13
Israel, G. 210–11
Italia–Francia 57
Italian Committee for the Study of Population (*Comitato italiano per lo studio dei problemi della popolazione*) 100
Italian Congress of Eugenics and Genetics (*Congresso Italiano di Genetica ed Eugenica*)
  first 157
  second 157, 167
  third 131, 216
Italian eugenics 88–9
  biological philosophy of Facism 95
  biotypological healthcare programme 93–5
  Fascist corporatism 91
  health and physical fitness 98–100
  Mussolini's population policy 97–8
  promotion of scientific moderation and research into human heredity 90
  safeguard of motherhood, women and children 97
  see also French Eugenics Society (*Société Française d'Eugénique*)
Italian racial movement 212
Italian Society of Genetics and Eugenics (*Società Italiana di Genetica ed Eugenica*) 87
Izquierdo, J. J. 79, 138

Jackson, J. 219
Jarnot, S. 217
Jayle, F. 47
Jennings, É. 219
Jonas, R. 7
Jordan, É. 113–14, 161, 183
Jorge, L. 163, 197

Kehl, R. 108–9, 144–9, 187, 197
  see also eugenic sterilization
King, D. 13
Kligman, G. 244
Knight, A. 137
Knudsen, E. R. 215
Kühl, S. 9, 125, 167, 199

La Difesa della Razza 211, 212, 213, 227
La Raza Latina 16
Lackerstein, D. 216
Lafuente, E. 75
L'alliance nationale pour l'accroissement de la population française (National Alliance for the Increase of the French Population) 54
Lamarck, J. B. 28
Lamas, C. S. 135
Landau, E. 78
Landouzy, L. 47
Landra, G. 205, 209–12, 214, 226–8, 233
Landry, A. 35, 42, 169, 172, 183, 245
Langeron, A. 221–3
Larbiou, B. 208, 217
Latin America 2
Latin eugenic movement 9, 11
Latin eugenic programme 10, 249
Latin eugenics
  communicable diseases, control measures for 5
  compulsory mass education, spread of 5
  'cutting-edge' genetic research 191–2
  defined 1
  distinctive characteristics of 8–9
  foreign and domestic immigration, dealing with 5–6
  health and hygiene programmes, development of 10
  history of 1–2, 13
  homiculture and 37–8
  intra-European dimension of 2
  in modernization of nation-states 4–5
  national unity and economic progress, role in establishing 3
  political relationships and international cooperation, role in developing 3–4, 9, 16
  post 1945 242–9
  scientific and cultural developments, role in 4, 9–10
  social and reproductive functions, role in 3, 10
  see also Argentina, eugenic developments in; Brazil, eugenic developments in; Cuba, eugenic developments in
Latin International Federation of Eugenic Societies 186
Latin race, decline of 6–7, 15
  causes 16–24
  proposals for revival 24–8
Laughlin, H. 105, 107, 151, 154, 157–63, 172
Laugier, H. 141, 209, 245
Lavrin, A. 36, 131, 134
Law for the Protection of the Family (*Decret Lege pentru Ocrotirea familiei*) 232
Law of Family Relations (*Ley sobre Relaciones Familiares*) 53–4
Law Relating to the Protection of Maternity and Infancy (*Loi relative á la protection de la maternité et de la première enfance*) 223–4
Lázaro, L. M. 74
Le Dantec, F. 28
Le Gall, P. 46
League for the Rights of Mothers and Children (*Liga para los Derechos de la Mujer y del Niño*) 56
Leclercq, J. 87
Lefaucheur, N. 33
Leon, S. M. 103, 120
Léonard, J. 48
Leonida, I. 32, 105–6
Leriche, R. 220–1
Leroy-Beaulieu, P. 54
Lesné, E. 187
Letard, E. 186
Levasseur, É. 18
Levi della Vida, G. 187

Levine, A. 29
Levine, P. 7
Lévy-Solal, E. 64
L'Hardy, G. 57
L'hygiène de la race (The Hygiene of the Race) 226
Lima, N. T. 143, 145–8, 161, 163
Livingstone, M. A. 11, 210
Lohse, F. 172
Lombrosian criminology 21, 23
Lombroso, C. 21–3, 85
López, O. V. 133
López-Cordón, M. V. 39
López-Guazo, L. S. 54, 137, 139–41
Loria, A. 42–4, 91
Löscher, M. 120
Loyo, G. 141
Lucamante, S. 22
Lucassen, L. 53
Luchaire, J. 59

Maciel, M. E. 143
Madrazo, E. D. 24
Maio, M. C. 143, 145–8
Mangiagalli, L. 52
Manias, C. 6
Manicatide, M. 178, 189
Manliu, I. 72, 104, 113, 225
Manouvrier, L. 47
Mantovani, C. 5, 22, 60, 77
Manuilă, S. 170, 172, 176, 178–9, 191, 229–31, 247, 251
Manuila, V. 193
Marañón, G. 75–6, 109, 133, 140, 197
March, L. 29, 42, 46–7, 49–51, 56, 62, 69, 78–82, 88, 166, 169–70
Margarit, O. 242
Margineanu, N. 243
Marín, R. C. 26
Marinescu, G. 24, 95, 125, 177–80, 183, 185–7, 189, 191, 193–5
Marques, V. R. B. 143
Marro, A. 42–3, 52
Martial, R. 172, 186–7, 207–9, 217, 221, 224, 232
maternal care, eugenic development related to
  Italian eugenics 97
  neo-Lamarckism 31–2
  Romania 179
maternidad consciente 75, 76, 109, 139
Mayer, J. 112
Mayr, E. 29
McDougall, M. L. 35
Mckiernan-González, J. 5
McLaren, A. 63
Mediterranean race 8
Mendes Correia, A. 17, 22, 29, 76, 84, 108–9, 202–4
Méry, H. 47
Mexico, eugenic developments in 136–43
  biotypology 139, 142
  child and maternal care 138
  eugenic sterilization and castration 138–40
  ideas of puériculture and pronatalism 139
  importance of education 138–9
  notion of 'conscious maternity' 139
  population management 142
  prenuptial medical certification 139
  puériculture and public health programmes 141, 147–8
  reproduction and socialization 139
Michaelis, M. 212
Michels, R. 42
Minkowska, F. 189
Miranda, M. A. 2, 130–1, 133, 136, 145, 198, 248
Mjöen, A. 77, 123
modernization, process of
  benefits 4
  biological racism and 8
  engineering of national communities 7
  Latin eugenics, role of 4–5
  racial decline and 20–1
modern nation-state 4
Moldovan, I. 7, 71–2, 106, 177, 179, 191, 231–2, 243
Monacelli, M. 97
Moncorvo Filho, A. 144
Montalvo, R. 27
Morcillo, A. G. 206
Morselli, E. 42, 45–6, 204
Mosso, A. 26
Mucchielli, L. 23
multi-cellular organism, organicist perspective 7
Munareto, G. D. 144

Murard, L. 24
Mussolini, B. 10–11, 54, 59, 65–6, 88, 90–1, 95–100, 124, 131, 140, 170, 200, 204–6, 210–16, 234–5

Nari, M. A. M. 134
Nash, M. 75, 206
Nasso, I. 124
National Agency for the Protection of Maternity and Infancy (*L'opera nazionale per la protezione della maternità e infanzia*) 97
National Congress of Population Science (*Congresso Nacional de Ciências da População*) 197, 203
National Congress of the Association of Christian Marriage (*Congrès National de l'Association du Mariage Chrétien*) 114
National Eugenics Office (*Office National d'Eugénique*) 69
National Institute for Demographic Studies (*Institut National d'Études Démographiques*) 245
National Institute for Maternology and Puériculture (*Instituto Nacional de Maternología y Puericultura*) 36
Nazi racial programme *see* German racism, and militarism
Nazi Sterilization Law of 1933 108, 122–8
  Belgian response 124
  French response 122–3
  Italian response 124
  Romanian response 125
  Spanish response 124–5
  US response 126
neo-Lamarckism 10–11, 15, 28–33, 38
  anthropology and 29–30
  concordance with Catholicism's religious paradigm 30–1
  environment management 31–2
  improvement of living conditions 22, 30
  influence of French medicine 29
  protection and supervision of mother during pregnancy 33
  social and biological management of women and children 31–2
  theory of adaptive and environmentally driven hereditary change 28–9
  transformation of Rio de Janeiro, example 30
neo-organicism 7
Nestor, I.-M. 243
Niceforo, A. 19, 42, 44–5, 51–2, 169–70
Nicollet, P. 83
Nisot, M.-T. 68–71, 77, 79–80, 107, 130, 138–9, 144
Noguera, E. 74, 109, 181
Novoa, A. 29
Nye, R. A. 32

Odobescu, G. 105–6, 178
Offen, K. 18, 31
Office for Racial Study and Propaganda (*Ufficio studi e propaganda sulla razza*) 211
Orbegoso, A. 2, 163
Ortega y Gasset, J. 27
Orzaes, C. C. 36
Osborn, F. 184
Otaola, J. M. 109
Ottaviani, A. 117

Padovan, C. 187
Palma, H. A. 133
Pan-American Conference on Eugenics and Homiculture (*Conferencia de Eugenesia y Homicultura*)
  first 154
  second 134, 154, 160, 174
  third 158, 161
Pan-American Office on Eugenics and Homiculture (*Oficina Panamericana de Eugenesia y Homicultura*) 157
Pan-Latinism 9
Papillault, G. 95
Parhon, C. I.
Pasteau, O. 120–1
Paté, H. 63–4
Paulo, L. F. 246
Pedersen, J. E. 32
Peláez, R. Á. 2, 26–7, 37, 67, 74, 109, 151, 153–8, 161–2, 205

Pende, N. 10, 91–5, 101, 131–3, 145, 170, 180–4, 200, 212–16, 232, 244, 246
Penrose, L. S. 192
Pérez, L. A. 7
Perrier, E. 29, 47, 78, 80
Perrier, R. 80
Persell, S. M. 29
Peruvian Conference on Eugenics (*Jordana Peruana de Eugenesia*)
  first 161, 163
  second 163
Pestalozza, E. 87, 89–90, 167–8
Petrovici, E. 193
Petruccelli, J. L. 20
Pieraccini, G. 5
Piéri, J. 114–15
Pimentel, I. F. 67, 77
Pinard, A. 29, 33–5, 42, 47, 60, 64, 69, 73, 78, 82–4, 131, 150–2, 154, 172, 190, 224
Piot, E. 18
Pitt-Rivers, G. H. L. F. 170
Ploetz, Alfred 17, 48, 80
Pogliano, C. 81
Poisson, G. 58
Popovici, G. 73
Portuguese eugenic programme 76–7
Portuguese Society for Eugenic Studies (*Sociedade Portuguesa de Estudos Eugénicos*) 195
Pozzi, L. 119
Preda, V. 243
premarital medical examination, debate on 77–87
  condemnation of eugenic sterilization 78
  French eugenicists 78–80
  French vs Italian eugenicists 80–5
  Italian eugenicists 80–1
  Romanian eugenicists 78
Presidents of the Latin International Federation of Eugenic Societies (1935–39) 196
Primer Curso Eugénico Español 74
Primeras Jornadas Eugénicas Españolas 109
Primo de Rivera, M. 74
Pronatalism 46, 56, 63, 78, 110, 136, 139, 170, 183, 218, 235, 244
'Protection of Infants' (*Protección de la Primera Infancia*) 56
Puig i Sais, H. 56, 110, 176
Punnett, R. C. 29

Querton, L. 42–3, 68
Quine, M. S. 2, 6, 22, 54, 96, 170, 197, 219
Quinlan, S. M. 6

racial cartography, process of 16
racial decline, causes of
  alcohol consumption and 32–3
  appearance of Teutonic people 16–17
  biological explanations 21
  Chamberlain's ideas 16–17
  Colajanni's views 24–5
  cultural incompatibilities of races and 19–20
  economic transformation and 20
  immigration and 20–1
  infant-mortality rates and decline in birth rates 21
  modernization and urbanization, due to 20–1
  'orthodox' deprecation of contemporary Italians and Spaniards 16
  predisposition to vigorous sexual activity 25–6
  racial degeneration in late-nineteenth-century Italy 18–19
  Sergi's measures to prevent degeneration 22
  theories of degeneration 24–6
  'Völkerchaos' (Chaos of Peoples) 16
racial hierarchies 8
racism, and eugenics
  African races, attitude towards 201
  Aryan race 210–11, 214
  'degenerate modernism' 205–6
  French eugenicists, views of 201
  German racial hygiene 207–17
  Gini's objections to 216
  immigration issues 204
  Italian racial discourse 205, 210
  Mediterranean race 211–12
  miscegenation between whites and non-whites 199–200, 204–5

normalization of race in Romania 225–35
Pende's endorsement of biotypology 214–16
'Portuguese race' 202–3
principle of human selection 202
racial ethnology ('ethnologie raciale') 217
racial immutability and incompatibility between superior and inferior races 208
racial purity and racial survival 204
radical eugenic programme of human breeding 201
superiority of Italians 200
'superior race,' notion of 201–2
Vallejo-Nágera's conservative racial hygiene 206–7
under Vichy era 217–24
see also racial decline, causes of
Radulescu-Motru, C. 243
Radzinowicz, L. 85
Ramacciotti, K. I. 133
Râmneanțu, P. 181, 184, 227, 231, 243–4
Ramos, A. P. 36
Ramos, D. F. 37, 79, 135, 150–64
Reggiani, A. H. 2, 6, 62, 131–3, 135, 223, 245
Reis, J. R. F. 145
religion and eugenics 238
Catholic Church on sterilization 119–22
Catholic eugenics, status of 113
Christian morality and eugenic ideas 114–16
correlation between Latinity and Christianity 117
in Latin countries 113
neo-Malthusian movement and 113
Protestant and Catholic attitude towards eugenics 112
Roman Catholic Church's attitude 113, 116
science of genetics and religion 121
Renshaw, J. L. 56
Revista de igienă socială (Review of Social Hygiene) 225
Rey, E. 17–18, 39
Ricci, M. 213
Richards, M. 2, 26, 75, 206
Richet, C. 35, 47, 78, 85, 201–2, 204, 207
Richter, I. 112
Rodriguez, J. 19, 23, 37
Rodríguez-Ocaña, E. 67
Rogier, H. 187
Roll-Hansen, N. 2
Romania, eugenic development in 3–4
biopolitics 7
cross-generational relationship within the eugenic community 193
eugenic sterilization 195–6
importance of 'environmental hygiene and education' 177
improvement of sanitary conditions 179
Latin eugenics and 12
normalization of race in Romania 225–35
promotion of biotypology 180–1
protection of women 179
relationship between individual and collective health 181–2
research on heredity and population health 177, 179, 183
Romanian eugenic programme 70–4
Romanian-Italian Institute of Racial and Demographic Studies (*Istituto Italo-Romeno di Studi Demografici e Razziali*) 228
Romanian Royal Society of Eugenics and Heredity (*Societea Regală Română pentru Eugenie și Studiul Eredității*) 178
Roquette-Pinto, E. 147, 149
Rosen, C. 112
Rosental, P.-A. 2, 4, 245–6
Rossi, A. 131–3
Roussy, G. 185, 187
rural biology 73

Saffiotti, F. U. 52
Sánchez-Albornoz, N. 21
Sand, R. 68–70, 81, 124
Sanger, M. 169
Sarmiento, D. F. 19
Sauvage, P. 219
Savorgnan, F. 62, 169, 172, 211, 247
Schell, P. A. 37, 153
Schneider, W. H. 29, 33, 36, 48, 79, 81–3, 95, 117, 204, 207, 209

Schraenen, W. 69, 87
Schreiber, G. 42, 49, 54, 78, 80–4, 108, 122–4, 151, 185, 187
Schreider, E. 95, 184, 209, 245
Scott, J. C. 96
Second Congress of the International Federation of Latin Eugenic Societies (*Al II-lea Congresal Federațiunii Societăților Latine de Eugenie*) 196–7
Second International Congress of Catholic Physicians (*II. Internationalen Kongress katholischer Ärzte*) 119
Second Pan-American Congress on Eugenics and Homiculture (*Segunda Conferencia Panamericana de Eugenesia y Homicultura de las Repúblicas Americanas*) 134
Sergi, G. 16–17, 22, 24, 51–2, 170, 211
Sinclair, A. 75
Sippial, A. T. 5
Skidmore, T. 143
Smith, J. S. 77
Sobral, J. M. 19
social and biological degeneration, signs of 3
social hygiene organizations 36
Society of Anthropology (*Sociedad Española de Antropología*) 74
Society of Biotypology (*Société de Biotypologie*) 95
Society for Puériculture and Pediatrics, Venezuela (*Sociedad Venezolana de Puericultura y Pediatría*) 163
Soffia, J. C. 32
Solano, S. 163, 197
Soldán, C. E. P. 129, 135, 143, 145, 154, 156, 163
Solonari, V. 230, 233
Somogyi, S. 97–8, 172
Sòrgoni, B. 205
Spanish eugenic programme 74–6
Speckman-Guerra, E. 27–8, 32
Spektorowski, A. 13
Spencer, H. 25
Spengler, J. J. 18
Stan, L. 229
Stepan, N. L. 2, 32, 37–8, 47, 52, 131, 140, 143, 145, 149–50, 156
Sterian, E. 104
Stern, A. M. 2, 21, 137, 139–42
Stoichita, I. 242–3
Strauss, P. 35, 55
Strauss Law of 1913 35
Suárez-Diaz, E. 142, 248
Sutter, J. 245

Taguieff, P.-A. 25
Tamagnini, E. 195–6
Tanner, J. 240
Tauro, G. 190
Tedesco, L. 22, 212
Thooris, M. A. 188
Thorp, R. 21
Timm, A. F. 77
Tissier, P. 60
Torrubiano, J. 110
Toscano, M. 214
Traetta, L. 95
Treaty of Frankfurt 15
Treves, A. 96
Tumblety, J. 223–4
Turda, M. 5–7, 12–13, 58, 70, 182, 228
Turpin, R. 108, 187, 189, 224

Vacher de Lapouge, G. 25, 79, 207–8, 217, 221
Vaida-Voevod, A. 73
Vallejo, G. 2, 5, 130–1, 133, 136, 248
Vallejo-Nágera, A. 111–12, 124, 134, 206–7
Vandellós, J. A. 110, 197
Van der Aa, J. S. 126–8
Vandervelde, E. 69
Variot, G. 35
Vasconcelos, J. 137
Vasilescu-Bucium, I. 193–5
Verano, A. F. 131
Verdicchio, P. 22
Vervaeck, L. 68, 80, 85–6, 121, 192
Vestemeanu, M. 193
Vicencio, A. 36
Vignes, H. 81–2, 95, 107–8, 185, 224
Viola, G. 145, 183–4
Vivó, V. 47, 55
Vlossak, E. 60
Voina, A. 71–3
Vornica, Gh. 213

Wadsworth, J. E. 144
Wanrooij, B. P. F. 77
Ward, L. F. 28
Wattiaux, H. 115
Waxweiler, E. 68
Weber, E. 5
Wegner, R. 2, 146, 149
Weinberg, D. 185
Weismann, A. 25, 29
Weiss, S. F. 48
Wishnia, J. 60
Wolff, E. C. 11
Woltmann, L. 17

Zimmermann, E. 19, 23–4, 130
Zola, É. 15
Zollschan, I. 184
Zuccarelli, A. 53
Zulawski, A. 5
Zylberman, P. 5, 24